An Introduction to Synoptic-Dynamic Meteorology

Gabriel Williams

ELSEVIER
ACADEMIC PRESS
An imprint of Elsevier

Academic Press is an imprint of Elsevier
125 London Wall, London EC2Y 5AS, United Kingdom
50 Hampshire Street, 5th Floor, Cambridge, MA 02139, United States

ISBN: 978-0-443-24858-0

For information on all Academic Press publications visit our website at https://www.elsevier.com/books-and-journals

Publisher: Candice Janco
Senior Acquisition Editor: Rakhshan Rizwan
Editorial Project Manager: Mason Malloy
Senior Project Manager: Nayagi Anandan
Cover Designer: Greg Haris

Typeset by TNQ Tech

Printed in the Great Britain
Last digit is the print number: 10 9 8 7 6 5 4 3 2 1

"The heavens declare the glory of God,
and the sky above proclaims His handiwork."
—Psalm 19:1

"For from Him, and through Him, and to Him are all things.
To Him be the glory forever. Amen."
—Romans 11:36

Preface

Introduction to Synoptic-Dynamic Meteorology

Historically, synoptic meteorology focused on observing large-scale weather systems, such as surface high- and low-pressure systems, jet streams and streaks, ridges and troughs, and surface fronts. However, the physical mechanisms that explain the structure and evolution of synoptic-scale systems are grounded in dynamic meteorology, which is a branch of fluid dynamics concerned with atmospheric motions. Dynamic meteorology applies the laws of classical mechanics, thermodynamics, and fluid dynamics to atmospheric phenomena.

Thanks to many technological advances, meteorologists in the 21st century benefit from insights deduced by these physical laws (typically associated with dynamic meteorology) and the wealth of observational data obtained from weather charts, satellite imagery, and other meteorological instruments (usually associated with synoptic meteorology). By integrating these insights, the modern meteorologist can:

1. Observe a synoptic-scale phenomenon and describe its essential characteristics,
2. Analyze the physical reasons for its existence, and
3. Predict its behavior over time.

For these reasons, synoptic meteorology and dynamic meteorology have merged into a unified field known as *synoptic-dynamic meteorology.*

In addition to advancements in observational technology, monumental progress in computing technology has made numerical weather prediction an essential component of meteorological analysis and forecasting. As a result, modern meteorologists must possess a solid foundation in atmospheric sciences, as well as well as proficiency in computer modeling and modern observational systems. The interdisciplinary character of meteorology has been recognized since the field's inception. In 1934, Carl-Gustaf Arvid Rossby - one of the founders of modern meteorology - wrote:

"The principal task of any meteorological institution of education and research must be to bridge the gap between the mathematician and practical man, that is, to make the weather man realize the value of a modest theoretical education, and to induce the theoretical man to take an occasional glance at the weather map."

The philosophy behind Rossby's statement remains as relevant today as it was in 1934. Meteorologists who can synthesize theoretical concepts with both conceptual and numerical observations are best equipped to contribute to operational forecasting and scientific progress in atmospheric science.

Purpose and Outline of This Textbook

In the 21st century, there has been rapid growth in interest and expertise in numerical weather prediction. However, as noted by the American Meteorological Society, education in modern synoptic-dynamic meteorology has steadily declined over the past decade. This trend may be attributed to several factors, including a growing interest in climate science. One significant factor is how synoptic-dynamic meteorology has been taught. Atmospheric dynamics is widely recognized as a challenge for both students and instructors, and many students find the transition from introductory meteorology to upper-level synoptic-dynamic meteorology especially difficult.

Based on my experience as a professor teaching meteorology and physics across all levels of the undergraduate curriculum, I believe students benefit from a well-designed bridge that allows them to apply their introductory physics and calculus knowledge to practical meteorology. This is the purpose of this textbook.

This textbook serves as a bridge between introductory meteorology at the freshman level and synoptic-dynamic meteorology at the senior level. It originated from course notes used in an undergraduate synoptic analysis course taught at the College of Charleston. The textbook assumes that students have completed their first year of calculus and introductory calculus-based physics, covering Newtonian mechanics, introductory thermodynamics, and introduction electromagnetism.[1] Various topics from introductory physics have been integrated throughout the text, with specific meteorological applications, to prepare students for advanced coursework in atmospheric dynamics.

The overarching approach of this textbook is to synthesize the observational emphasis of traditional synoptic meteorology with the theoretical foundations provided by dynamic meteorology.

This textbook is divided into three parts. Part I introduces the tools used to measure meteorological data and the techniques for analyzing and interpreting them.

- Chapter 1 introduces the four fundamental observational variables that recur throughout the text—pressure, temperature, water vapor, and wind—along with related concepts from mechanics and thermodynamics, including Archimedes' principle, buoyancy, temperature scales, and vector analysis. Derived variables such as virtual temperature and dew point are also presented.
- Chapter 2 describes modern instruments for atmospheric measurement, particularly those in automated surface observation systems.
- Chapter 3 explains common meteorological data formats (e.g., METAR and radiosonde codes) and teaches how to decode and interpret station plots.

[1]Some example textbooks include *Physics for Scientists and Engineers* by Raymond Serway and John Jewett and *Principles of Physics* by David Halliday, Robert Resnick, and Jearl Walker.

- Chapter 4 introduces the analysis of weather maps, emphasizing manual isopleth analysis and the identification of synoptic weather features such as fronts, troughs, ridges, and jet streaks.

Part II covers the essential principles of atmospheric dynamics and thermodynamics needed to interpret meteorological observations.

- Chapters 5 and 6 examine the dominant forces driving atmospheric motion. These chapters review Newton's laws, linear momentum, and work, and introduce mathematical tools often overlooked in introductory calculus courses—such as the scalar and vector products, and the gradient operator.
- Chapter 7 applies the equations of motion to geostrophic and gradient flows, culminating in a discussion of thermal wind balance and geostrophic adjustment. These chapters not only derive the relevant equations but also demonstrate how these balance conditions appear on weather maps.
- Chapter 8 examines thermodynamic processes that influence atmospheric motion, reviewing the first law of thermodynamics and introducing the concept of atmospheric stability.

Part III applies the foundational concepts from Part II to specific synoptic-scale systems.

- Chapter 9 explores the structure and evolution of upper-tropospheric jets and jet streaks, emphasizing their critical role in generating vertical motion.
- Chapter 10 focuses on the structure and development of Rossby waves, introducing key concepts such as vorticity and potential vorticity.
- Chapter 11 examines the structure and evolution of surface fronts and the physical mechanisms of frontogenesis.
- Chapter 12 concludes the textbook by synthesizing material from the preceding chapters into a discussion of midlatitude cyclones, framed by the Norwegian cyclone model and its subsequent modern refinements.

Although this textbook emphasizes conceptual understanding conceptual understanding and the physical governing synoptic-scale phenomena, it does not extensively cover certain advanced topics. Subjects like quasigeostrophic theory, formal scale analysis of motion, isentropic coordinates, and baroclinic instability are intentionally excluded to maintain accessibility and focus. Rather than beginning with complex mathematics, this text builds a conceptual framework first, introducing formal derivations and real-world illustrations as students progress. This approach is intended to make complex ideas more intuitive, engaging, and rewarding to learn.

Companion Student Resources

An important complement to this textbook is the companion student site, which provides practical resources to reinforce key concepts. The site includes Jupyter notebooks (using MetPy) and MATLAB scripts, each designed to correspond directly to material in the text.

- Chapter 3: Jupyter notebooks for accessing surface and upper-air observation data.
- Chapter 4: Jupyter notebooks for creating contoured and uncontoured weather charts.
- Chapters 5 and 8: MATLAB scripts and Jupyter notebooks examining thickness and thermal wind balance.
- Chapter 7: MATLAB scripts simulating balanced flow and geostrophic adjustment.
- Chapters 9 and 10: MATLAB scripts evaluating idealized jet streaks and troughs.
- Chapters 11 and 12: Jupyter notebooks analyzing surface fronts and frontal cyclones.

MATLAB is included for its accessibility to most university students, while Python- given its growing prominence - serves as an excellent complementary tool. No prior experience with either language is required; small modifications are sufficient to tailor scripts for individual use. It is my hope that these resources inspire students to develop their own scripts and analyses as they deepen their understanding of synoptic-dynamic meteorology.

Each chapter concludes with a series of problems—over 150 in total—varying in difficulty and scope. Many involve weather map interpretation, so printing the maps available on the companion site is recommended. Complete solutions are available in a separate instructor manual provided by the publisher.

Acknowledgments

After more than 2 years of writing, illustrating, and editing, completing this project is both a celebration and a moment of reflection. Although this has been a solo writing effort, many individuals have contributed both directly and indirectly to this work. Professionally, I am grateful to those who inspired me to pursue an academic career in atmospheric physics. Dr. Wayne H. Schubert first encouraged me to explore atmospheric dynamics and showed how my physics background could be applied to meteorology. Dr. Susan C. van den Heever, my first instructor in synoptic-dynamic meteorology, provided clear and rigoroug explanations of synoptic phenomena that have left a lasting impression.

Special mention is reserved for my family. My parents, Gabriel Sr. and Linda Williams, made significant sacrifices to ensure I could pursue academia as a vocation. My wife, Alicia Williams, and my daughters, Alexandria and Victoria, have provided me with unwavering love and support throughout this entire endeavor.

Contents

CHAPTER

Introduction

1

Chapter outline

To study the dynamics of synoptic-scale phenomena such as surface fronts, jets, and cyclones, we must first understand the fundamental properties of the atmosphere. Physically, the atmosphere behaves as a continuously evolving fluid system. To describe how it changes over time, we need to identify the variables that define its state. In this chapter, we will describe the four basic variables commonly used to specify the state of the atmosphere—namely pressure, temperature, water vapor, and wind.

1.1 Pressure

We often think of atmospheric pressure intuitively as the weight of the air above a given location. To develop a more complete understanding of atmospheric

An Introduction to Synoptic-Dynamic Meteorology. https://doi.org/10.1016/B978-0-443-24858-0.00001-0

pressure, we must revisit the concepts of stress and strain from introductory physics.

1.1.1 Stress and strain

When external forces act on an object (whether it is a solid, liquid, or gas), they can cause changes in the object's shape or size. This behavior is described using the concepts of stress and strain. **Stress** is defined as the external force acting on an object per unit area of its cross-section, while **strain** refers to the resulting deformation from this stress. In general, there are three types of stress that can be applied on an object, as shown in Fig. 1.1. When an external force is applied perpendicular to the cross-section (as shown in Fig. 1.1A), the object is under **tensile stress**, and the resulting change in length is known as a **tensile strain**. When an external force is applied parallel to one of the faces of the object (as shown in Fig. 1.1B), the object is under **shear stress**, and the resulting change in the sheared face is known as a **shear strain**. When external forces are applied perpendicular over the entire surface of the object (as shown in Fig. 1.1C), the object is under **bulk stress**, and the resulting change in volume is called a **bulk strain.**

We can distinguish between solids, liquids, and gases by examining how they respond under these three types of stress. When any stress (whether tensile, shear, or compressive) is applied to a solid object, internal molecular forces *resist* the deformation. This behavior supports the commonly held view that solids tend to maintain their shape and size. In contrast, when a shearing or tensile stress is applied to a fluid (ie, a liquid or gas), it flows in response. This leads to the common understanding that fluids do not have a definite shape.

Fluids cannot withstand shear or tensile stresses, but they can sustain bulk stresses. Liquids and gases can be distinguished by how they respond to bulk stress. Liquids possess strong internal cohesive forces that provide significant resistance to compression, making them nearly incompressible (meaning their density is

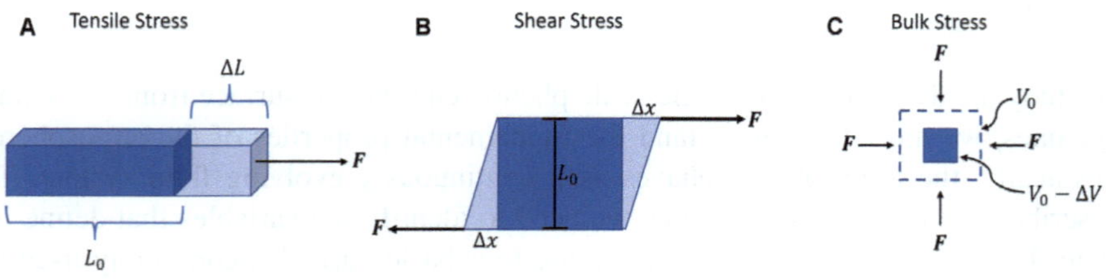

FIGURE 1.1 Objects under stress and strain

(A) An object under tensile stress. The amount by which the length of the object changes due to the applied force is ΔL. (B) A cross-section of an object under shear strain. The shear stress causes the top face of the object to move to the right and the bottom face to move to the left by Δx. (C) A cross-section of an object under bulk stress. The object undergoes a change in volume ΔV without a change in shape.

No Permission Required

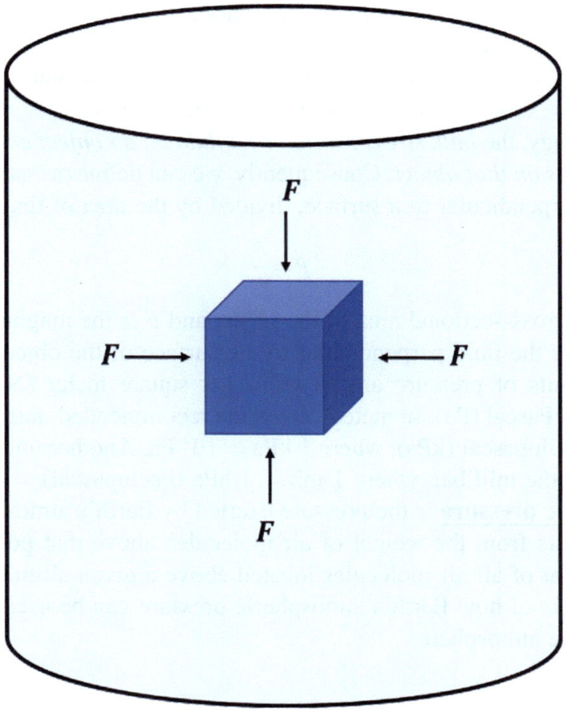

FIGURE 1.2 An object submerged in a fluid

A schematic of an object fully submerged in a fluid. This is analogous to the behavior of any object submerged within the atmosphere. Notice that forces are exerted perpendicular to each face of the object.

No Permission Required

approximately constant). In contrast, gases lack such cohesive forces, so their density varies significantly with applied bulk stress. For this reason, gases are classified as compressible fluids.

In a static fluid,[1] the only stress that can be exerted on a submerged object is a bulk stress, since fluids cannot sustain shear and tensile stress. To understand why this is the case, let's consider an object submerged in a fluid, as shown in Fig. 1.2. According to Newton's Third Law of motion, the force exerted by the fluid on the object must be equal and opposite to the force exerted by the object on the fluid. Because a static fluid cannot support shear stress, any force parallel to the object's surface would cause the fluid to accelerate, violating the condition of static equilibrium. In simpler terms, when an object is submerged in a static fluid, the fluid

[1] A static fluid is a fluid that is in static equilibrium, implying that the net force acting on the fluid is zero.

will rearrange itself until all parallel forces along the object's surface cancel out. Therefore, *at any point on the surface of the object, the force exerted by the fluid is perpendicular to that surface*. This is why the effect of a static fluid on its surroundings can be described entirely in terms of bulk stress.

In meteorology, the *bulk stress exerted by a fluid on an object corresponds to the pressure it exerts on that object*. Consequently, we can define pressure as the force a fluid exerted perpendicular to a surface, divided by the area of that surface:

$$P = \frac{F}{A} \tag{1.1}$$

where A is the cross-sectional area of the object and F is the magnitude of the total force exerted by the fluid perpendicular to the surface of the object. Based on Eq. (1.1), the SI units of pressure are Newtons per square meter (N m^{-2}) which is also known as a Pascal (Pa). In meteorology, the recommended unit for atmospheric pressure is the kilopascal (kPa), where 1 kPa $= 10^3$ Pa. Another unit that remains in common use is the millibar, where 1 mb = 1 hPa (hectopascal) = 100 Pa.

Atmospheric pressure is the pressure exerted by Earth's atmosphere at a given altitude. It results from the weight of air molecules above that point - that is, the combined weight of all air molecules located above a given altitude. Box 1.1 provides an example of how Earth's atmospheric pressure can be used to estimate the total mass of the atmosphere.

1.1.2 Archimedes' principle

As discussed in the previous section, atmospheric pressure results from the weight of the air above a given location. Consequently, atmospheric pressure is greatest at Earth's surface and decreases with increasing altitude. Because gases are

Box 1.1 Estimating the total mass of Earth's atmosphere

If we know the average surface atmospheric pressure on Earth, we can provide an estimate of the total mass of Earth's atmosphere.

Treating Earth's atmosphere as a sphere in which Earth's average radius is approximately 6371 km, we note that Earth's surface area is

$$A = 4\pi r^2 = 4\pi\left(6.37 \times 10^6 \text{ m}\right)^2 = 5.10 \times 10^{14} \text{ m}^2$$

The pressure from the entire atmosphere is felt at Earth's surface, which has an average value of 101.325 kPa = 1.01325×10^5 Pa. Therefore, the force exerted by the atmosphere must be

$$F = PA = \left(1.01325 \times 10^5 \text{ Pa}\right)\left(5.10 \times 10^{14} \text{ m}^2\right) = 5.17 \times 10^{19} \text{ N}$$

The force exerted by Earth's atmosphere is due to the weight of all the air molecules above the surface. This weight can be written as Mg, where M is the total mass of the atmosphere and g is the free-fall acceleration at Earth's surface. Therefore, the estimated mass of the atmosphere is:

$$M = \frac{F}{g} = \frac{5.17 \times 10^{19} \text{ N}}{9.80 \frac{\text{m}}{\text{s}^2}} = 5.28 \times 10^{18} \text{ kg}$$

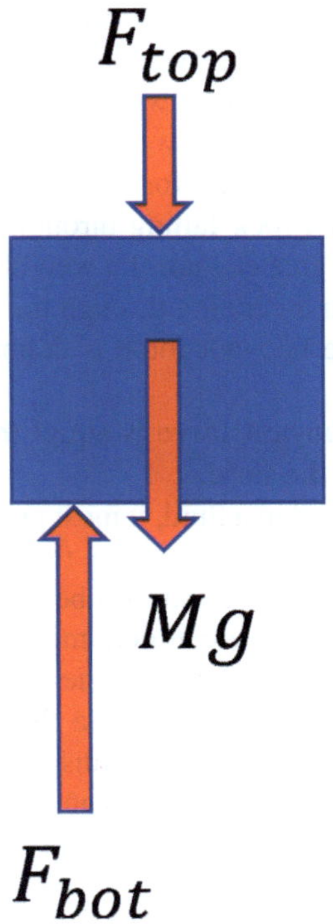

FIGURE 1.3

Free body diagram of a sample of air. A free body diagram of a sample of air where F_{top} is the force applied on the top surface due to air pressure, F_{bot} is the force applied on the bottom surface due to air pressure, and Mg is the weight of the air sample.

No Permission Required

compressible fluids, air is most compressed near the surface, where the pressure is highest. As atmospheric pressure decreases with height, air density also decreases. In Ch. 5.4, we will demonstrate that both atmospheric pressure and density decrease approximately exponentially with height.

To understand the effects of decreasing atmospheric pressure with height, consider a sample of air within the atmosphere, as shown in Fig. 1.3. Three forces act on this air sample: (1) the weight of the air sample, represented by Mg where M is the mass of the air sample and g is the acceleration due to gravity; (2) the force F_{top}, resulting from atmospheric pressure acting on the top face of the sample; and (3) the force F_{bot}, resulting from atmospheric pressure acting on the bottom face of the sample. If the atmosphere is in static equilibrium, the net force on the sample must be zero. According to Newton's Second Law, this condition can be expressed as $F_{bot} - F_{top} - Mg = 0$, which indicates that $F_{bot} > F_{top}$. In other words, because atmospheric pressure decreases with height, there must be a net upward force on the air sample. This net upward force is known as the **buoyant force**.

The argument stated above regarding Earth's atmosphere applies to any object that is fully submerged in a fluid. Every such object experiences a buoyant force, $\boldsymbol{F_B}$, due to vertical pressure differences within the fluid. Whether an object sinks or rises depends on the relationship between the magnitude of the buoyant force and the object's weight. If the buoyant force is less than the weight of the object, the object will sink - such as a rock falling through the atmosphere. On the other hand, if the buoyant force is greater than the weight of the object, the object will rise - such as a helium balloon ascending through the air. According to introductory physics, the magnitude of the buoyant force is defined by Archimedes' principle, which states that:

The magnitude of the buoyant force is equal to the weight of the fluid displaced by the object within the fluid.

When an object is submerged in a fluid, it must push a certain amount of fluid out of its way. According to Archimedes' principle, the buoyant force acting on the object is equal to the weight of the fluid that has been displaced by the object.

Using Archimedes' principle, we can determine whether an object will rise or sink in a fluid by comparing the densities of the fluid and the object. To see why, consider an object that is fully submerged in a fluid. In this case, the volume of the object is equal to the volume of the fluid it displaces. According to Archimedes' principle, an object will sink if its weight is greater than the buoyant force acting on it. This can be written mathematically as:

$$F_g > F_B \Rightarrow m_{obj}g > m_f g$$

where m_{obj} is the mass of the object and m_f is the mass of displaced fluid. Because density is defined as $\rho = M/V$ where M is the mass and V is the volume, we can rewrite the above expression as:

$$\rho_{obj}V_{obj} > \rho_f V_f \Rightarrow \rho_{obj} > \rho_f$$

where the volume of the object V_{obj} is equal to the volume of the displaced fluid V_f for a fully submerged object. Therefore, an object will sink in a fluid if its density is greater than the density of the surrounding environment. Conversely, an object will rise in a fluid if its density is less than the surrounding environment. Therefore, according to Archimedes' principle, vertical motion within the atmosphere is governed by the density of the surrounding atmosphere. We have already established that atmospheric density depends on atmospheric pressure. In the following sections, we will show that atmospheric density also depends on temperature and the amount of water vapor present. We begin by examining the concept of temperature.

1.2 Temperature

We often think of temperature in intuitive terms - as a measure of how "hot" or "cold" an object feels. However, to provide a scientific definition of temperature, we must first revisit the concept of thermal equilibrium from introductory physics.

1.2.1 Temperature and thermal equilibrium

Consider two objects, B and C, that are initially isolated from each other, with object B being "hotter" than object C, as shown in Fig. 1.4. When these objects are brought into thermal contact, energy is exchanged between them in the form of heat. Empirical observation shows that heat flows from the hotter object to the colder one until the net energy transfer stops. At this point, object B is said to be in **thermal equilibrium** with object C. To define this condition more precisely, we must introduce a third object A. If object A is in thermal equilibrium with object B, and also in thermal equilibrium with object C, then we can conclude that object B *must* be in thermal equilibrium with object C. This principle is known as the **0th law of thermodynamics**, which establishes that thermal equilibrium is characterized by a single measurable property - temperature. Therefore, as illustrated in Fig. 1.4, object A functions as a thermometer, providing a standardized definition of temperature.

To define temperature quantitatively based on the concept of thermal equilibrium, we need identify properties of matter that respond consistently and reliably to heating and cooling. These properties are known as **thermometric properties**. To create a standardized definition of temperature, one thermometric property from a specific substance must be chosen as a reference. Historically, the standard substance has been an ideal gas (which will be precisely defined in Ch. 1.3), and the chosen thermometric property is the pressure of the ideal gas at a fixed volume.

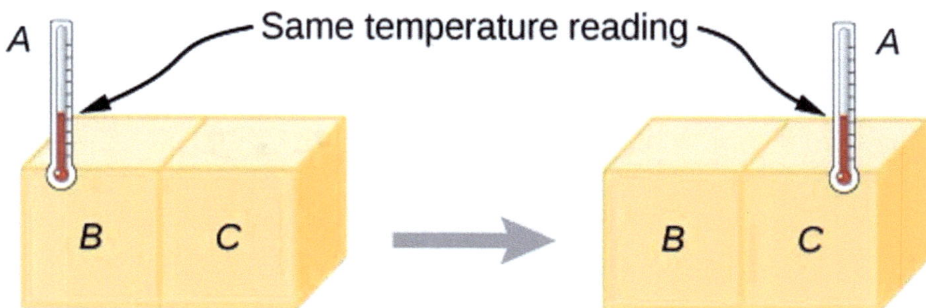

FIGURE 1.4 Schematic of the 0th law of thermodynamics

If thermometer *A* is in thermal equilibrium with object *B*, and object *B* is in thermal equilibrium with object *C*, then thermometer *A* is also in thermal equilibrium with object *C*. As a result, the reading on thermometer *A* remains unchanged when it is moved to make contact with object *C*.

Access for free at https://openstax.org/books/university-physics-volume-2/pages/1-introduction.

A thermometer constructed using this method is called a **constant volume gas thermometer**, and we will explore this procedure in the next section.

1.2.2 Temperature and temperature scales

Fig. 1.5 shows a schematic of a constant-volume gas thermometer, which features a U-shaped tube containing a mercury reservoir. When the system reaches equilibrium, the mercury level remains horizontally aligned, indicating that the pressure at point A (exerted by the gas in the bulb) is equal to the pressure at point B (due to the combined effect of atmospheric pressure and the mercury column). As the pressure in the bulb changes (whether from variations in temperature or atmospheric conditions), the pressure at point B will adjusts accordingly, resulting in a change in the mercury column height, denoted by h.

To function as a thermometer, the device must be calibrated using two reproducible physical phenomena that occur at well-defined temperatures. Traditionally, the constant-volume gas thermometer is calibrated using the freezing point and the boiling point of water. First, the gas bulb is placed in an ice-water mixture, and the mercury column height h is recorded after thermal equilibrium is reached. Next, the bulb is placed in boiling water, and the height h is measured again after equilibrium is established. These two height measurements correspond to distinct gas pressures, which allow the construction of a linear temperature scale. Historically, the Celsius scale was established by *defining* the

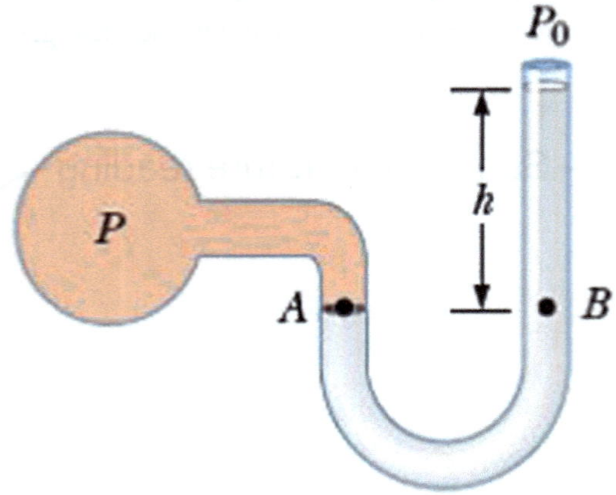

FIGURE 1.5 Schematic of an open-end manometer. Schematic of a constant volume gas thermometer known as an open-end manometer. The left side consists of a low-density gas in a bulb of fixed volume with pressure *P*. The U-shaped tube has one arm that is connected to the gas to be measured and another end open to the atmosphere with pressure P_0. This device contains mercury that can rise or fall based on the pressure of the gas.

Access for free at https://openstax.org/books/university-physics-volume-2/pages/1-introduction.

temperature at the freezing point of water as 0°C and by *defining* the temperature at the boiling point as 100°C. By plotting the measured pressures against these two temperatures, a straight line is obtained, serving as a calibration curve for determining unknown temperatures. To use the thermometer, the gas bulb is brought into thermal contact with the substance of interest, and the resulting change in gas pressure alters the height of the mercury column. This height indicates the pressure of the gas, which is then be used to determine the temperature from the calibration curve.

Now suppose temperatures are measured using various gas thermometers filled with different low-density gases at varying initial pressures. If we plot the pressure vs temperature curves for each thermometer, we obtain a set of of calibration curves, as illustrated in Fig. 1.6. Extrapolating these linear curves to zero pressure reveals that they all intersect at a fixed temperature of $T = -273.15°C$. Since pressure cannot be negative, this temperature is considered the lowest possible temperature attainable, known as **absolute zero**. This fundamental temperature forms the basis of the absolute temperature scale, commonly referred to as the **Kelvin scale**. The Kelvin scale sets −273.15°C as its zero point; therefore, the conversion factor between the Celsius scale and the Kelvin scale is:

$$T_C = T - 273.15 \tag{1.2}$$

where T_C is the temperature measured in Celsius and T is the temperature measured in Kelvin.

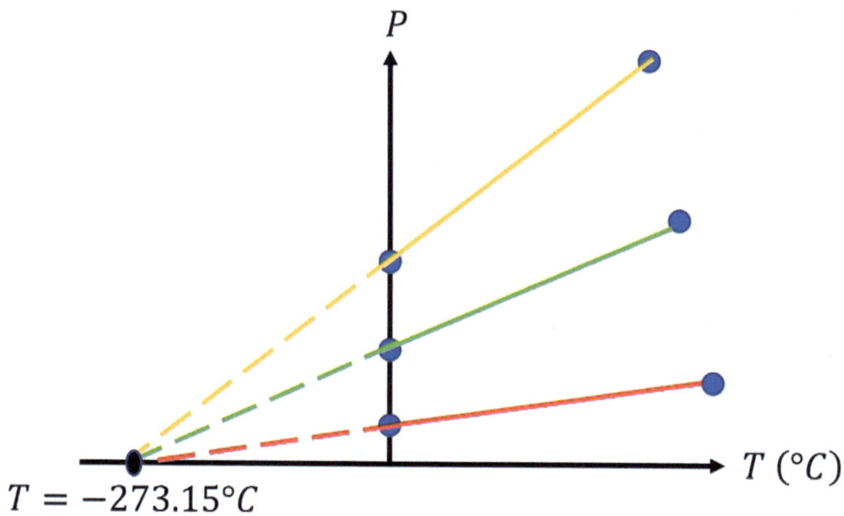

FIGURE 1.6 Establishing the Kelvin scale. Pressure vs temperature for three low-density gases. Note that for all gases, the pressure extrapolates to zero at the temperature −273.15°C.

No Permission Required

Because the Kelvin scale is the absolute temperature scale, *it should be used for all scientific computations involving temperature*. Moreover, since it is a linear scale, *any other linear temperature scale can be defined relative to the Kelvin or the Celsius scale*. Box 1.2 illustrates how this can be done.

The Kelvin scale is primarily used for scientific contexts, while the Fahrenheit and Celsius scales are more common in operational analysis and forecasting. Typically, upper-air weather charts display temperatures in degrees Celsius, whereas surface weather charts in the United States use degrees Fahrenheit. It is helpful to memorize key threshold temperature values in degree Celsius and the corresponding values in degree Fahrenheit. Table 1.1 provides a collection of common temperatures for both scales.

1.3 Properties of low-density gases

In our discussion on temperature and its scales, we defined temperature using the observed behavior of low-density gases. In this section, we will review the basic properties of low-density gases.

1.3.1 The ideal gas law

In introductory physics and chemistry, it is shown that the properties of low-density gases can be summarized by four experimental laws:

1. **Boyle's law**, which is named after British chemist Robert Boyle, states that for a fixed amount of gas held at a constant temperature, the pressure of a low-density gas is inversely proportional to its volume. Thus, as a gas is compressed, its pressure increases. Conversely, as the gas expands, its pressure decreases. Boyle's law can be written mathematically as $P \propto V^{-1}$.
2. **Charles's law**, which is named after French scientist Jacques Charles, states that for a fixed amount of gas held at a constant pressure, the volume of a low-density gas is directly proportional to its temperature. Hence, Charles's law explains how gases tend to expand when heated or contract when cooled. Charles's law can be written mathematically as $V \propto T$.
3. **Gay-Lussac's law**, named after French chemist Joseph Louis Gay-Lussac, states that for a fixed amount of gas held at a constant volume, the pressure of a low-density gas is directly proportional to its temperature.[2] Therefore, the pressure of a gas in a fixed container increases when heated and decreases when cooled. This property of low-density gases was used in constructing the constant

[2] It should be noted that the French chemist Guillaume Amontons was the first to empirically establish the relationship between the pressure and the temperature of a gas, whereas Joseph Gay-Lussac determined the relationship more precisely. For this reason, this law is also referred to as Amontons's law.

Box 1.2 Conversion formulas between the Fahrenheit, Celsius, and Kelvin scales

The Fahrenheit temperature scale is a linear temperature scale defined so that the freezing point of water is 32°F and the steam point of water is 212°F. Based on this information, we can derive the conversion formulas between the Fahrenheit, Celsius, and Kelvin scales.

Because the Celsius and the Fahrenheit scales are both linear, there should be a linear relationship between the two scales. We can write this linear scale as:

$$T_F - T_{0,F} = \alpha(T_C - T_{0,C})$$

where $T_{0,F} = 32°F$ is the freezing point of water in Fahrenheit, $T_{0,C} = 0°C$ is the freezing point of water in Celsius, and α is the slope of our scale. Substituting these values into our above expression gives:

$$T_F - 32 = \alpha T_C$$

To determine the slope of our scale, we use our two data points, which are the freezing point of water and the steam point of water. This gives

$$\alpha = \frac{T_{steam,F} - T_{freeze,F}}{T_{steam,C} - T_{freeze,C}} = \frac{212 - 32}{100 - 0} = \frac{9}{5}$$

Therefore, the linear scale, which converts from the Celsius scale to the Fahrenheit scale is given by

$$T_F - 32 = \frac{9}{5}T_C$$

$$\Rightarrow T_F = \frac{9}{5}T_C + 32 \tag{1.3}$$

Substituting Eq. (1.2) into Eq. (1.3) gives the conversion formula from the Kelvin scale to the Fahrenheit scale

$$T_F = \frac{9}{5}(T - 273.15) + 32$$

$$\Rightarrow T_F = \frac{9}{5}T - 459.67 \tag{1.4}$$

where T is the temperature measured in Kelvin. To obtain the Fahrenheit to Celsius relation, we can solve Eq. (1.3) for T_C to obtain

$$T_C = \frac{5}{9}(T_F - 32) \tag{1.5}$$

To obtain the Fahrenheit to Kelvin relation, we can solve Eq. (1.4) for T to obtain

$$T = \frac{5}{9}(T_F + 459.67) \tag{1.6}$$

Table 1.1 Common temperatures for meteorological forecasting

°C	−40	−30	−20	−10	−5	0	5	10	15	20	25	30	40
°F	−40	−22	−4	14	23	32	41	50	59	68	77	86	104

No Permission Required

volume-gas thermometer in Ch. 1.2. Gay-Lussac's law can be written mathematically as $P \propto T$.

4. **Avogadro's law**, named after Italian scientist Amedeo Avogadro, states that for a gas at fixed temperature and pressure, the volume of a low-density gas is directly proportional to the number of moles[3] of gas. This can be written mathematically as $V \propto n$.

Combining these four experimental laws leads to the relationship $PV \propto nT$ or $PV/nT =$ constant. This means that the thermal properties of all low-density gases can be uniquely defined based on their pressure, volume, temperature, and the number of moles of gas present. Modern experiments on the thermal properties of low-density gases have confirmed this relationship, and the constant of proportionality has been measured to be $R = 8.314\ J\ mol^{-1}\ K^{-1}$ (U.S. Standard Atmosphere, 1976). Because this proportionality constant is independent of the type of low-density gas used, R is referred to as the **universal gas constant**. Therefore, the thermal behavior of low-density gases can be summarized by a single equation:

$$PV = nRT \tag{1.7}$$

Eq. (1.7) is known as the **ideal gas law**. It is derived from experimental observations and is valid for low-density gases, in which the average distance between molecules is much greater than the size of the molecules themselves. For all common atmospheric gases near room temperature and atmospheric pressure, the average molecular spacing is roughly 10 times the size of a molecule. Therefore, the ideal gas law provides a sufficiently accurate description for our purposes.

For atmospheric science, it is useful to rewrite the ideal gas law in terms of the density ρ of the gas. We can rewrite Eq. (1.7) in terms of the pressure, which gives

$$P = \left(\frac{n}{V}\right)RT$$

Because $\rho = m/_V$ where m is the total mass of the gas and V is the total volume of the gas, we can write this expression as

$$P = \rho\left(\frac{n}{m}\right)RT = \rho\left(\frac{R}{M_{mol}}\right)T$$

where M_{mol} is known as the **molar mass** of the substance, which is defined as the mass per unit mole of the substance. Therefore, Eq. (1.7) can be written as

$$P = \rho R^* T, \quad R^* = \frac{R}{M_{mol}} \tag{1.8}$$

The **specific gas constant**, denoted as R^*, varies depending on the type of low-density gas. To calculate the specific gas constant, you need to know the molar mass

[3] The mole is the SI unit of measurement for the amount of substance. It is defined as particles, which may be atoms or molecules.

Box 1.3 Calculating the specific gas constant of water vapor

Let's calculate the specific gas constant for water vapor R_v. The molar mass of a gas can be determined from the atomic mass of each element used to construct water vapor. From the periodic table, we read that hydrogen has an atomic mass of 1.008 g/mol, and oxygen has an atomic mass of 15.999 g/mol. Therefore, the molar mass of water vapor H_2O is:

$$M_{mol} = \left(2 \times 1.008 \frac{\text{g}}{\text{mol}}\right) + 15.999 \frac{\text{g}}{\text{mol}} = 18.015 \frac{\text{g}}{\text{mol}}$$

Therefore, the specific gas constant of water vapor is:

$$R_v = \frac{R}{M_{mol}} = \frac{8.314 \frac{\text{J}}{\text{mol}\cdot\text{K}}}{18.015 \frac{\text{g}}{\text{mol}}} = \frac{8.314 \frac{\text{J}}{\text{mol}\cdot\text{K}}}{0.018015 \frac{\text{kg}}{\text{mol}}} = 461.5 \frac{\text{J}}{\text{kg}\cdot\text{K}}$$

of the gas, which can be obtained from the periodic table. Box 1.3 illustrates how to determine the specific gas constant for water vapor.

1.3.2 The kinetic theory of gases

The previous section discussed the properties of low-density gases based on experimental observations. However, these properties can also be understood more deeply by examining gases from a molecular perspective. The theory that provides a molecular model for low-density gases is called the **kinetic theory of gases**. Although a complete mathematical description of the kinetic theory of gases is outside the scope of this text, we will provide a qualitative overview of this theory.

Consider a collection of n moles of an ideal gas in a container of volume V. The ideal gas law applies to low-density gases, where the volume occupied by gas molecules is much smaller than the distance between them, as illustrated in Fig. 1.7. Therefore, for a low-density gas, we can safely assume that each molecule is sufficiently far from the others, allowing us to neglect any intermolecular attractive forces. This assumption has two important implications:

1. The only molecular interactions that need to considered are the collisions between gas molecules and the collisions of gas molecules with the container walls.
2. The total energy of the gas arises solely from the kinetic energy of its molecules.

The kinetic theory of gases also makes two additional assumptions about molecular motion:

1. Gas molecules are in continuous, random motion, meaning there is no preferred direction for their movement, as indicated by the velocity vectors in Fig. 1.7.
2. The only interactions among the molecules are due to elastic collisions, which conserve kinetic energy.

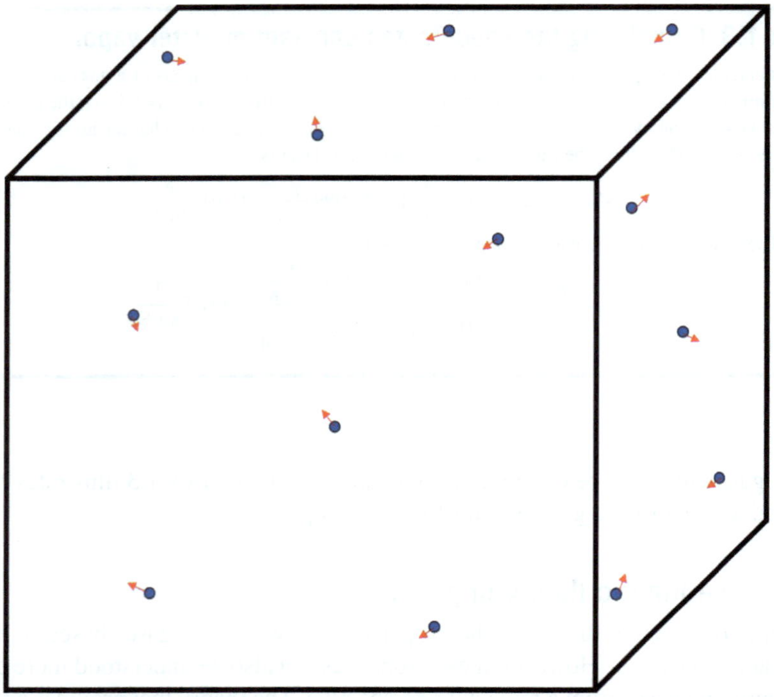

FIGURE 1.7 Molecular model of an ideal gas. A schematic of a low-density gas within a volume *V*. The red vector arrows depict the velocity of each molecule. Because the gas container is fixed in space, it is assumed that the gas molecules are in continuous, random motion.

No Permission Required

This theory provides a relationship between the pressure exerted by a gas and the molecular kinetic energy of its molecules. To understand this, let's consider a single molecule and then generalize to all molecules in the gas.

Suppose a molecule is located a distance *d* from the wall of its container. When this molecule strikes the wall, it undergoes an elastic collision, transferring kinetic energy and exerting a force on the wall. Faster-moving molecules will exert greater forces than slower ones, indicating that *the force on the wall depends on the molecule's kinetic energy.*

Extending this idea to all gas molecules in the container, the total number of collisions with the container walls results in a net force, which we observe as pressure. This pressure depends on the average molecular speed (since faster molecules would imply a greater force) and the density of the gas (since more molecules result in more collisions).

In summary, *the pressure exerted by the gas is directly related to its molecular kinetic energy and density.* The pressure arises from the momentum transferred during molecular collisions with the walls of the container.

The kinetic theory of gases also clarifies how temperature relates to molecular kinetic energy. Suppose we heat the gas in a fixed container, as shown in Fig. 1.7. Because heat is a form of energy transfer, adding heat increases the molecular kinetic energy. Thus, an increase in the temperature of gas corresponds to an increase in the average kinetic energy of the gas molecules. Therefore, for a low-density gas, *temperature can be interpreted as a direct measure of the molecular kinetic energy of a gas.*

To summarize, the kinetic theory of gases provides a simple model of the thermal behavior of low-density gases. It describes an ideal gas as a collection of molecules in constant, random motion. Measurable quantities like temperature and pressure are directly linked to the molecular motion of the ideal gas.

1.4 Properties of moist air

In the analysis of an ideal gas presented in the previous section, we assumed that the gas consisted of a single species. However, Earth's atmosphere is composed of multiple types of gases. Table 1.2 lists the primary constituents of the atmosphere near Earth's surface. The concentration of permanent gases—nitrogen, oxygen, and argon—remains approximately constant over time and across Earth's atmosphere. In contrast, the concentration of variable gases—such as water vapor, carbon dioxide, and methane—varies over time and across Earth's atmosphere. Fortunately, the ideal gas law can be generalized to a homogeneous mixture of low-density gases. In this section, we will extend the ideal gas model to include a mixture of gases, which will allow the examination of the thermodynamic properties of moist air.

1.4.1 Dalton's law of partial pressures

The **partial pressure** of a gas in a mixture is the pressure it would exert if it *alone* occupied the entire volume occupied by the mixture. The total pressure exerted by

Table 1.2 Mean composition of the atmosphere

Permanent Gases			Variable Gases		
Gas	**Chemical symbol**	**Percent (by volume) of dry air**	**Gas**	**Symbol**	**Percent by volume**
Nitrogen	N_2	78.08	Water vapor	H_2O	0–4
Oxygen	O_2	20.95	Carbon dioxide	CO_2	0.038
Argon	Ar	0.930	Methane	CH_4	0.00017

No Permission Required

the gas mixture is determined by Dalton's law of partial pressures, which states the following:

The total pressure exerted by a homogeneous mixture of gases is equal to the sum of the partial pressures of its individual components.

To apply Dalton's law to a mixture of ideal gases, consider a mixture of N ideal gases, which occupies a volume V at temperature T. The partial pressure of each gas constituent satisfies Eq. (1.7):

$$P_j = \frac{n_j RT}{V}$$

where n_j is the number of moles of the jth gas constituent and P_j is the partial pressure of the jth gas constituent. By Dalton's law, the total pressure exerted by the mixture is:

$$P_{total} = \sum_{j=1}^{N} P_j = \sum_{j=1}^{N} n_j \frac{RT}{V} = \frac{RT}{V} \sum_{j=1}^{N} n_j = \frac{nRT}{V}$$

We can use Dalton's law of partial pressures to determine the specific gas constant for dry air, as shown in Box 1.4.

Box 1.4 Calculating the specific gas constant for dry air

To determine the specific gas constant for dry air, we first will need to determine the molar mass of dry air. The molar mass of any gas mixture (including dry air) can be written as

$$M_d = \sum_{i=1}^{N} \chi_i M_i$$

where M_i is the molar mass of the *ith* gas constituent and χ_i is the mole fraction of the *ith* gas constituent, which is defined as

$$\chi_i = n_i / n$$

where n_i is the number of moles of the *ith* gas constituent and n is the total number of moles in the gas mixture. Because the temperature and pressure of the gas are fixed, Avogadro's law indicates that the mole fraction is equal to the volume fraction for each gas constituent. Based on Table 1.2, the dominant contributions to dry air are nitrogen, oxygen, and argon. Using the periodic table, the molar mass of each of these gas constituents is:

$$M_{N_2} = \left(2 \times 14.007 \frac{\text{g}}{\text{mol}}\right) = 28.014 \frac{\text{g}}{\text{mol}} \quad M_{O_2} = \left(2 \times 15.999 \frac{\text{g}}{\text{mol}}\right) = 31.998 \frac{\text{g}}{\text{mol}} \quad M_{Ar} = \left(1 \times 39.948 \frac{\text{g}}{\text{mol}}\right) = 39.948 \frac{\text{g}}{\text{mol}}$$

Based on Table 1.2, the molar mass of dry air is given by

$$M_d = \sum_{i=1}^{N} \chi_i M_i = \chi_{N_2} M_{N_2} + \chi_{O_2} M_{O_2} + \chi_{Ar} M_{Ar}$$

$$= (0.7808)\left(28.014 \frac{\text{g}}{\text{mol}}\right) + (0.2095)\left(31.998 \frac{\text{g}}{\text{mol}}\right) + (0.00930)\left(39.948 \frac{\text{g}}{\text{mol}}\right) = 28.95 \frac{\text{g}}{\text{mol}}$$

Therefore, the specific gas constant for dry air is given by

$$R_d = \frac{R}{M_d} = \frac{8.314\,\text{Jmol}^{-1}\text{K}^{-1}}{28.95\,\text{gmol}^{-1}} = \frac{8.314\,\text{Jmol}^{-1}\text{K}^{-1}}{0.02895\,\text{kgmol}^{-1}} = 287.2\,\text{Jkg}^{-1}\text{K}^{-1}$$

1.4.2 Virtual temperature and buoyancy

Dalton's law of partial pressures also helps us understand how water vapor affects the thermal properties of air. This can be demonstrated by deriving an expression for the ideal gas law as it applies to moist air.

Consider a volume $V$of moist air at temperature $T$and total pressure P, containing a mass m_d of dry air and mass m_v of water vapor. According to Dalton's law, the total pressure of the moist air is the sum of the partial pressure from water vapor (known as **vapor pressure**) and the partial pressure from dry air. Therefore, the total pressure of moist air can be written as

$$P_{total} = e + P_d$$

where e is the vapor pressure and P_d is the partial pressure of dry air. Because the temperature of the gas is fixed, Dalton's law also implies that the total density can be expressed as

$$\rho = \rho_d + \rho_v$$

where ρ_d is the density of dry air and ρ_v is the density of water vapor. Applying Eq. (1.8) to the total density gives:

$$\rho = \rho_d + \rho_v = \frac{P_d}{R_d T} + \frac{e}{R_v T}$$

where R_d and R_v are the specific gas constant for dry air and for water vapor, respectively. Because $P_{total} = e + P_d$, the total density can be written as:

$$\rho = \frac{P_{total} - e}{R_d T} + \frac{e}{R_v T}$$

$$\Rightarrow \rho = \frac{P_{total}}{R_d T}\left[1 - \frac{e}{P} + \frac{e}{P}\left(\frac{R_d}{R_v}\right)\right]$$

$$\Rightarrow \rho = \frac{P_{total}}{R_d T}\left[1 - \frac{e}{P}\left(1 - \frac{R_d}{R_v}\right)\right]$$

Therefore, the total pressure exerted by moist air can be expressed as

$$P_{total} = \rho R_d T_v, \ T_v = \frac{T}{1 - \frac{e}{P_{total}}\left(1 - \frac{R_d}{R_v}\right)} \tag{1.9}$$

In atmospheric science, T_v refers to the **virtual temperature**. It is defined as the temperature that dry air would need to have in order to match the pressure and density of a given sample of moist air. The difference between the virtual temperature and the actual ambient temperature, known as the **virtual temperature correction**, is shown in Fig. 1.8. It's important to note that the virtual temperature correction is always positive, which implies that $T \le T_v$. Furthermore, this correction is approximately a linear function of e/P_{total}.

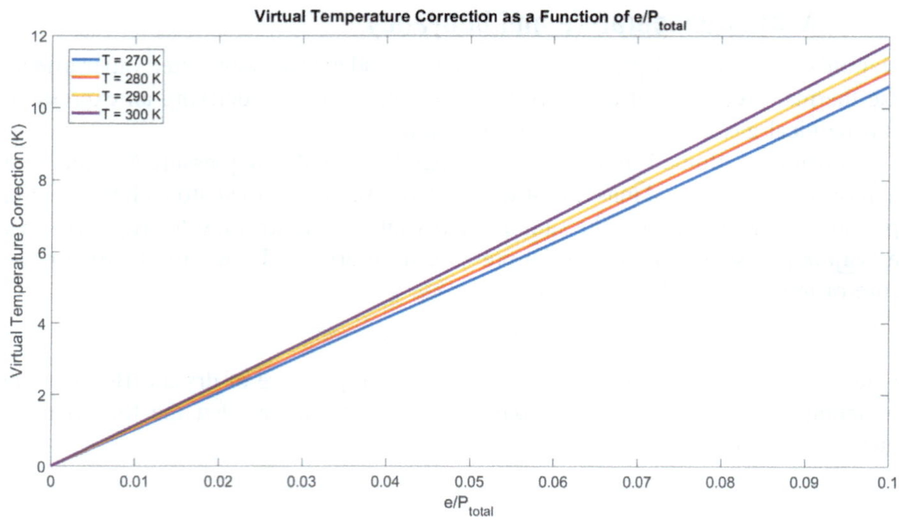

FIGURE 1.8 Virtual temperature correction vs vapor pressure. The graph shows the relationship between virtual temperature correction vs *e/P*$_{total}$ for *T* = 270 K, *T* = 280 K, *T* = 290 K, and *T* = 300 K.

No Permission Required

Furthermore, the virtual temperature correction increases as the ambient temperature rises. This means that the influence of water vapor on the thermal properties of moist air becomes more significant at higher temperatures. According to Eq. (1.9), density (ρ) is inversely proportional to virtual temperature (T_v), so as the virtual temperature increases, the density of moist air *decreases*. Therefore, *adding water vapor to air reduces its overall density.* Box 1.5 illustrates how water vapor affects the thermal properties of the air.

Another variable that can be used to assess the amount of moisture in the air is the **mixing ratio**, which is defined as:

$$r = \frac{m_v}{m_d}$$

where m_v is the mass of water vapor in the air sample and m_d is the mass of dry air in the sample. Although the mixing ratio is a dimensionless quantity, it is commonly expressed in grams of water vapor per kilogram of dry air, since water vapor comprises less than 5% of the total mass of air in Earth's atmosphere. Observations indicate that the mixing ratio ranges from 2 g/kg to approximately 15 g/kg in the mid-latitudes near the surface. Moreover, the mixing ratio decreases approximately exponentially with increasing height (Shi and Bates, 2005).

We can relate the mixing ratio to the vapor pressure by using Dalton's law of partial pressures. Note that

Box 1.5 The effects of water vapor on air density

To understand the effects of water vapor on air density, let's consider a sample of moist air that is composed of dry air and water vapor at a temperature of 20.0°C. Let's suppose that the density of dry air alone is 1.188 kg/m^3 and the density of water vapor alone is 4.445×10^{-3} kg/m^3.

To calculate the virtual temperature of the moist air sample using Eq. (1.9), we need to calculate the total pressure and the total density to determine the virtual temperature. From Dalton's law of partial pressures, the total density can be written as:

$$\rho = \rho_d + \rho_v = 1.188\,\frac{\text{kg}}{\text{m}^3} + 4.445 \times 10^{-3}\,\frac{\text{kg}}{\text{m}^3} = 1.192\,\frac{\text{kg}}{\text{m}^3}$$

The total pressure exerted by the moist air is equal to the sum of the vapor pressure and the partial pressure of dry air according to Dalton's law of partial pressures. The vapor pressure is given by

$$e = \rho_v R_v T = \left(4.445 \times 10^{-3}\,\frac{\text{kg}}{\text{m}^3}\right)\left(461.5\,\frac{\text{J}}{\text{kg·K}}\right)(293.15\,\text{K}) = 601.4\,\text{Pa}$$

where the temperature was converted from Celsius to Kelvin using Eq. (1.2). The partial pressure of dry air is given by

$$P_d = \rho_d R_d T = \left(1.188\,\frac{\text{kg}}{\text{m}^3}\right)\left(287\,\frac{\text{J}}{\text{kg·K}}\right)(293.15\,\text{K}) = 99951\ \text{Pa}$$

Therefore, the total pressure exerted by the moist air is

$$P_{total} = P_d + e = 99951\,\text{Pa} + 601.4\,\text{Pa} = 100552\,\text{Pa} = 100.552\,\text{kPa}$$

Using Eq. (1.9), the virtual temperature is given by

$$T_v = \frac{P_{total}}{\rho R_d} = \frac{100552\,\text{Pa}}{\left(1.192\,\frac{\text{kg}}{\text{m}^3}\right)\left(287\,\frac{\text{J}}{\text{kg·K}}\right)} = 293.9\,\text{K} = 20.8°\text{C}$$

where the temperature was converted from Kelvin to Celsius using Eq. (1.2).

This result indicates that a sample of dry air would need to have a temperature of 20.8°C to have the same density as this sample of moist air at a temperature of 20.0°C. This illustrates that because moist air is less dense than dry air at the same temperature and pressure, the virtual temperature is always greater than the actual temperature.

This example emphasizes an important point regarding air masses in the atmosphere: the temperature and the moisture content of air affect the density. In future chapters, we will see that adding moisture to the air increases the buoyancy of air masses, which will help to promote rising motion within the atmosphere.

$$e = \frac{e}{P_{total}} P_{total} = \left(\frac{e}{e + P_d}\right) P_{total} = \left(\frac{\rho_v R_v T}{\rho_v R_v T + \rho_d R_d T}\right) P_{total} \Rightarrow e = \left(\frac{\rho_v R_v}{\rho_v R_v + \rho_d R_d}\right) P_{total}$$

where $P_{total} = e + P_d$ from Dalton's law and the ideal gas law has been used. By using the definition of density, note that $\rho_d = m_d/V$ and $\rho_v = m_v/V$. Therefore, we can rewrite the above expression as:

$$e = \left(\frac{m_v R_v V}{m_v R_v V + m_d R_d V}\right) P_{total} = \left(\frac{r R_v}{r R_v + R_d}\right) P_{total} = \left(\frac{r}{r + \dfrac{R_d}{R_v}}\right) P_{total}$$

where the definition of the mixing ratio has been used. By substituting this expression into Eq. (1.9), it can be shown that the virtual temperature can be written in terms of the mixing ratio:

$$T_v \simeq (1 + 0.608r)T \tag{1.10}$$

where the mixing ratio is expressed in kg/kg (Wallace and Hobbs, 2006). Eq. (1.10) is a useful expression for obtaining the virtual temperature from the ambient temperature and the mixing ratio. According to Eq. (1.10), the virtual temperature correction is a linear function of the mixing ratio and the temperature, as implied in Fig. 1.8.

To summarize, Eqs. (1.9) and (1.10) provide an important foundation for understanding buoyancy in Earth's atmosphere. In Ch. 1.1, we demonstrated that vertical motion within the atmosphere is governed by the density of the surrounding air. According to Eqs. (1.9) and (1.10), the atmospheric density is influenced by ambient pressure, the temperature, and the amount of water vapor present.

In this section, we have shown that higher mixing ratio leads to higher virtual temperature, which implies increased buoyancy. Furthermore, because the virtual temperature correction is approximately a linear function of actual temperature, the effect of water vapor on the density of moist air becomes more significant at higher temperatures. In Ch. 8, we will explain how the thermal properties of moist air affect atmospheric stability within Earth's atmosphere.

1.5 Atmospheric motion

Up to this point, we have shown that temperature, pressure, and moisture content are essential for describing the state of the atmosphere. However, because the atmosphere is a fluid system, atmospheric motion can alter these conditions. To fully characterize the state of the atmosphere, we must also define the variables that describe atmospheric motion.

1.5.1 Vectors and vector components

In introductory physics, physical variables are typically classified into two categories: scalar and vector quantities. **Scalar quantities** have magnitude but no directional information, such as distance, time, mass, and energy. In contrast, **vector quantities** have both magnitude and direction, which include velocity and acceleration.

So far, the variables introduced in this chapter, such as pressure, temperature, density, and moisture, are all scalar quantities. However, to accurately describe atmospheric motion, both wind speed and wind direction must be specified, indicating

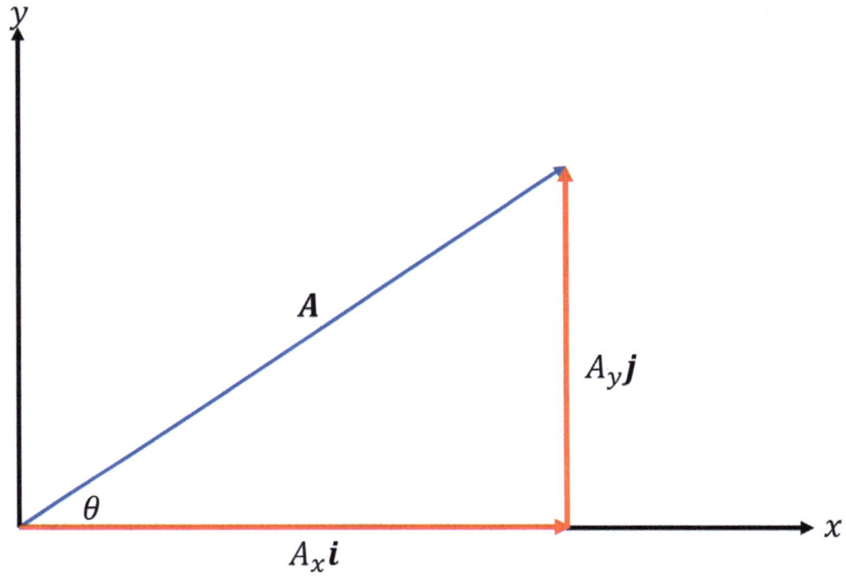

FIGURE 1.9 Schematic of a vector in unit-vector notation. Schematic of a vector *A* with vector components A_x and A_y. The unit-vector *i* corresponds to the positive *x*-direction, and the unit vector *j* corresponds to the positive *y*-direction. Using unit-vector notation, we can express this vector as: $\boldsymbol{A} = A_x\boldsymbol{i} + A_y\boldsymbol{j}$.

No Permission Required

that wind is a vector quantity. Therefore, before providing atmospheric motion in detail, it is important to review the mathematical properties of vectors as covered in introductory physics.

To perform mathematical calculations with vectors, we must understand the concept of vector components. A vector component represents the projection of a vector along a specific axis. Consider a vector **A** lying in the *xy*-plane and forming an angle θ with the positive *x*-axis, as shown in Fig. 1.9.[4] The component A_x represents the projection of **A** along the *x*-axis, and the vector component A_y represents the projection of **A** along the *y*-axis.

These components can be either positive or negative: A_x is positive if the projection of **A** lies on the positive *x*-axis, and A_y is positive if the projection lies along the positive *y*-axis.

The vector components form two sides of a right triangle, with a hypotenuse of length A corresponding to the magnitude of **A**. Based on the definitions of sine and cosine, we can express the components as follows:

[4] In this text, vector quantities will be denoted in bold type.

$$A_x = A\cos\theta, A_y = A\sin\theta \quad (1.11)$$

This formulation allows us to calculate the components of any vector given its magnitude and the angle with respect to the positive x-axis. Furthermore, the signs of the components A_x and A_y depend on the quadrant in which the angle θ lies. Conversely, the magnitude and direction of $\boldsymbol{A}$ can be determined from its components using the relations:

$$A = \sqrt{A_x^2 + A_y^2}, \theta = \tan^{-1}\left(\frac{A_y}{A_x}\right) \quad (1.12)$$

Thus, a vector can be fully specified either by its components or by its magnitude and direction.

To express a vector in terms of its components, we use unit vectors. A **unit vector** has a magnitude of exactly 1 and is used to indicate a specific direction in space. Hence, unit vectors provide a convenient way to describe direction in space. In this text, we use $\boldsymbol{i}, \boldsymbol{j}$, and $\boldsymbol{k}$ to represent unit vectors pointing in the positive x, y, and z directions, respectively. Using unit vector notation, we can express vector $\boldsymbol{A}$ as follows:

$$\mathrm{A} = A_x\mathrm{i} + A_y\mathrm{j} + A_z\mathrm{k}$$

With these definitions of vector components and unit vectors, we can now add or subtract vectors. For example, to perform the vector operation $\boldsymbol{R} = \boldsymbol{A} \pm \boldsymbol{B}$, we simply add or subtract the components.

$$\mathrm{R} = \mathrm{A} \pm \mathrm{B} = \left(A_x\mathrm{i} + A_y\mathrm{j} + A_z\mathrm{k}\right) \pm \left(B_x\mathrm{i} + B_y\mathrm{j} + B_z\mathrm{k}\right) = (A_x \pm B_x)\mathrm{i} + \left(A_y \pm B_y\right)\mathrm{j} + (A_z \pm B_z)\mathrm{k}$$

This indicates that the components of the vector $\boldsymbol{R}$ are

$$R_x = A_x \pm B_x, R_y = A_y \pm B_y, R_z = A_z \pm B_z$$

Following the same reasoning, the scalar multiplication of a vector can also be done in component form:

$$\boldsymbol{R} = a\boldsymbol{A} = a\left(A_x\boldsymbol{i} + A_y\boldsymbol{j} + A_z\boldsymbol{k}\right) = aA_x\boldsymbol{i} + aA_y\boldsymbol{j} + aA_z\boldsymbol{k}$$

Box 1.6 demonstrates how to perform vector addition and subtraction in unit vector notation.

1.5.2 Description of the wind field

The motion of a given object can be described using three variables—position, velocity, and acceleration. The position of any object is given by its coordinates (x, y, z), which can be represented by a position vector:

$$\boldsymbol{r} = x\boldsymbol{i} + y\boldsymbol{j} + z\boldsymbol{k}$$

In synoptic meteorology, motion in the $\boldsymbol{i}$ direction corresponds to the zonal direction (east-west), motion in the $\boldsymbol{j}$ direction corresponds to the meridional direction

Box 1.6 Performing vector addition and subtraction

Suppose that $\boldsymbol{A}$ has a magnitude of 20.0 with a direction 45.0° and $\boldsymbol{B}$ has a magnitude of 25.0 with a direction of 300.0°. Assume that the direction of each vector is measured from the positive x-axis.

First, let's calculate the vector $\mathbf{C} = \mathbf{A} - \mathbf{B}$ and express the answer in unit-vector notation. First, we find the components by using Eq. (1.11):

$$A_x = A\cos\theta = 20.0\cos(45.0°) = 14.1, A_y = A\sin\theta = 20.0\sin(45.0°) = 14.1$$
$$B_x = B\cos\theta = 25.0\cos(300.0°) = 12.5, B_y = B\sin\theta = 25.0\sin(300.0°) = -21.7$$

Now, we subtract the vectors in unit-vector notation.

$$\boldsymbol{C} = \boldsymbol{A} - \boldsymbol{B} = (A_x - B_x)\boldsymbol{i} + (A_y - B_y)\boldsymbol{j} \Rightarrow \boldsymbol{C} = (14.1 - 12.5)\boldsymbol{i} + (14.1 - (-21.7))\boldsymbol{j} = 1.6\boldsymbol{i} + 35.8\boldsymbol{j}$$

Second, let's calculate the vector $\boldsymbol{D} = \boldsymbol{A} + 2\boldsymbol{B}$ and express the answer in terms of the magnitude and direction. Writing $\boldsymbol{D}$ in unit-vector notation gives:

$$\boldsymbol{D} = \boldsymbol{A} + 2\boldsymbol{B} = (A_x + 2B_x)\boldsymbol{i} + (A_y + 2B_y)\boldsymbol{j} \Rightarrow \boldsymbol{D} = [14.1 + 2(12.5)]\boldsymbol{i} + [14.1 + 2(-21.7)]\boldsymbol{j} = 39.1\boldsymbol{i} - 29.3\boldsymbol{j}$$

Based on the above expression, we note that $D_x = 39.1$ and $D_y = -29.3$. We can now obtain the magnitude and the direction of $\boldsymbol{D}$ using Eq. (1.12)

$$D = \sqrt{D_x^2 + D_y^2} = \sqrt{(39.1)^2 + (-29.3)^2} = 48.9$$
$$\theta = \tan^{-1}\left(\frac{D_y}{D_x}\right) = \tan^{-1}\left(\frac{-29.3}{39.1}\right) = -36.8°$$

Since $D_y < 0$ and $D_x > 0$, then we expect that $\boldsymbol{D}$ will point in the fourth quadrant. Therefore, the direction of $\boldsymbol{D}$ relative to the positive x-axis will be $-36.8° + 360° = 323°$.

(north-south), and motion in the $\boldsymbol{k}$ direction corresponds to the vertical direction. In this coordinate system, $\boldsymbol{i}$ is positive toward the east, $\boldsymbol{j}$ is positive toward the north, and $\boldsymbol{k}$ is positive upward.

The velocity vector is defined as the rate of change of the position vector. There are two common forms of velocity: average velocity, which describe the change in position over a finite time interval, and the instantaneous velocity, which describes the rate of change at a specific moment. These two vectors are defined as:

$$\boldsymbol{V}_{avg} = \frac{\Delta \boldsymbol{r}}{\Delta t}, \boldsymbol{V} = \frac{d\boldsymbol{r}}{dt}$$

respectively. In unit-vector notation, the instantaneous velocity vector can be written as:

$$\boldsymbol{V} = u\boldsymbol{i} + v\boldsymbol{j} + w\boldsymbol{k}$$

Here, u is the zonal (east-west) velocity component, v is the meridional (north-south) velocity component, and w is the vertical velocity component. These are often referred to as the **zonal wind**, **meridional wind**, and **vertical wind**, respectively. A positive zonal wind ($u > 0$) indicates wind blowing from west to east, a positive meridional wind ($v > 0$) indicates wind blowing from south to north, and a positive vertical wind ($w > 0$) indicates rising motion.

Atmospheric wind is typically reported in terms of wind speed and direction. In vector terms, wind speed is the magnitude of the instantaneous velocity vector. Because weather charts display motion in the horizontal *xy*-plane, we focus primarily on horizontal wind. As a vector, horizontal wind can be expressed in unit vector notation to capture both magnitude and direction. Eq. (1.13) expresses the horizontal wind speed and direction in terms of their components:

$$V_H = \sqrt{u^2 + v^2}, \quad \theta = \tan^{-1}\left(\frac{v}{u}\right) \tag{1.13}$$

Box 1.7 provides examples of how to express horizontal wind in unit-vector notation as well as in terms of magnitude and direction.

Acceleration is another key concept for understanding motion. It describes how quickly the velocity vector of an object changes. As with velocity, there are two types of acceleration vectors: the average acceleration vector over a time interval and instantaneous acceleration. These are defined as:

$$\boldsymbol{a}_{avg} = \frac{\Delta \mathbf{v}}{\Delta t}, \boldsymbol{a} = \frac{dv}{dt}$$

Wind can accelerate in two fundamental ways. First, if the air speeds up or slows down, the magnitude of the instantaneous velocity vector (i.e. the wind speed) changes, which is one form of acceleration. Second, when the air changes direction, the direction of the instantaneous velocity vector also changes, which also constitutes acceleration.

Box 1.7 Expressing the horizontal wind in vector notation

Typically, the wind direction is expressed by the direction *from* which it comes. Let's suppose that the components of the horizontal wind are $u = 3.00$ m/s and $v = 4.00$ m/s. To determine the direction from which the wind is blowing, we first determine the direction of the horizontal wind using Eq. (1.13):

$$\theta = \tan^{-1}\left(\frac{v}{u}\right) = \tan^{-1}\left(\frac{4.00\,\text{m/s}}{3.00\,\text{m/s}}\right) = 53.1^\circ$$

This states that the wind is blowing in the direction given by 53.1° (which is to the northeast). This implies that the wind is blowing from 180° + 53.1° = 233° (which is from the southwest).

We can also determine the components of the horizontal wind if we know the horizontal wind speed and the direction of motion. Suppose that we have a 5.00 m/s wind blowing from the southwest. To determine the horizontal wind vector in unit-vector notation, we first recognize that a wind blowing *from* the southwest is blowing *toward* the northeast, which corresponds to $\theta = 45.0^\circ$ using the standard Cartesian system. Using Eq. (1.11), the velocity components are given by

$$u = V_H \cos\theta = \left(5.00\,\frac{\text{m}}{\text{s}}\right)\cos(45.0^\circ) = 3.54\,\frac{\text{m}}{\text{s}}$$

$$v = V_H \sin\theta = \left(5.00\,\frac{\text{m}}{\text{s}}\right)\sin(45.0^\circ) = 3.54\,\frac{\text{m}}{\text{s}}$$

In both cases, wind undergoes acceleration due to by unbalanced forces acting on it, as described by Newton's Laws of Motion. We will explore the implications of this further in Ch. 5–7.

Questions

1. Use the concept of stress and strain to explain how vertical wind shear can be present in the atmosphere.
2. Use the kinetic theory of gases to provide a molecular interpretation of Boyle's law, Charles's law, Gay-Lussac's law, and Avogadro's law.
3. On a certain day, the temperature in Buffalo, NY, is 40.0°F, whereas the temperature in Miami, FL, is 80.0°F. Does this mean that the temperature in Miami is "twice as hot" as the temperature in Buffalo? Defend your answer.
4. Explain why $R_v > R_d$ must imply that water vapor is less dense than dry air at the same temperature and at the same pressure.
5. The Rankine temperature scale (abbreviated °R) uses the same size degrees as Fahrenheit, but it is measured up from absolute zero like Kelvin.
   a. Determine the conversion formula between Rankine and Fahrenheit. Use your formula to calculate 80.0°F on the Rankine scale.
   b. Determine the conversion formula between Rankine and Kelvin. Use your formula to calculate 300.0 K on the Rankine scale.
6. Determine whether the following statements are true or false. Defend your answer in three sentences or less:
   a. It is possible for the wind speed to increase while having a negative instantaneous acceleration.
   b. It is not possible for the wind to have zero instantaneous velocity and negative instantaneous acceleration at any point in its motion.
   c. If the average wind speed is the same as the instantaneous wind velocity at all times within a time interval, then the wind must have zero acceleration.
7. Consider a column of air that is above a horizontal area of 1.00 m^2. If the bottom of the air column has a pressure of 1000 kPa and the top of the air column has a pressure of 500 kPa, calculate the mass of air within the air column.
8. Consider an air mass with initial temperature of 20.0°C and a density of 1.15 kg/m^3.
   a. Calculate the new density of the air if the temperature of the air mass increases to 30.0°C without any changes in pressure.
   b. Determine the new density of the air if the pressure of the air mass changes to 20.0 kPa while the temperature of the air mass remains at 30.0°C.
9. Answer the following questions regarding virtual temperature:
   a. Calculate the virtual temperature correction for moist air at 25.0°C that has a mixing ratio of 15 g/kg.

b. Calculate the vapor pressure and the virtual temperature if an air mass has a total pressure of 980 hPa, a temperature of 13.5°C and a mixing ratio of 2.30 g/kg.

10. Vector $\boldsymbol{A}$ has a magnitude of 10.0 with a direction of 25.0°. Vector $\boldsymbol{B}$ has a magnitude of 15.0 with a direction of 160.0°. If the direction of each vector is measured from the positive x-axis, then calculate the vector $\boldsymbol{C} = \boldsymbol{B} + 2\boldsymbol{A}$ and express your answer in unit-vector notation.

11. Vector $\boldsymbol{A}$ has a magnitude of 20.0 with a direction of 45.0°. Vector $\boldsymbol{B}$ has a magnitude of 25.0 with a direction of 300.0°. If the direction of each vector is measured from the positive x-axis, then calculate the magnitude and the direction of the vector $\boldsymbol{D} = \boldsymbol{B}/2 - 4\boldsymbol{A}$.

12. Suppose an air mass moves with a position vector given by $\boldsymbol{r} = \left[(10.0\,\text{m}) + \left(2.50\frac{\text{m}}{\text{s}}\right)t + \left(-3.50\frac{\text{m}}{\text{s}^2}\right)t^2\right]\boldsymbol{k}$

a. Calculate the initial velocity and the initial acceleration of the air mass. Express your answer in unit-vector notation.

b. At what time will the air mass have zero instantaneous velocity?

c. In words, describe the motion of the air mass over the first 5 seconds of its motion.

13. Suppose an air mass moves with a position vector given by $\boldsymbol{r} = \left[\left(-2.50\frac{\text{m}}{\text{s}^3}\right)t^3 + \left(3.50\frac{\text{m}}{\text{s}}\right)t\right]\boldsymbol{i} + \left[\left(5.00\frac{\text{m}}{\text{s}^2}\right)t^2 - \left(3.50\frac{\text{m}}{\text{s}}\right)t + (5.00\,\text{m})\right]\boldsymbol{j}$

a. Calculate the initial velocity of the air mass. Express your answer in terms of magnitude and in direction.

b. Calculate the initial acceleration vector of the air mass. Express your answer in unit-vector notation.

c. At what time will the zonal wind be zero? At what time will the meridional wind be zero?

d. In words, describe the motion of the air mass over the first 10 seconds of its motion.

14. As discussed in Ch. 1.1, the density of Earth's atmosphere is not uniform with height. Let's demonstrate that this must be the case:

For the sake of argument, let's suppose that the density of Earth's atmosphere was constant at 1.30 kg/m^3. Assuming that the pressure at the top of Earth's atmosphere is 0 kPa and that the pressure at Earth's surface is 101.325 kPa, determine the height of the atmosphere under these conditions.

What would be the height of the atmosphere if the air density decreased linearly to zero with height? Assume that at sea level the air pressure is 101.325 kPa and the air density at sea level is 1.30 kg/m^3.

15. Suppose that a volume of moist air contains 1.96% water vapor at 30.0°C. If the total pressure of the moist air is 1.014×10^5 Pa, determine the density of the moist air sample.

References

Shi, L., Bates, J., 2005. 5th AMS Annual meeting San Diego, CA American Meteorological Society Boston, MA Unpublished Content Distribution and Variability of Atmospheric Temperature and Water Vapor Derived from HIRS Measurement.

United States Committee on Extension to the Standard Atmosphere 1976 National Oceanic and Atmospheric Administration. National Aeronautics and Space Administration, United States Air Force, Washington D.C. Unpublished content U.S. Standard Atmosphere, 1976.

Wallace, J., Hobbs, P., 2006. Atmospheric Science: An Introductory Survey, second ed. Academic Press. https://doi.org/10.1016/C2009-0-00034-8.

CHAPTER

Weather instrumentation and observations

2

Chapter outline

The previous chapter explained that pressure, temperature, water vapor, and the wind are the four essential variables needed to define the state of the atmosphere. To understand how the atmosphere evolves over time, it is important to continuously monitor these variables. In the United States, weather observations at the surface are recorded at least once every hour using a version of the **Automated Surface Observation System (ASOS)**, while observations above the surface are collected at least twice daily using weather balloons (also known as balloons). In this chapter, we will examine how these atmospheric variables are measured by providing an overview of the instruments commonly used to collect meteorological data, with a focus on the equipment commonly used by the U.S. National Weather Service (NWS).

An Introduction to Synoptic-Dynamic Meteorology. https://doi.org/10.1016/B978-0-443-24858-0.00002-2

2.1 Temperature measurements

As mentioned in Ch. 1, thermometers are designed based on properties of matter that consistently and reliably respond to heating and cooling (known as thermometric properties). Many thermometric properties can be used to construct thermometers; however, in this section, we will focus on the two most common ones used for temperature measurements from surface observation systems.

2.1.1 Thermal expansion

The most common thermometric property used in thermometer design is **thermal expansion** - the tendency of matter to change its shape and/or size in response to temperature changes. This phenomenon is rooted in the molecular structure of matter. We can think of matter as a collection of atoms held together by molecular bonds, which behave like simple harmonic oscillators, similar to masses connected by springs in introductory physics. According to the molecular model of matter, the kinetic energy of molecules is associated with the vibration of atoms. As temperature increases, the atoms vibrate with greater amplitude, resulting in a net expansion of the material.

To mathematically quantify thermal expansion, consider heating an object so that its temperature changes by dT and its volume changes by dV. The fractional change in volume can be expressed as:

$$\frac{dV}{V} = \beta \quad dT \tag{2.1}$$

where β is known as the **volume expansion coefficient**. This coefficient measures the fractional change in volume per unit change in temperature.

Values of the volume expansion coefficients for common liquids and solids are given in Table 2.1. Notice that the volume expansion coefficients for liquids are typically about 100 times larger than those for solids. This indicates that liquids undergo much more noticeable volume changes when heated or cooled. This property forms the basis of the classic liquid-in-glass thermometers, as shown in Figure 2.1.

These thermometers consist of a glass bulb attached to a sealed, graduated tube approximately 25 cm long, with a very narrow capillary opening extending from the

Table 2.1 The volume expansion coefficients for common solids and liquids

Solids	β (K^{-1})	Liquids	β (K^{-1})
Brass	56×10^{-6}	Benzene	1.24×10^{-4}
Copper	51×10^{-6}	Ethyl alcohol	1.12×10^{-4}
Glass	27×10^{-6}	Mercury	1.82×10^{-4}
Iron or steel	35×10^{-6}	Water	2.10×10^{-4}

No Permission Required

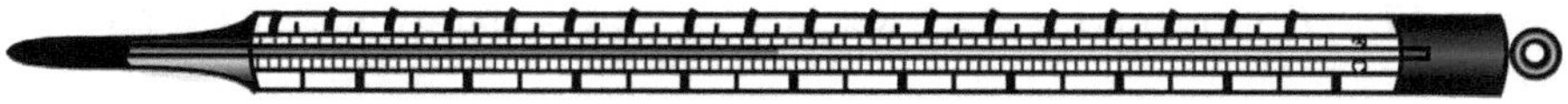

FIGURE 2.1 Mercury thermometer

Schematic of a mercury thermometer in which the narrow tube is a vacuum.

-No Permission Required

bulb to the end of the tube. The bulb contains a liquid, usually mercury or red-colored ethyl alcohol, that is free to move up the tube. As the air temperature increases, the liquid in the bulb expands and rises within the tube. Using the same calibration procedure discussed in Ch. 1.2 for the constant volume gas thermometer, the height of the liquid column can be calibrated to represent the air temperature.

2.1.2 Electronic thermometers

Automated observation stations often use electronic thermometers to measure temperature. A common type of electronic thermometer used in meteorology, including by the NWS, is the **resistive temperature device** (RTD), also known as a **resistance thermometer**.

To understand how a RTD device works, it's helpful to recall a concept from introductory physics: the electrical resistance R between two points on a conductor is defined by the formula:

$$R = \frac{\Delta V}{I}$$

where ΔV is the potential difference (or voltage) between the points and I is the current flowing through the conductor. Since the current represents the flow rate of conduction electrons, resistance quantifies how much the conductor impedes this flow.

At the microscopic level, resistance arises from collisions between conduction electrons and the fixed atoms (or ions) that make up the conductor. As a result, resistance depends on both the conductor's geometry and its temperature. As temperature increases, the frequency of these collisions rises, which in turn increases the resistance.

Let's suppose we measure the resistance of a conductor at a reference temperature T_0, and then determine the resistance at a different temperature. From introductory calculus, we know that the resistance as a function of temperature can be approximated using a Taylor series expansion about T_0:

$$R(T) = R(T_0) + \left(\frac{dR}{dT}\right)_{T=T_0} \Delta T + \frac{1}{2!}\left(\frac{d^2R}{dT^2}\right)_{T=T_0} (\Delta T)^2 + \ldots$$

where $\Delta T = T - T_0$. If ΔT is much less than T_0, we can approximate this expression as:

$$R(T) \approx R_0 + \alpha \Delta T, \alpha = \left(\frac{dR}{dT}\right)_{T=T_0}$$

where R_0 is the resistance of the conductor at the reference temperature T_0, and α is the **temperature coefficient of resistance**.

Experiments show that this linear approximation is effective over a wide range of temperatures, particular those encountered in atmospheric science. Therefore, if an instrument is calibrated using a material's known resistance at 0°C, the temperature of the conductor can be determined by measuring its resistance.

A schematic of a typical RTD is shown in Figure 2.2. In this design, a platinum wire is wound around a ceramic core. After winding, the assembly is sealed with a coating of molten glass. The sealing process ensures the long-term stability of the RTD and helps limit the thermal expansion of the platinum at high temperatures.

ASOS weather stations use an RTD to measure ambient surface temperature. The device has a reported operating range of −62°C to 54°C with a resolution of 0.06°C. The maximum error ranges from ±1°C to 2°C, depending on the specific measurement within that range (Nadolski, 1998).

2.2 Pressure measurements

In atmospheric science, the standard instruments used to measure atmospheric pressure are called **barometers**. In general, barometers measure the pressure difference between atmospheric pressure on one side of the sensor and a reference pressure on the other side. In this section, we will explore this process for two types of barometers.

2.2.1 Mercury barometers

Figure 2.3 shows a basic mercury barometer. This instrument consists of a long tube, closed at one end, filled with mercury, and then inverted into a dish of mercury,

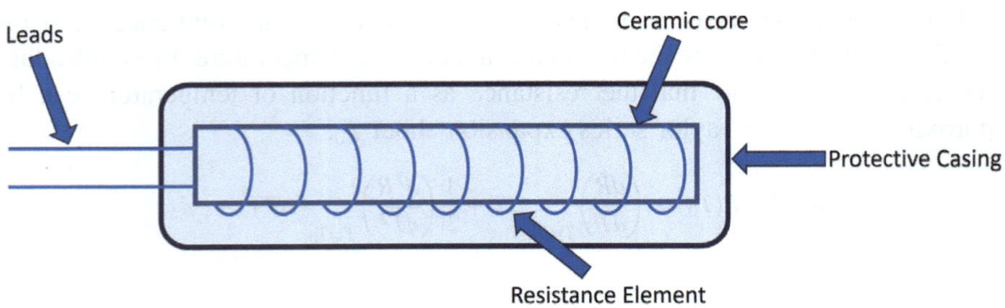

FIGURE 2.2 Resistive temperature device

A schematic of a resistive temperature device (RTD). Typically, the leads are made of platinum because of its long-term stability and because platinum can be used over a wide range of temperatures.

-No Permission Required

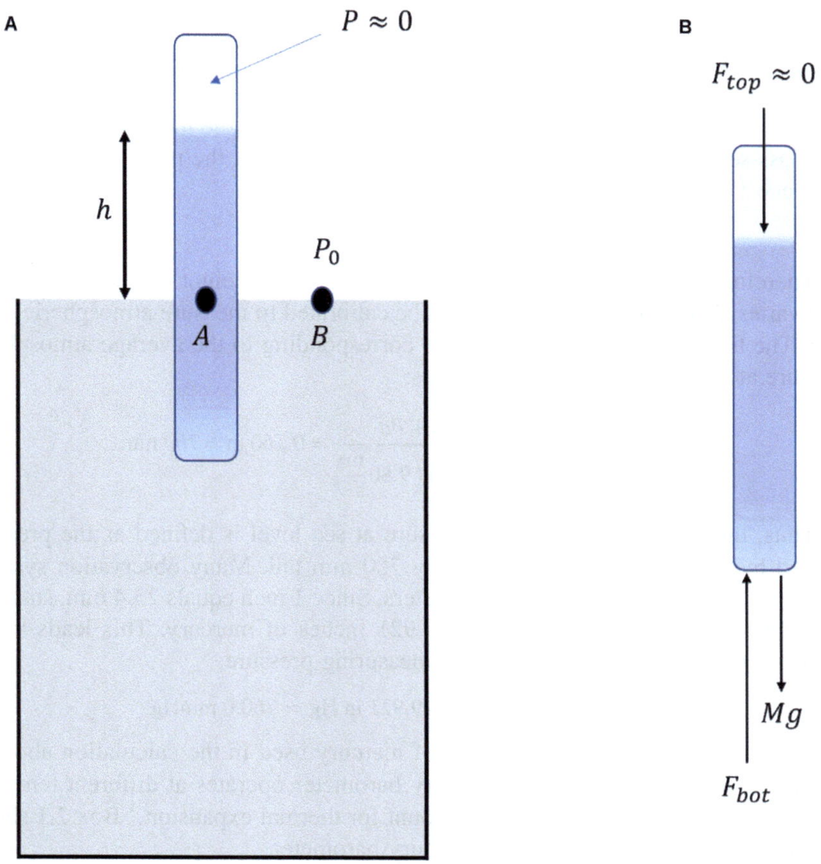

FIGURE 2.3 Mercury barometer

(A) Schematic of a basic mercury barometer. (B) Free-body diagram of the mercury column of height *h*.

-No Permission Required

forming a vacuum near the top. Because the mercury level is horizontal and stationary, the pressure at point A (due to the mercury column) must equal the pressure at point B (from the surrounding atmospheric pressure). Therefore, if we can determine the pressure at point A, we can determine the atmospheric pressure.

We can derive a relationship between the pressure at point A and the height (h) of the mercury column. In Figure 2.3B, three forces act on the mercury column: the weight of the mercury column (Mg), the force due to the pressure of the mercury below point A, and the force due to the space above the mercury column. Since the space above the mercury contains only mercury vapor, whose pressure is negligible at ordinary temperatures, we can assume that the force from the top (F_{top}) is zero. Given that the mercury column is in static equilibrium, we have:

$$Mg = F_{\text{bot}}$$
$$\Rightarrow \rho V g = P_A A$$

where ρ is the density of the mercury, V is the volume of the mercury column, A is the cross-sectional area of the mercury column, and P_A is the pressure at point A.

Since $V = Ah$, then we have

$$\rho(Ah)g = P_A A \quad \Rightarrow P_A = \rho g h$$

Therefore, as the atmospheric pressure changes, the height of the mercury column varies accordingly. This height can be calibrated to measure atmospheric pressure. The height of the mercury column corresponding to the average atmospheric pressure at sea level can be calculated as:

$$h = \frac{P_{sfc}}{\rho g} = \frac{1.01325 \times 10^5 Pa}{\left(13.6 \times 10^3 \frac{\text{kg}}{\text{m}^3}\right)\left(9.80 \frac{\text{m}}{\text{s}^2}\right)} = 0.760 \text{ m} = 760 \text{ mm}$$

Thus, the average atmospheric pressure at sea level is defined as the pressure exerted by a column of mercury exactly 760 mm tall. Many observation systems use inches of mercury instead of millimeters. Since 1 inch equals 25.4 mm, standard atmospheric pressure at sea level is 29.921 inches of mercury. This leads to the commonly used conversion factors for measuring pressure:

$$1 \text{ atm} = 101.325 \text{ kPa} = 29.921 \text{ in Hg} = 760.0 \text{ mmHg}$$

It should be noted that the density of mercury used in the calculation above is applicable only at 0°C. When a mercury barometer operates at different temperatures, adjustments must be made to account for thermal expansion.[1] Box 2.1 shows how thermal expansion affects the mercury barometer.

2.2.2 Electronic barometers

Automated observation stations often use electronic barometers to measure atmospheric pressure. A widely used type of electronic barometer in meteorology, including by the NWS, is the **capacitive pressure transducer**. This device converts changes in pressure into an electrical output signal.[2] To understand how it works, it is helpful to review the properties of capacitors from introductory physics.

A parallel-plate capacitor, as shown in Figure 2.4, consists of two conductors connected to a battery. When the plates are connected to the battery with voltage ΔV, electrons are transferred between them. The plate connected to the positive terminal becomes positively charged, while the plate connected to the negative

[1] Each mercury barometer has an instrument error caused partly by the surface tension of the mercury against the glass tube.

[2] An electrical transducer is a device that converts a mechanical signal into an electrical output.

Box 2.1 Effects of thermal expansion on mercury barometers

Let's determine the height of the mercury column for a mercury barometer if the temperature of mercury was measured at $-5.00°C$.

First, we must examine the change in volume due to thermal expansion. Because the cross-sectional area of the mercury column does not change, we can infer the change in height of the mercury by examining the volume expansion of the mercury. Because we know the height of the mercury column at 0°C, we can perform a Taylor series of the volume of the mercury column centered at 0°C:

$$V(T) = V(T_0) + \left(\frac{dV}{dT}\right)_{T=T_0} \Delta T + \frac{1}{2!}\left(\frac{d^2V}{dT^2}\right)_{T=T_0} (\Delta T)^2 + \ldots$$
$$= V(T_0) + \beta V(T_0)\Delta T + \frac{1}{2!}\left(\frac{d^2V}{dT^2}\right)_{T=T_0} (\Delta T)^2 + \ldots$$

where $\Delta T = T - T_0$ and $\beta V = dV/dT$ from Eq. (2.1). Since the volume expansion coefficient for mercury is $\beta = 1.82 \times 10^{-4}\ K^{-1}$ (according to Table 2.1) and $\Delta T = -5.00$ K, then we can use a linear approximation to obtain the volume

$$V(T) \approx V(T_0) + \beta V(T_0)\Delta T \Rightarrow V \approx V(T_0)[1 + \beta \Delta T]$$
$$\Rightarrow \frac{\Delta V}{V(T_0)} \approx \beta \Delta T = \left[1.82 \times 10^{-4} K^{-1}\right](-5.00K) = -9.10 \times 10^{-4}$$

Since the cross-sectional area remains approximately constant, then

$$\frac{\Delta V}{V(T_0)} = \frac{\Delta h}{h(T_0)}$$

Since the height of the mercury column is 760 mm at 0°C, then the height of the mercury column at $-5°C$ is given by

$$h = h(T_0) + \Delta h = h(T_0)\left(1 + \frac{\Delta h}{h_0}\right) = (760\ \text{mm})\left(1 - 9.10 \times 10^{-4}\right) = 759.3\ \text{mm}$$

Although a height difference of 0.700 mm appears to be small, if this is converted into kPa, we obtain

$$0.700\ \text{mmHg} \times \frac{101.325\ \text{kPa}}{760.0\ \text{mmHg}} = 0.0933\ \text{k}Pa = 0.933\ \text{mb}$$

Thus, a 5°C difference in temperature can cause an error of approximately 0.9 mb in pressure. For this reason, mercury barometers require temperature corrections to account for the thermal expansion of the fluid. In other words, to obtain accurate pressure reading unaffected by temperature, all mercury barometers are corrected as if they were measured at the same temperature.

terminal becomes negatively charged. This process creates an electric field directed toward the negatively charged plate, with a magnitude given by $E = \Delta V/d$ where d is the distance between the plates.

Experiments with capacitors demonstrate that the amount of charge stored on the capacitor is linearly proportional to the potential difference between its conductors. This relationship can be expressed as $Q = C\Delta V$, where the proportionality constant C is known as the **capacitance**. The capacitance depends on the geometric shape of the conductors and the separation between the plates. This quantity is significant because it indicates how much electric charge (and, consequently, how much electrical energy) a device can store.

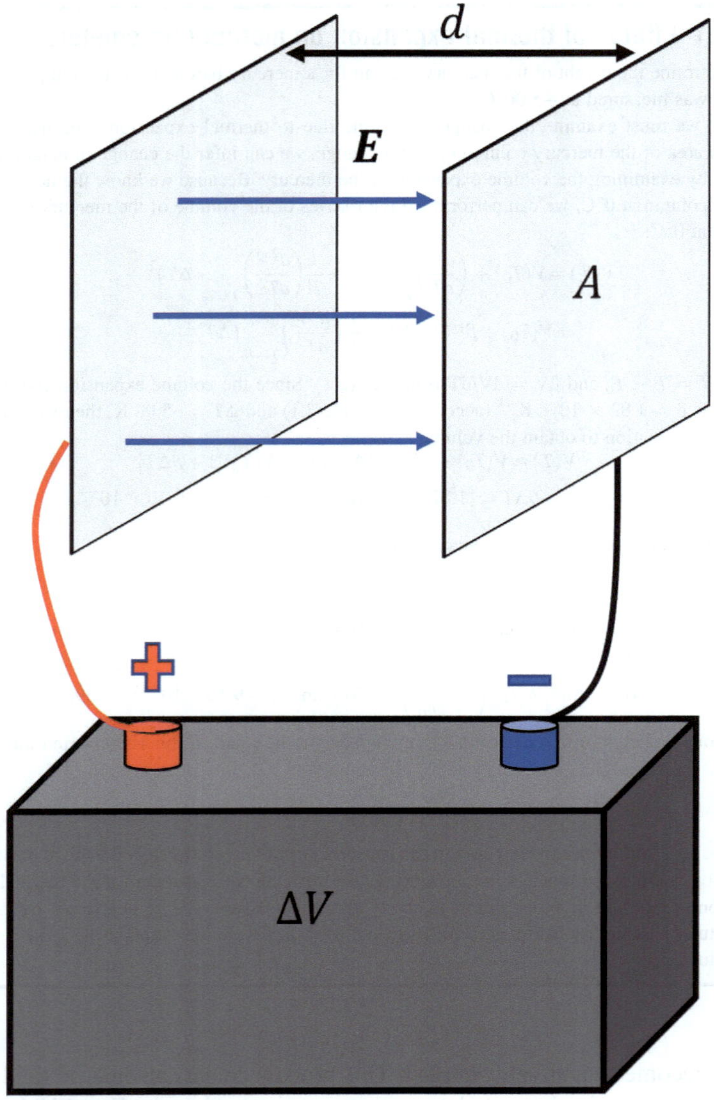

FIGURE 2.4 Parallel plate capacitor

A schematic of a parallel plate capacitor in which each conducting plate has a cross-sectional area and the conducting plates have a separation distance *d*. When the capacitor is connected to the battery with voltage ΔV, electrons are transferred between the plates and the wires so that the conducting plates become charged.

-No Permission Required

For a parallel-plate capacitor, as depicted in Figure 2.4, the capacitance depends on the area A of the plates and the separation distance d between them. Conceptually, we expect that the capacitance increases with plate area, since larger conducting plates can hold more charge. Conversely, capacitance decreases as the plate separation increases. Because the voltage across the plates is fixed by the battery, and $E = \Delta V/d$, increasing the distance between the plates increases the electric field strength, which results in a larger accumulation of charge on the plates.

Changes in capacitance caused by variations in plate separation can be used to measure atmospheric pressure. Figure 2.5 shows a schematic of a capacitive pressure

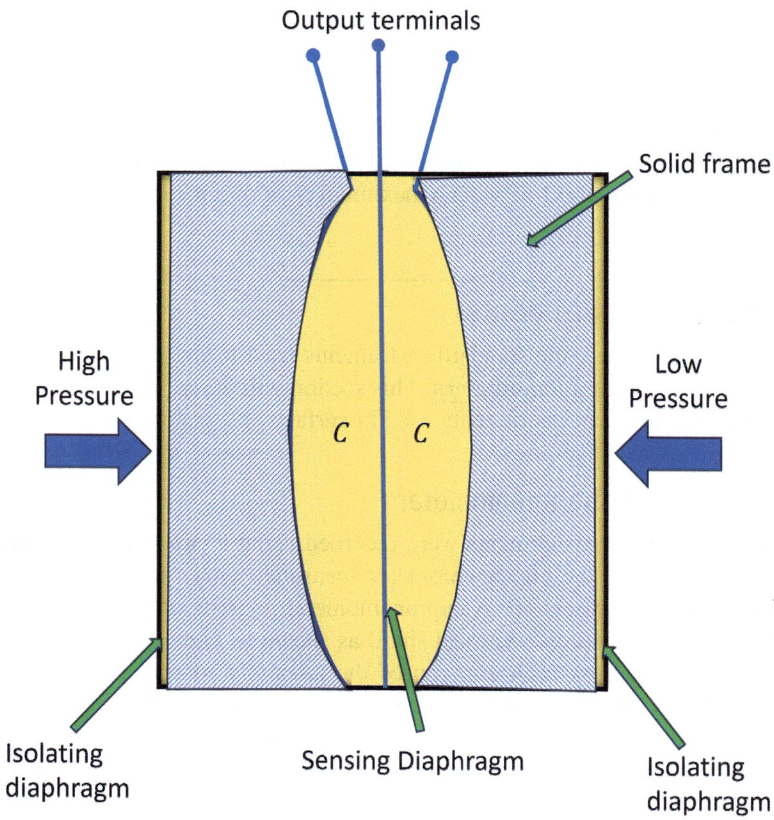

FIGURE 2.5 Capacitive pressure transducer

Schematic of a capacitive pressure transducer. In this design, the sensing element is a metal sensing diaphragm located between two stationary metal surfaces, comprising three plates for a pair of capacitors. The insulating fluid between the plates helps to amplify the capacitance of the system.

-No Permission Required

transducer, which uses a metal sensing diaphragm[3] placed between two stationary conducting plates, effectively forming two capacitors. Pressure differences across the diaphragm cause it to bend toward the lower pressure side, altering the distance between the capacitor plates. As the diaphragm moves toward lower pressure, the capacitance on the low-pressure side increases while the capacitance on the high-pressure side also changes. Therefore, the total capacitance of the system varies in response to changes in atmospheric pressure.

The capacitor plates are connected to electrodes that detect changes in current and voltage as pressure changes. Because both voltage and current are affected by changes in capacitance, capacitive pressure transducers can convert pressure variations into an electrical output signal.

The pressure sensors used at ASOS weather stations operate on principles similar to those of the capacitive pressure transducers described above. One side of each sensor is permanently evacuated to create a vacuum, making them well-suited for measuring atmospheric pressure. Accurate pressure measurements are essential for both aviation[4] and meteorology, which is why most ASOS stations use three independent pressure sensors. A pressure report is issued only when at least two of the sensors agree within acceptable limits. The standard working range of these pressure sensors is 572 to 1066 hPa, with a resolution of 0.1 hPa and a maximum error of ±0.7 hPa (Nadolski, 1998).

2.3 Wind measurements

In atmospheric science, the standard instruments used to measure wind speed and direction are known as **anemometers**. This section will describe the two most common wind measurement instruments used in surface observation systems.

2.3.1 Cup and vane anemometer

Traditionally, wind measurements were recorded using a **cup and vane anemometer**. In this system, the cup anemometer measures wind speed while the wind vane measures wind direction. A cup anemometer consists of at least three hemispherical cups mounted on a vertical shaft, as shown in Figure 2.6. The difference in wind pressure across each cup causes the assembly to spin around the shaft. From introductory physics, we know that the angular speed ω of an object is related to its tangential speed by the formula: $\omega = v/r$. Thus, the rotation rate of the cup anemometer is directly proportional to wind speed. Older ASOS stations used an

---

[3] A diaphragm is a thin membrane that isolates one side of a medium from another, while allowing pressure to be transmitted with minimal loss.

[4] Atmospheric pressure is the most important surface observation for aviation as it helps determine the altitude of an aircraft above the ground. Because pilots cannot directly observe atmospheric pressure, it has always been carefully measured. Consequently, pressure sensors are routinely calibrated against a reference standard to ensure accuracy.

FIGURE 2.6 Wind vane and cup anemometer

-No Permission Required

automated version of the cup anemometer, in which the rotation of the cups drives an electric generator, and the generator's output is calibrated in knots (where 1 knot = 0.514 m/s = 1.15 mph) (Nadolski, 1998).

Most wind vanes consist of a long arrow with a tail that moves freely around a vertical post, as shown in Figure 2.6. The arrow always points into the wind, allowing it to indicate the wind direction. It's important to note that the meteorological convention for wind direction differs from the vector-based convention described in Ch. 1. For example, horizontal wind directions reported by a wind vane are based on compass bearings, where 0° corresponds to the north and degrees increase clockwise up to 360°, as shown in Figure 2.7B. This contrasts with the mathematical

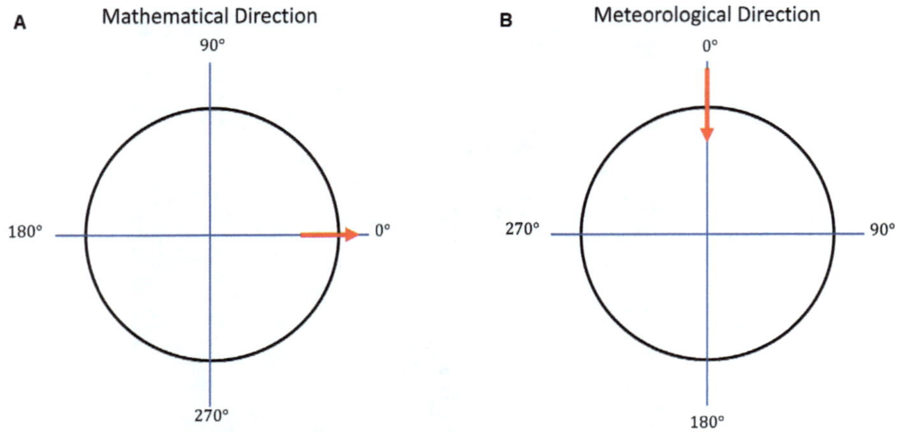

FIGURE 2.7 Meteorological coordinate system

(A) The standard Cartesian coordinate system for defining mathematical direction. Notice that it is to the east with degrees increasing in a counterclockwise direction. The vector arrow indicates the direction of the wind in which the mathematical direction is zero. (B) The standard coordinate system for defining meteorological direction. Notice that it is *from* the north with degrees increasing in a clockwise direction. The vector arrow indicates the direction of the wind in which the meteorological direction is zero.

-No Permission Required

convention, in which 0° is defined as east and the degrees increase counterclockwise 360°, as shown in Figure 2.7A. Additionally, recall that winds are typically named by the direction *from* which they originate, whereas in mathematics, angles indicate the direction toward which an object moves. Box 2.2 demonstrates how to determine the direction of the wind vector using the meteorological coordinate system.

2.3.2 Sonic anemometer

Within the NWS, the traditional cup and vane anemometer has largely been replaced by a modern electronic anemometer known as the **sonic anemometer**. The primary components of a sonic anemometer are ultrasonic sound emitters and receivers positioned at opposite ends of a sampling volume, as shown in Figure 2.8. Sonic anemometers operate by measuring the time it takes for a sound wave to travel between two transducers. When the sound wave travels in the same direction as the wind, its speed increases; conversely, when it moves against the wind, its speed decreases.

Each sonic anemometer typically has three equally spaced transducers, enabling it to measure the time required for sound waves to travel along multiple paths. In practice, six travel-time measurements are taken (one in each direction along the

Box 2.2 Determining the direction of the wind vector

To derive a mathematical expression for the meteorological direction, there are two cases to consider: when $u > 0$ and when $u \leq 0$. For the first case where $u > 0$, let's consider a wind from the northwest. According to (b), a wind from the northwest has a meteorological direction of 315°. Since a wind from the northwest points to the southeast such that $v = -u$, this implies that it will have a mathematical direction of

$$\theta = \tan^{-1}\left(\frac{v}{u}\right) = \tan^{-1}(-1) = -45^\circ$$

Now, let's consider a wind from the southwest. According to (b), a wind from the southwest has a meteorological direction of 225°. Since a wind from the southwest points to the northeast such that $u = v$, this implies that it will have a mathematical direction of

$$\theta = \tan^{-1}\left(\frac{v}{u}\right) = \tan^{-1}(1) = +45^\circ$$

These two results suggest that for $u > 0$, the meteorological direction can be written as

$$\alpha = 270^\circ - \tan^{-1}\left(\frac{v}{u}\right)$$

where it is assumed that the range of arc tangent is between -90° and $+90^\circ$.

Using the same procedure as above, it can be shown that for $u \leq 0$, the meteorological direction can be written as

$$\alpha = 90^\circ - \tan^{-1}\left(\frac{v}{u}\right)$$

where it is assumed that the range of arc tangent is between -90° and $+90^\circ$. In conclusion, the wind direction in terms of the meteorological direction can be written as

$$\alpha = \begin{cases} 90^\circ - \tan^{-1}\left(\frac{v}{u}\right), & u \leq 0 \\ 270^\circ - \tan^{-1}\left(\frac{v}{u}\right), & u > 0 \end{cases} \tag{2.2}$$

three paths). These data are processed using an algorithm to determine both wind speed and wind direction.[5]

Using sonic anemometers, ASOS measures wind direction and speed every second. This high-frequency monitoring allows ASOS stations to distinguish effectively between sustained winds and wind gusts. Sustained wind speed is defined as the average wind speed over the most recent 2-minute period. To reduce variability, this 2-minute average is computed from a series of 5-second averages, which in turn are derived from 1-second measurements. Thus, the 5-second averages serve as the fundamental data units used to calculate all reportable wind values, and this smoothing process ensures a stable and representative measure of the sustained wind.

To track changes in sustained wind speed over time, the 2-minute average wind speed is updated every 5 seconds. These 5-second averages are also used to identify wind gusts. A wind gust is recorded by the NWS when rapid fluctuations in wind

[5] It should be noted that the speed of sound wave propagation also depends upon temperature. However, the sonic anemometer detects two opposite directions on two channels so that the effect of temperature on sound wave speed can be ignored.

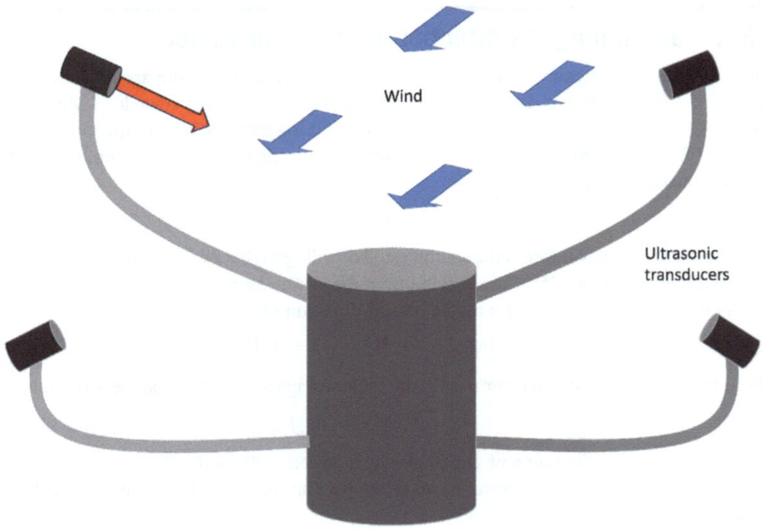

FIGURE 2.8 Sonic anemometer

Schematic of a sonic anemometer, which consists of two ultrasonic transducers for each measurement direction. An ultrasound pulse is generated by one of the transducers, while the other transducer receives the signal and measures the time delay between the emitted and received pulses.

-No Permission Required

speed occur, with a variation exceeding 10 knots between peaks and lulls during the 10 minutes preceding the observation. The reported gust is the maximum 5-second wind speed observed during this period. Thus, gusts represent brief, transient increases in wind speed that stand out against surrounding values on the 5-second scale.

The ASOS anemometer has a working wind speed range of 0 to 125 knots, with a resolution of 1 knot and an accuracy of ± 2 knots. For wind direction, the range is $0°$ to $359°$, with a resolution of $1°$ and an accuracy of $\pm 5°$ (Nadolski, 1998).

2.4 Humidity and precipitation measurements

In Ch. 1, we explored how the presence of water vapor directly affects the thermodynamic state of the atmosphere. In this section, we will introduce the most common instruments used to measure humidity and precipitation at surface observation stations.

2.4.1 Measuring water vapor

Before discussing instruments for measuring atmospheric moisture, it is important to understand the concept of saturation.[6] Consider a volume of air at temperature T and pressure P that is in contact with liquid water, as shown in Figure 2.9A. Water molecules in the liquid are in constant motion, and some molecules travel fast enough to escape the surface and enter the air as vapor, leading to evaporation. At the same time, some water vapor molecules in the air lose energy and return to the liquid phase, leading to condensation.

The air is considered **unsaturated** when the evaporation rate exceeds the condensation rate, as shown in Figure 2.9A. If evaporation continues in a closed

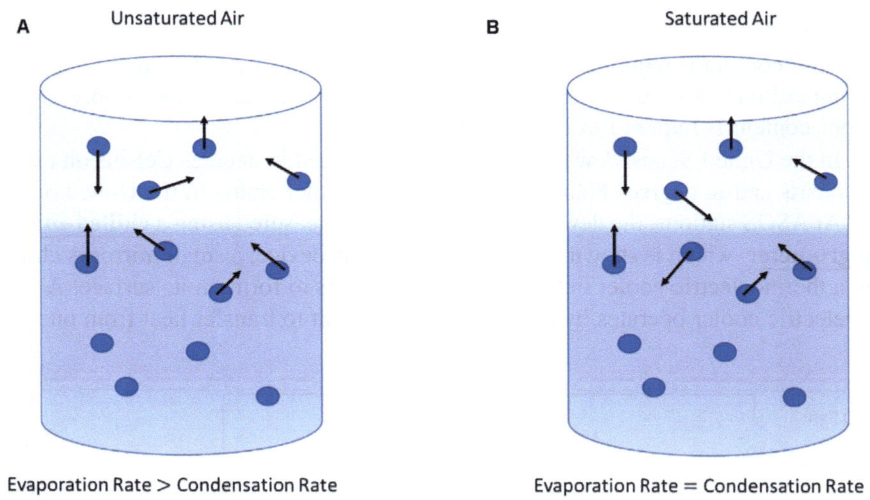

FIGURE 2.9 Schematic of unsaturated vs saturated air

(A) Unsaturated air. Since more molecules are evaporating than condensing, net evaporation is occurring. (B) Saturated air. Since the number of water molecules leaving the liquid (i.e., evaporation) balances those water molecules that are entering the liquid (i. e., condensation), the air above the liquid is saturated.

-No Permission Required

[6] The term "saturated" is widely used within meteorology and atmospheric science, but it can be somewhat misleading. It may suggest that air behaves like a sponge that can "hold" a certain amount of water, which is not accurate. In reality, saturated air refers to a state in which water vapor is in equilibrium with liquid water. For more information, see Schroeder, Daniel V., An Introduction to Thermal Physics (Oxford, 2021; online edn, Oxford Academic, 18 Mar. 2021), https://doi.org/10.1093/oso/9780192895547.001.0001 Bohren, Craig, and Bruce Albrecht, Atmospheric Thermodynamics, 2nd edn (Oxford, 2023; online edn, Oxford Academic, 28 Sept. 2023), https://doi.org/10.1093/oso/9780198872702.001.0001

container, the number of water vapor molecules in the air will increase until the condensation rate equals the evaporation rate. At this point, the air becomes saturated, as shown in Figure 2.9B.

The evaporation rate is influenced by the humidity - that is, the amount of moisture already present in the air. As the humidity increases, the evaporation rate decreases until saturation is reached. Temperature also plays a critical role: higher temperatures cause water molecules to move more rapidly, increasing the likelihood that they will escape the liquid surface. Conversely, when the air cools, water vapor molecules slow down, increasing the likelihood of condensation.

From this discussion, we can see that air becomes saturated in two primary ways: (1) by increasing the amount of water vapor in the air, or (2) by lowering the air temperature. The second mechanism leads to an important concept known as the **dew point temperature** (T_d), or simply dew point. The dew point is defined as the temperature to which air must be cooled, at constant pressure, for saturation to occur. Because cooling is required for saturation, it follows that $T_d \leq T$, with $T = T_d$, indicating saturated air. Lower values of T_d correspond to drier air, as they imply a lower vapor content is required to reach saturation.

In the United States, dew point is typically reported in degrees Celsius on upper-air charts and in degrees Fahrenheit on surface weather charts in the United States.

At ASOS stations, the dew point temperature is measured using a **chilled-mirror hygrometer**, which is shown in Figure 2.10. In this device, a small mirror is chilled by a thermoelectric cooler until condensation begins to form on its surface. A thermoelectric cooler operates by using electrical current to transfer heat from one side

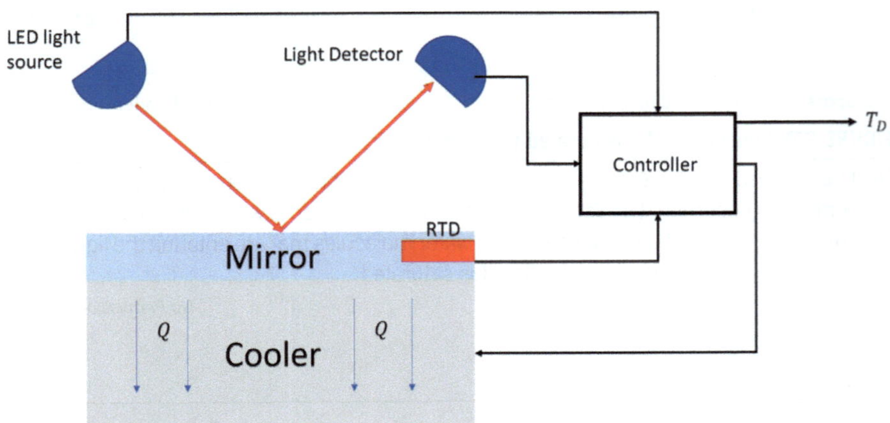

FIGURE 2.10 Chilled mirror hygrometer

A chilled mirror hygrometer. When current flows through the thermoelectric cooler, heat is transferred away from the mirror so that the temperature of the mirror decreases. The controller receives the reflected light from the light detector, and the controller controls the mirror's temperature.

-No Permission Required

of the device to the other. This cooling lowers the temperature of the mirror until the air near its surface reaches saturation, allowing the dew point to be measured. The appearance of dew on the mirror marks the dew point temperature.

A beam of light, typically from an LED source, is directed at the mirror surface, and a light detector monitors the reflected light. As the gas sample flows over the chilled mirror, dew droplets form on its surface, causing the reflected light to scatter. As the amount of reflected light decreases, the output from the light detector also decreases, which in turn controls the mirror's temperature. The body of the mirror contains a RTD to monitor ambient temperature. Thus, the temperature of the mirror when dew droplets begin to form on its surface is recorded as the dew point temperature.

Once the air and dew point temperatures are measured, the relative humidity can be calculated using the formula:

$$RH = \frac{r}{r_s} \times 100\%$$

where r is the mixing ratio (defined in Ch. 1) and r_s is known as the **saturation mixing ratio**, which represents the mixing ratio of the air when it is saturated. Because the evaporation rate depends on temperature, it follows that the saturation mixing ratio must also be a function of temperature. Therefore, r_s represents the maximum possible mixing ratio for a given air mass at a given temperature.

Moreover, because r_s depends on temperature, the relative humidity is *not* a direct measure of moisture content. Instead, it is influenced by both temperature and moisture content. Thus, the relative humidity of the air can change for two reasons: either by changing the amount of water vapor in the air or by changing the temperature of the air.

Because r_s is a function of temperature and r is a function of the dew point temperature, the relative humidity can be calculated by using both the dew point temperature and the ambient temperature.

The device used to measure both dew point and ambient temperature is called a **hygrothermometer**. The ASOS hygrothermometer has a reported operating range −80°F to +86°F with a resolution of 0.1°F. The accuracy of the hygrothermometer depends on the **dew point depression**, defined as the difference between the ambient temperature and the dew point temperature. When the dew point depression nears 0°F (indicating nearly saturated air), the maximum error is ±2°F to 5°F. Conversely, when the dew point depression nears 60°F (indicating air that is almost at 0% relative humidity), the maximum error increases to ±8°F to 14°F (Nadolski, 1998).

For weather forecasting, it is important to consider not only the presence of water vapor but also the occurrence of water condensate. In the following section, we will explore the instruments commonly used to measure cloud cover and precipitation.

2.4.2 Measuring cloud cover

One method for measuring cloudiness near the surface is to assess visibility. Figure 2.11A shows a schematic of a **visibility sensor** commonly used at ASOS stations. The sensor's transmitter emits a light beam, and the visibility is calculated by

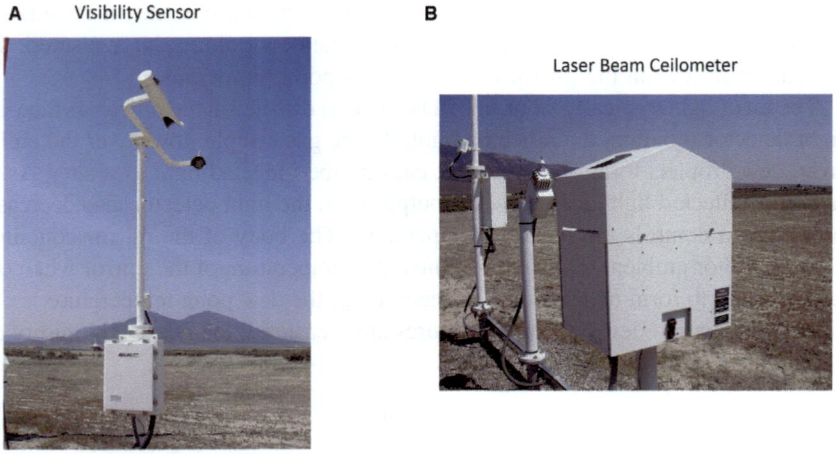

FIGURE 2.11 Visibility sensor and laser beam ceilometer

(A) A visibility sensor and (B) a laser beam ceilometer from the Automated Surface Observation System (ASOS) at Ely Airport in Ely, NV.

measuring how much light reaches the receiver. It is important to note that the sensor only samples a small volume of air - about 0.75 cubic feet - between the transmitter and the receiver. As a result, the sensor may not accurately represent visibility conditions across the broader environment.

To measure cloud cover at higher altitudes, ASOS stations typically use a **laser beam ceilometer**, as shown in Figure 2.11B. This device emits light pulses and calculates cloud height by measuring the time between transmission and reception of the reflected signals. However, the laser ceilometers has some limitations. Like the visibility sensor, it can only detect clouds directly overhead. To estimate conditions elswehere in the sky, an automated algorithm extrapolates from local observations. For this reason, the laser ceilometer may report misleading cloud conditions if the environment about it is not representative of the broader environment. Additionally, falling precipitation increases infrared light scattering, reducing the accuracy of cloud height measurements.

2.4.3 Measuring precipitation

To measure precipitation accurately, it is necessary to determine the precipitation type, precipitation intensity, and total accumulated precipitation. The NWS uses a **Light Emitting Diode Weather Identifier** (LEDWI) to distinguish between rain and snow and to assess precipitation intensity. As shown in Figure 2.12A, the LEDWI features an infrared light transmitter and receiver mounted on a crossarm approximately 10 feet above the platform's base.

The transmitter emits an infrared beam toward the receiver, and the LEDWI measures the amount of light scattering caused by falling precipitation. Because frozen precipitation is denser than liquid precipitation, it produces greater scattering. Based on the degree of scattering, the LEDWI can determine both the type and intensity of the precipitation.

To differentiate between freezing rain and standard rain, the NWS uses a **freezing rain sensor**, as shown in Figure 2.12B. This sensor consists of a small cylindrical probe that vibrates at a known frequency, along with a feedback coil that monitors changes in vibration. When ice accumulates on the probe, the added mass causes a decrease in the vibrational frequency. By analyzing this change, the freezing rain sensor can effectively distinguish freezing rain from standard rain.

The simplest instrument for measuring precipitation accumulation is the **tipping bucket rain gauge**. As illustrated in Figure 2.13A, this device consists of a cylindrical container that collects rain and channels it into a small bucket mounted on one side of a pivoting lever. When the bucket fills, it tips the lever, empties its contents, and positions an empty bucket from the opposite side beneath the funnel. Each tip is recorded electronically, and the frequency of tips per hour provides a measure of the rainfall rate.

Although the tipping bucket gauge is a simple tool for measuring precipitation, it has some well-known limitations. One major issue arises during heavy rainfall events (i.e., precipitation rates exceeding 1.80 inches per hour) when the gauge cannot keep up with the water flow, leading to an underestimation of total accumulation. The ASOS software typically corrects for this bias during most heavy rainfall events; however, during extremely heavy rainfall events (i.e., precipitation rates exceeding 10 inches per hour), the rain gauge may still under-report total accumulation.

A Light Emitting Diode Weather Identifier

B Freezing Rain Sensor

FIGURE 2.12 LEDWI and freezing rain sensor

(A) A Light Emitting Diode Weather Identifier (LEDWI) and (B) a freezing rain sensor from the Automated Surface Observation System (ASOS) at Ely Airport in Ely, NV.

FIGURE 2.13 Precipitation Gauges

(A) A tipping bucket precipitation gauge from Tonopah Airport in Nevada. (B) An all-weather precipitation accumulation gauge from Eureka Airport in Nevada.

Another common issue with tipping bucket gauges occurs during snow events. To measure frozen precipitation, the gauge melts the snow and funnels the resulting water into the bucket. However, accurately measuring snowfall and snow depth is more challenging, as these measurements are significantly influenced by local environmental factors. For example, wind conditions have a greater impact on snowfall accuracy than on rainfall measurements, making lighter snow events especially prone to errors (Rasmussen et al., 2012).

Because of these difficulties in measuring frozen precipitation with tipping bucket gauges, the **all-weather precipitation accumulation gauge** (AWPAG) was developed, as shown in Figure 2.13B. Unlike the tipping bucket gauge, the AWPAG does not rely on a tipping mechanism. Instead, it continuously collects precipitation, and records the weight as precipitation increases. The AWPAG also uses internal algorithms to account for evaporation, local wind changes, and other short-term fluctuations caused by non-precipitating events.

2.5 Radiosondes

Before the modern meteorological era, upper-air observations were expensive and infrequent. However, the invention of the **radiosonde** revolutionized our ability to measure key atmospheric variables, such as temperature, humidity, and pressure, throughout the vertical profile of the atmosphere.

The radiosonde is housed within a small white box, as shown in Figure 2.14, and includes a GPS receiver, a temperature sensor, a pressure sensor, a humidity sensor, and an antenna. The temperature sensor operates as an RTD, which measures temperature changes based on variations in electric resistance as the radiosonde ascends. The pressure sensor is a capacitive pressure sensor that detects pressure changes as the radiosonde ascends. The humidity sensor works on a similar principle to the RTD; it measures the electrical resistance, which varies with the amount of water vapor content. This allows for the calculation of the water vapor mixing ratio. A built-in transmitter then relays the radiosonde's data back to a ground-based receiving system in real time.

To obtain upper-air observations, radiosondes are attached to weather balloons, as shown in Figure 2.15. As the balloon ascends, the radiosonde can reach altitudes over 35 km, experiencing temperatures as cold as −130°F (−92°C) and atmospheric pressures below 10 hPa. Vertical wind profiles are inferred by tracking the drift of the balloon using GPS.[7] As the weather balloon rises into regions of lower pressure, it expands to equalize its internal pressure with the surrounding environment. Once the balloon exceeds its elastic limit—typically when it reaches a diameter of 20 to 25 feet in diameter—it bursts, and the radiosonde returns to Earth by a small parachute. The parachute slows the descent, reducing the risk to life or property on the ground.

Since atmospheric pressure can be reliably measured by the radiosonde and since atmospheric pressure consistently decreases with height, radionsonde data such as temperature, relative humidity, wind speed, and wind direction are typically reported as functions of pressure rather than height. As we will explore in further in Ch. 5, the

FIGURE 2.14 Radiosonde

A radiosonde instrument that is lifted into the atmosphere via a large hydrogen-filled balloon, courtesy of the National Oceanic and Atmospheric Administration.

-No Permission Required

[7] When a radiosonde is tracked so that winds aloft are provided in addition to the pressure, temperature, and humidity data, it is called a rawinsonde observation.

FIGURE 2.15 Weather balloon launch

A radiosonde ascending a few minutes after release of the balloon, courtesy of the National Atmospheric and Oceanic Administration. The radiosonde is the tiny box approximately 85 feet (26 meters) below the balloon. The orange object is the parachute.

-No Permission Required

relationship between pressure and altitude is a nonlinear function of virtual temperature. Therefore, by measuring temperature and humidity as a function of pressure, we can estimate the corresponding altitudes at which the data were recorded. For this reason, the pressure levels are often accompanied by height estimates in upper-level data sets. Now that we have examined the instruments used to collect meteorological data, we are ready to analyze the data they produce, which is the subject of the next chapter.

Questions

1. A bimetallic thermometer uses a coiled strip made of two different metals laminated together. Using the concept of thermal expansion, explain how this device can be used as a thermometer.
2. For a classic glass-in-bulb thermometer that is heated from below, explain why the height of the mercury initially *decreases* before it increases.
3. Explain why it would be impractical to replace a mercury barometer with water.
4. In your own words, explain why the saturation vapor pressure must be a function of temperature and independent of atmospheric pressure under normal conditions.
5. As mentioned in Ch. 2.1, the sealing process of the RTD helps to limit the thermal expansion of the platinum. What would happen to the accuracy of the RTD if there was no restriction to the thermal expansion of the platinum? You may assume that the volume expansion coefficient of platinum is different than the ceramic bobbin. [Hint: Consider the stress that would be placed on the wire.]
6. Explain how dew, frost, or heavy fog can cause "false tips" in a standard rain gauge.
7. In your own words, explain how the hygrothermometer in an ASOS measures temperature and dew point.
8. According to the Glossary of Meteorology, drizzle is a form of liquid precipitation that is characterized by very small, numerous, and uniformly dispersed water drops that fall to the ground, and drizzle is frequently accompanied by low visibility and fog. Explain why it should be difficult to detect drizzle using only the present weather sensor.
9. As a radiosonde rises within the atmosphere, there is a time delay between the recording of data and the vertical ascent of the radiosonde. Explain how this can lead to errors in radiosonde measurements.
10. Suppose that the approach of a storm causes the height of a mercury barometer to drop by 20 mm from the normal height. What is the atmospheric pressure?
11. Determine the wind speed (in knots) and the wind direction (measured relative to north) if the zonal wind component is 10.0 knots and the meridional wind component is 15.0 knots.
12. Determine the wind speed (in knots) and the meteorological direction of the sustained wind:
    a. if the zonal wind component is 10 mph and if the meridional wind component is 8 mph.
    b. if the zonal wind component is –15 mph and if the meridional wind component is 12 mph.

13. Determine the sustained wind (in miles per hour) in unit-vector notation using two different methods: (1) applying Eq. (2.2) and (2) using a standard Cartesian system:

a. if the wind speed is 10 knots and the meteorological wind direction is from 170°.

b. if the wind speed is 20 knots and the meteorological wind direction is from 330°.

14. Suppose that V is the wind speed measured by a cup anemometer and α is the meteorological direction measured by a wind vane. Show that the components of the wind vector are given by

$$u = -V \sin(\alpha), v = -V \cos(\alpha). \tag{2.3}$$

15. In Box 2.1, we use a Taylor series approximation to derive a linear approximation for the volume given by given by $V(T) \approx V(T_0)[1 + \beta(T - T_0)]$. Derive the same linear approximation by directly integrating Eq. (2.1). You may assume that β is constant over the same temperature range. [Hint: For x << 1, $e^x \approx 1 + x$.]

References

Nadolski, V., 1998. Automated surface observing system (ASOS) user's guide. Available from: https://www.weather.gov/media/asos/aum-toc.pdf.

Rasmussen, R., Baker, B., Kochendorfer, J., Meyers, T., Landolt, S., Fischer, A., Black, J., Theriault, J., Paul, K., Gochis, D., Smith, C., Nitu, R., Hall, M., Ikeda, K., Gutmann, E., 2012. How well are we measuring snow: the NOAA/FAA/NCAR winter precipitation test bed. Bull. Am. Meteorol. Soc. 93 (6), 811–829. https://doi.org/10.1175/BAMS-D-11-00052.1.

CHAPTER

Analyzing weather observations

3

Chapter outline

The meteorological instruments discussed in the previous chapter provide continuous and reliable observational data around the world. Under the oversight of the World Meteorological Organization (WMO), they form a crucial component of Global Observing System (GOS).[1] These data are used in various meteorological applications, including weather forecasting, meteorological analysis, and climate monitoring.

In this chapter, we will explore how meteorological data are processed and decoded. We will give an overview of the common formats for reporting surface meteorological data, such as METAR code, and upper-air meteorological data, such as radiosonde code. We will also explain how to decode and interpret surface station plots on weather charts and upper-air station plots on isobaric charts.

3.1 Surface observations

There are approximately 17,500 stations and platforms worldwide that make surface weather observations, typically recorded on an hourly basis. In the United States,

[1] There are many components of the GOS, such as surface observations, upper-air observations, marine observations, aircraft-based observations, satellite observations, and weather radar observations. For this text, we will focus our surface observations and upper-air observations.

An Introduction to Synoptic-Dynamic Meteorology. https://doi.org/10.1016/B978-0-443-24858-0.00003-4

most surface observations are recorded at airports, which primarily use a version of the **Automated Surface Observation System (ASOS)**, as shown in Figure 3.1.

As discussed in Ch. 2, an ASOS station contains the following instruments:

- A wind tower equipped with a wind vane and a sonic anemometer to measure wind speed and direction.
- A hygrothermometer, consisting of a resistive temperature device (RTD) and a chilled mirror hygrometer, to measure temperature and dewpoint.
- A tipping bucket gauge or an all-weather precipitation gauge to measure total accumulated precipitation.
- A Light Emitting Diode Weather Identifier (LEDWI) and a freezing rain sensor to measure precipitation intensity and identify precipitation type.
- A laser beam ceilometer to measure cloud base heights above the station, along with a visibility sensor to measure local visibility.
- A central processing unit (CPU) that houses three barometers to report station pressure, as well as the ASOS computer and power supply.

FIG. 3.1 ASOS Station in Lubbock, TX

An Automated Surface Observing System (ASOS) in Lubbock, TX. (Courtesy National Oceanic and Atmospheric Administration.)

-No Permission Required

In the United States and Canada, the most common format for reporting surface observation data is known as **Meteorological Terminal Aviation Routine** (or METAR) code. This section provides an overview of the common rules associated with a METAR report.[2]

3.1.1 Body of METAR

The body of a METAR report contains a station identifier; date and time; report modifier; prevailing visibility and runway visual range; present weather; sky condition; temperature and dew point; and altimeter setting. An example of the body of a METAR report is provided in Table 3.1.

Let's break down each portion of the surface observation.

The first column in Table 3.1 describes the report type. METAR is the standard observation, issued at the end of each hour. During significant or severe weather events (such as low visibility, low clouds, frozen precipitation, or thunderstorms) a report may be issued more frequently, typically every 20 minutes. In such cases, the report type is designated as *SPECI* (short for "special").

The second column in Table 3.1 describes the station identifier, which is a four-letter code based on the International Civil Aviation Organization (ICAO). The first letter represents the subcontinent or region, while the remaining three letters specify the station location. For example, the letter **K** indicates a station in the continental United States, and **CHS** is the three-letter code for Charleston, specifically Charleston International Airport.[3]

The third column in Table 3.1 indicates the date and time of the METAR report. In this column, the first two digits correspond to the day of the month, the following four digits indicate the time of the observation, and Z refers to Zulu time. For example, **12** represents the 12th day of the month, while **0821Z** indicates the time of the observation (8:21 Z). Zulu time, also known as **Universal Time Coordinate** (UTC) time, is the time and date in Greenwich, England, without accounting for daylight saving time rules. To convert from Zulu to local time, you must account for your time zone offset from Greenwich and whether daylight saving time is in effect.

Table 3.2 provides time zone conversions for the continental United States. For example, since Charleston is in the Eastern Time Zone, 0821Z corresponds to 3:21 AM Eastern Standard Time (EST).

Although the specific dates of Daylight saving time changes from year to year, it generally begins in early March and ends in early November. When daylight saving

[2] A complete description of all the rules associated with a METAR report can be found in the Federal Meteorological Handbook, No. 1: Surface Weather Observations and Reports. This manual is publicly available, and can be found here: https://www.icams-portal.gov/resources/ofcm/fmh/FMH1/fmh1_2019.pdf.

[3] The full list of US station IDs can be found online at: https://www.cnrfc.noaa.gov/metar.php.

Table 3.1 An example of the body of a METAR report from Charleston, SC

Report type	Station identifier	Time and date of report	Report modifier	Wind data	Visibility	Present weather	Sky condition	Temperature and dew point	Altimeter	Remarks section
METAR	KCHS	120821Z	AUTO	05014G20KT	10SM	-SHRA	BKN009 BKN014 OVC100	17/06	A2980	RMK A02

No Permission Required

time begins, clocks are moved forward 1 hour. Therefore, to convert from UTC to local daylight saving time, you add 1 hour to the standard time. Consequently, **0821Z** corresponds to 4:21 AM Eastern Daylight Time (EDT).

The fourth column in Table 3.1 describes any modifications to the METAR report. There are two possibilities:

- AUTO — the observation was taken by an ASOS with no human adjustments.
- COR — one or more elements of the report were corrected by a human observer.

In Table 3.1, the Charleston observations were recorded solely by an ASOS.

The fifth column in Table 3.1 displays the wind data recorded by the wind vane and anemometer. The first three digits represent the meteorological direction of the winds (in degrees), while the next two digits report the sustained wind speed in knots, always reported to the nearest knot. In some cases, there may be code that begins with **G**, which represents the wind gusts in knots. For example, **05014G20KT** indicates sustained winds of 14 knots from 50° (i.e., from the northeast) with wind gusts up to 20 knots. If the winds are light and variable, the code begins with **VRB**, meaning that the wind direction is variable and not steady. For instance, **VRB005KT** indicates light and variable winds of 5 knots.

The sixth column in Table 3.1 shows the prevailing visibility recorded by the visibility sensor, typically reported in statute miles in North America. For example, **10SM** means 10 statute miles. While visibility can occasionally be as high as 20 or 30 statute miles, it usually ranges from less than ¼ statute mile to 10 statute miles.

The seventh column in Table 3.1 presents current weather information recorded by the LEDWI. This data shows the type and intensity of precipitation, along with other weather phenomena that may obscure visibility. Table 3.3 provides a list of common items found in the METAR present weather group. The code always follows this order: precipitation intensity, precipitation description, precipitation type, and obscuration. Light precipitation is indicated by a minus (−) sign, moderate precipitation has no symbol, and heavy precipitation is denoted with a plus (+) sign. If precipitation is occurring within 5 to 10 statute miles of the observation station, the code begins with **VC** (vicinity). It is important to note that the intensity qualifier

Table 3.2 Time zone conversion for the Continental United States

Time zone	Hours behind Z	00Z	12Z
Eastern Standard Time (EST)	5	7 pm	7 am
Central Standard Time (CST)	6	6 pm	6 am
Mountain Standard Time (MST)	7	5 pm	5 am
Pacific Standard Time (PST)	8	4 pm	4 am

is always linked to the precipitation type. For example, −**SHRA** should be read as "light rain shower."

The eighth column in Table 3.1 contains information on sky coverage, cloud base heights, and any special cloud types recorded by the laser beam ceilometer. The format consists of a three-letter sky coverage amount (as shown in Table 3.4) followed by a three-digit cloud base height expressed in hundreds of feet. For example, a code of **BKN009** indicates broken clouds at 900 feet. Significant cloud types [such as Towering Cumulus (TCU) or Cumulonimbus (CB)] or occurrences of a shower or thunderstorm are added at the end of the code. For example, the code **SCT035TCU** indicates scattered clouds at 3,500 feet with towering cumulus present.

Multiple cloud layers are reported with multiple codes. For example, **BKN009 BKN014 OVC100** means broken clouds at 900 feet, broken clouds at 1400 feet, and overcast clouds at 10,000 feet. The ceiling, which is important for the aviation industry, is defined as the lowest layer of overcast or broken clouds. For example, **BKN009 BKN014 OVC100** indicates that the ceiling is 900 feet. If the sky is obscured and the ceiling cannot be determined (often due to fog), the station will report **VV** (vertical visibility), which indicates the maximum height that can be seen upward.

The ninth column in Table 3.1 gives the temperature and dew point recorded by the hygrothermometer at the ASOS station in degrees Celsius, rounded to the nearest whole degree. The first two digits indicate the temperature, while the last two digits represent the dew point. For example, a reading of **17/06** indicates a temperature of 17°C (or 63°F) and a dew point of 6°C (or 43°F). If either the temperature or dew point falls below 0°C, an "M" will precede the value. For example, the code **03/M02** indicates a temperature 3°C (37°F) and a dew point of −2°C (28°F).

The tenth column in Table 3.1 reports the altimeter setting, which is the pressure value often mentioned in TV weather reports. The altimeter setting is the station

Table 3.3 Present weather code

Intensity or proximity	Description of precipitation	Precipitation type	Obscuration to visibility	Other
Light (–) Heavy (+) In the vicinity (VC)	Blowing (BL) Freezing (FZ) Low Drifting (DR) Shallow (MI) Partial (PR) Patches (BC) Shower (SH) Thunderstorm (TS)	Drizzle (DZ) Hail (GR) Ice Crystals (IC) Ice Pellets (PL) Rain (RA) Small Hail and/or Snow Pellets (GS) Snow (SN) Snow Grains (SG) Unknown Precipitation (UP)	Fog (FG) Haze (HZ) Mist (BR) Sand (SA) Spray (PY) Smoke (FU) Volcanic Ash (VA) Widespread Dust (DU)	Dust/Sand Whirls (PO) Funnel Cloud (FC) Sandstorm (SS) Squalls (SQ)

No Permission Required

Table 3.4 Cloud cover indicator

Descriptor	Meaning
FEW	Few clouds (1/8 to 1/4 cloud coverage)
SCT	Scattered clouds (3/8 to 1/2 cloud coverage)
BKN	Broken clouds (5/8 to 7/8 cloud coverage)
OVC	Overcast condition (Full coverage)

No Permission Required

pressure (from the CPU pressure sensors) adjusted to sea level using a standard atmospheric chart. This adjustment removes the effect of station elevation on pressure readings, and it is essential for setting aircraft altimeters. In North America, the altimeter setting is always denoted by a leading **A** and is followed by a four-digit pressure value expressed in inches of mercury. For example, **A2980** means 29.80 inches of mercury. This can be converted to kPa by recalling that 760 mmHg = 101.325 kPa, which gives

$$29.80\ \text{inHg} \times \frac{25.4\ \text{mm}}{1\ \text{in}} \times \frac{101.325\ \text{kPa}}{760\ \text{mm Hg}} = 100.914\ \text{kPa} = 1009.1\ \text{mb}$$

Following the altimeter reading, the three-letter code **RMK** appears. This marks the end of the standard METAR observations in the body of the report and begins the remarks section. The contents of the remarks section will vary by location and weather situation. The following section introduces the most common remarks used in North America.

3.1.2 METAR remarks section

Consider the METAR report from Monroe, Louisiana (KMLU) given below:

KMLU 171153Z AUTO 03007KT 6SM RA BR BKN009 BKN014 OVC025 09/08 A3001 RMK AO2 SLP163 P0013 60033 70033 T00940079 10167 20094 51007

The remarks section of this METAR report can be subdivided into sections, as shown in Table 3.5.

In the remarks section of Table 3.5, the first column indicates the type of automated element that follows **RMK**. In this case, A02 signifies that the ASOS station is equipped with a precipitation sensor capable of determining precipitation type, such as the LEDWI. In contrast, older ASOS systems without this capability are coded as **A01**.

The second column of Table 3.5 details the peak wind indicator, which reflects the maximum three-second wind speed recorded over the past hour. This value is only reported when the wind speed exceeds 25 knots, so it may not appear in every METAR report. As in the main body of METAR report, the first three digits represent the meteorological wind direction, and the subsequent two or three digits

Table 3.5 An example of the remarks section of a METAR report from Monroe, Louisiana

Type of automated station	Peak wind and wind shift	Visibility group	Lightning group	Beginning and ending of precipitation and thunderstorms	Sea-level pressure	Precipitation group	Temperature group	Pressure tendency
AO2					SLP163	P0013 60033 70333	T00940079 10167 20094	51007

No Permission Required

correspond to the wind speed. Following the wind direction and speed, the report includes a four-digit timestamp showing when the peak wind occurred. For example, the code **PK WND 29042/0112** indicates that there was a peak wind of 42 knots at 290°, recorded at 0112Z.

The third column in Table 3.5 contains additional information on visibility. The most common **RMK** elements related to visibility are as follows:

1. **BINOVC** - *Breaks in overcast*: Indicates small clear patches within an otherwise overcast sky.
2. **CIG** - *Ceiling*: Refers to the height of the lowest broken (BKN) or overcast (OVC) layer.
3. **V** - *Variable visibility or ceiling***:** For example, **VIS 2V3** indicates visibility fluctuating between 2 and 3 statute miles, while **CIG025V030** indicates a variable cloud ceiling between 2,500 and 3,000 feet.

The fourth column in Table 3.5 provides information about the location and type of lightning when significant thunderstorms are nearby. The most common remarks regarding lightning include:

CG: Cloud to Ground
IC: Intracloud
CC: Cloud to Cloud
CA: Cloud to Air
OCNL: Occasional
FRQ: Frequent
CONS: Continuous

The fifth column in Table 3.5 includes remarks on the location, direction, and timing of thunderstorms. The most common remarks are the following:

1. **TSB:** This indicates the time when the thunderstorm began, based on the minutes past the hour. For example, the code **TSB24** means that the thunderstorm began 24 minutes past the hour.
2. **RAB:** This represents the time when rain began, expressed in the same format as **TSB.**
3. **LOC:** This indicates the location, usually expressed in terms of quadrants relative to the station, such as northeast, southwest, etc. The code **OHD** indicates that the storm is overhead, while **ALQDS** indicates storms in all quadrants.
4. **DIR:** This refers to the direction of the storm, also expressed in terms of quadrants relative to the station.

In our METAR example found in Table 3.5, no significant thunderstorm activity is present, so these fields are blank.

The sixth column in Table 3.5 contains remarks on sea-level pressure. This measurement differs from the altimeter setting: the altimeter setting is calculated using a standard atmospheric chart that adjusts for elevation, while the sea-level pressure in this remarks section is derived using temeprature data (discussed in Ch. 5.4). To distinguish it from the altimeter setting, we refer to it as **reduced sea-level pressure**. In general, pilots and the aviation industry use the altimeter setting, while meteorologists prefer reduced sea-level pressure, as it is often more accurate for forecasting. The three-digit code in this column (i.e., **SLP163** in our example) represents the last three digits of the pressure in hectopascals (hPa). From this, we determine that the sea-level pressure is 1016.3 hPa since a reading of 916.3 hPa would only occur in extremely intense hurricanes.

The seventh column in Table 3.5 contains information on precipitation accumulation. The **P** group reports the amount of precipitation that has fallen since the last standard METAR report, measured in hundredths of an inch. For example, in our METAR example, the code **P0013** indicates that 0.13 inches of precipitation has fallen in the past hour. Trace amounts of precipitation, defined as less than 0.01 inches, are typically coded as **P0000**. However, this should not be confused with a report of zero precipitation. If no precipitation has occurred, the **P** group will not be included in the METAR report.

The next group, which begins with **6**, reports either the 3-hour or 6-hour accumulated precipitation (also in hundredths of an inch), depending on the reporting time:

- At 03Z, 09Z, 15Z, and 21Z, this group reports the 3-hour precipitation total.
- At 00Z, 06Z, 12Z, and 18Z, this group reports the 6-hour precipitation total.

In our example, the report time is closest to **12Z**, so the code **60033** indicates that 0.33 inches of precipitation has fallen over the past 6 hours. The following group, which begins with **7**, reports the 24-hour accumulated precipitation, also in hundredths of an inch. In our example, the code **70033** indicates that 0.33 inches of precipitation has fallen over the past 24 hours.

The eighth column in Table 3.5 contains temperature remarks. The **T** group provides both the air temperature and the dew point to the nearest tenth of a degree Celsius, offering greater precision than the values in the main body of the METAR report. The digit immediately after **T** is **0** for positive values and **1** for negative values**,** followed by three digits specifying the temperature. The dew point is reported in the same format. For example, the code **T00940079** corresponds to a temperature of 9.4°C and a dew point temperature of 7.9°C. Additional groups provide temperature extremes. The group that begins with **1** reports the 6-hour maximum temperature, while the group that begins with **2** reports the 6-hour minimum temperature, using the same format as the **T** group. Thus, for our example, the code **10167** represents a 6-hour maximum temperature of 16.7°C, and the code **20094** represents a 6-hour minimum temperature of 9.4°C. These two groups only appear in routine METAR reports at 00Z, 06Z, 12Z, and 18Z but are valuable for forecast verification and climate records.

The tenth and final column in Table 3.5 provides the 3-hour sea-level pressure tendency, also called the **3-hour reduced sea-level pressure tendency**. The first digit after **5** describes the trend (using the classification in Table 3.6). The final three digits give the pressure change in tenths of hPa over the past 3 hours. For example, the code **51007** indicates that the reduced sea-level pressure changed by 0.7 hPa over the past 3 hours, with the trend showing an initial rise followed by a steady or slowly continuing rise. Significant pressure changes may also be explicitly reported using by **PRESRR** (pressure rising rapidly) or **PRESFR** (pressure falling rapidly), which appear after the **RMK A02** section.

3.2 Surface station plots

The previous section outlined the procedure for processing data from an individual surface weather station. In practice, however, weather observations are collected simultaneously from numerous weather stations around the world. To streamline the interpretation of surface observations across different regions, meteorologists use a shorthand notation known as a **station plot**. In this system, each observation station is represented by a symbol that summarizes the current weather conditions

Table 3.6 Pressure tendency table

Primary requirement	Pressure tendency description	Pressure tendency code
The current atmospheric pressure is higher than 3 hours ago	Increasing, then decreasing	0
	Increasing, then steady or Increasing, then increasing more slowly	1
	Increasing steadily	2
	Decreasing, then increasing or Steady, then increasing or Increasing, then increasing more rapidly	3
The current atmospheric pressure is the same as 3 hours ago	Increasing, then decreasing	0
	Steady	4
	Decreasing, then increasing	5
The current atmospheric pressure is lower than 3 hours ago	Decreasing, then increasing	5
	Decreasing, then steady or Decreasing, then decreasing more slowly	6
	Decreasing steadily	7
	Steady, then decreasing or Increasing, then decreasing or Decreasing, then decreasing more rapidly	8

No Permission Required

at that location. A schematic of the standard surface station plot is presented in Figure 3.2.

The temperature (**TT**) is located in the upper left of the station circle and is measured in degrees Fahrenheit. The dew point temperature (**T_dT_d**) is positioned in the lower left of the station circle, also measured in degrees Fahrenheit. The current weather (**ww**) symbol appears directly below the temperature. Figure 3.3 illustrates the common weather symbols used in station plots.

The precipitation recorded over the past 24 hours (**RR**) appears in the lower right and is measured in inches. Cloud cover (**N**) is shown by the amount of shading within the station circle. A completely unshaded circle indicates clear skies. If the circle is one-fourth shaded, it signifies few clouds (or partly cloudy conditions). A half-shaded circle indicates scattered clouds, whereas a three-fourths-shaded circle represents broken clouds. A fully shaded circle denotes overcast conditions.

Wind is represented by a wind barb showing both direction (**dd**) and speed (**ff**). The barb points in the direction *from which* the wind is blowing, and speed - measured in knots (kts) - is indicated by the number and type of flags on the barb (see Figure 3.4).

Sea-level pressure (**PPP**) is shown to the right of the station circle. Typical values for sea-level pressure range from 970 hPa to 1030 hPa, but the station plot only displays the last three digits. If **PPP** exceeds 500, add a "9" before the value and place a decimal between the last two digits. For example, the code **986** means 998.6 hPa. Conversely, if **PPP** is less than 500, add a "10" before the value, with the decimal in the same position. For example, the code **244** means 1024.4 hPa.

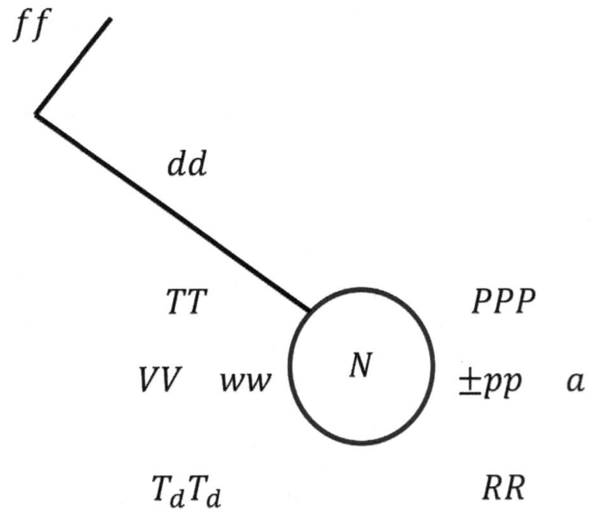

FIG. 3.2 Schematic of surface station plot. A description of each code is provided in the text.

-No Permission Required

FIG. 3.3 Common weather symbols on surface station plots.

(Courtesy of Weather Prediction Center [WPC] of the National Oceanic and Atmospheric Administration [NOAA].)

-No Permission Required

Pressure tendency is plotted to the right of the cloud cover symbol (**N**) and just below the sea-level pressure value (**PPP**). It consists of a magnitude (**±pp**) in tenths of hPa and a symbol (**a**) to indicate how the pressure has changed over the past 3 hours, as shown in Figure 3.5.

Box 3.1 provides an example (shown in Figure 3.6) illustrating how a surface station plot has been decoded.

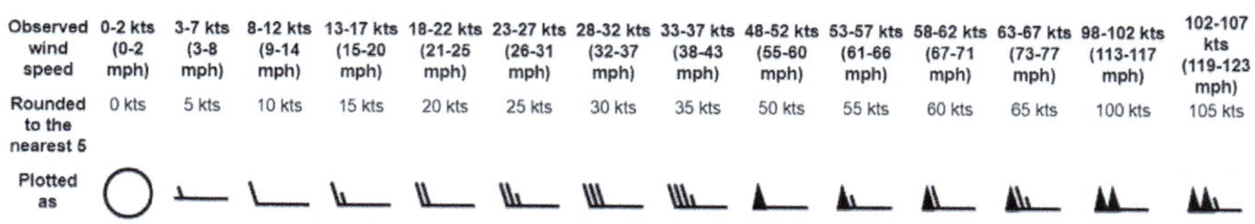

FIG. 3.4 Common wind speed symbols on surface station plots.

(Courtesy of Weather Prediction Center [WPC] of the National Oceanic and Atmospheric Administration [NOAA].)

-No Permission Required

3.3 Upper air observations

As mentioned in Ch. 2, radiosondes are the essential instruments for measuring upper-air observations. Approximately 1,000 upper-air stations, collectively known as the **global radiosonde network**, launch weather balloons equipped with radiosondes. These instruments measure atmospheric pressure, wind speed and direction, temperature, and humidity from the surface up to altitudes of about 30 km.

More than two-thirds of the stations, including all stations in the United States and Canada, launch weather balloons twice a day, at 00Z and 12Z. This schedule ensures that radiosonde data is collected at the same times each day for every station across the globe. During significant weather events, such as landfalling hurricanes or severe weather outbreaks, upper-air stations may also launch weather balloons at additional times, such as 06Z and 18Z.

In addition to these temporal limitations, the radiosonde network has spatial limitations. Oceans cover roughly two-thirds of Earth's surface, yet radiosonde observations over the ocean are taken by only about 15 ships equipped with Automated Shipboard Aerological Program (ASAP) facilities, most of which operate in the North Atlantic. On land, spatial coverage is uneven and largely depends on a country's resources, with less affluent regions generally having fewer stations than more developed areas.

FIG. 3.5

Pressure tendency symbols commonly used on surface weather plots.

(Courtesy of Weather Prediction Center [WPC] of the National Oceanic and Atmospheric Administration [NOAA].)

-No Permission Required

> **Box 3.1 Decoding a surface station plot**
>
> Consider the surface station plot shown in Figure 3.6. According to this plot, there is northwesterly wind flow at a speed of 30 knots. The current temperature is 27°F, with a dew point of 25°F. The weather conditions are overcast, with light rain reported at the station. The total accumulated precipitation over the past 24 hours is 0.22 inches. Additionally, the sea-level pressure is at 999.7 hPa, and this pressure has been continuously decreasing over the past 3 hours, with a pressure tendency of 1.1 hPa.

To help offset these limitations, aircraft-based observations are gathered from more than 4,000 aircraft in collaboration with the International Civil Aviation Organization (ICAO), the International Air Transport Association (IATA), and commercial airlines. These observations provide data on pressure, wind, temperature, and humidity. The Aircraft Meteorological Data Relay (AMDAR) system collects high-quality measurements of air temperature and wind speed and direction at cruising level, as well as at selected levels during ascent and descent. The volume of aircraft data has increased to an estimated 700,000 reports per day, corresponding to roughly 90,000 AMDAR profiles from 550 airports worldwide.

The **WMO TEMP** or **RAOB** format is the most common way to record radiosonde data. It is typically divided into 5 sections:

1. **TTAA** - Contains data from the *mandatory* pressure levels: surface, 1000 hPa, 925 hPa, 850 hPa, 700 hPa, 500 hPa, 400 hPa, 300 hPa, 250 hPa, 200 hPa, 150

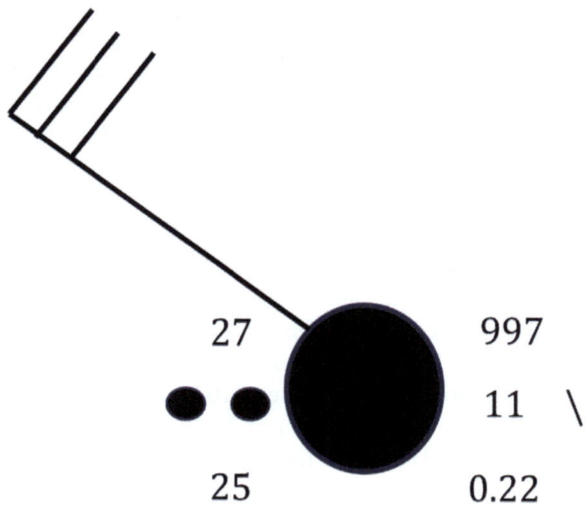

FIG. 3.6 An example of a surface station plot.

Station plot symbol.

-No Permission Required

hPa, and 100 hPa. This section reports temperature, dew point depression (defined as the air temperature minus the dew point temperature), height of each pressure level, wind direction, and wind speed.
2. **TTBB** - Contains data from *significant* levels, chosen where temperature or relative humidity changes noticeably with height. Sufficient points are included so the profile can be reconstructed accurately by connecting the points with straight lines, implying linear changes between them. This section reports temperatures, dew point depressions, and their corresponding pressure levels (in hPa).
3. **PPBB** - Presents wind direction and speed at significant levels, selected in portions of the sounding where wind changes nonlienarly with height. As with TTBB, the data allows linear interpolation between points.
4. **TTCC** - Similar to TTAA, but for mandatory levels below 100 hPa. The levels reported are 70, 50, 30, and 10 hPa.
5. TTDD - Contains significant-level data for pressures below 100 hPa.

We will begin by discussing mandatory level data.

3.3.1 Decoding mandatory level data

Let's start with an example of the TTAA section of a sounding as shown in Table 3.7.

The letters above each code in Table 3.7 identify a repeatable three-code pattern. The TTAA data is decoded as follows:

1. The superscript identifier **A** serves as a station identifier based on five-digit code from the World Meteorological Organization (WMO). In this example, **72518** is the station identifier for Albany, NY.
2. The superscript identifier **B** is used to identify the day and time of the radiosonde observation. To decode this portion, subtract 50 from the first two digits to obtain the day (e.g., in this case $81 - 50 = 31$). The next two digits indicate the time of the radiosonde observation (for this example, **00** means 00 UTC). The last digit signifies the highest level of wind data (in this instance, **1** denotes wind data up to 100 mb). Therefore, for this example, the radiosonde observation **81001** was taken at 00 UTC on the 31st day of the month.
3. The superscript identifier **C** indicates the mandatory pressure/height levels associated with the radiosonde launch. The first two digits identify the level, and the next three digits indicate either the surface station pressure (in hPa) or the height level (in meters). Table 3.8 shows the decoding method. In this example, the code **99030** corresponds to a pressure of 1030 hPa at the surface level, while the code **40736** corresponds to a height of 7,360 meters at the 400-hPa level.
4. The superscript identifier **D** refers to the temperature/dew point depression group for the preceding pressure/height level. The first three digits represent the tens, ones, and tenths digits for temperature. Note that if the tenths digit is an odd number, the temperature is negative. The final two digits are related to the dew point depression. If the dew point depression code is 50 or less, the digits give

Table 3.7 An example of the TTAA code for a radiosonde launch in Albany, NY

		B	**A**	**C**	**D**	**E**	**C**	**D**	**E**
	TTAA	81001	72518	99090	08863	33004	00263	10464	34016
C	**D**	**E**	**C**	**D**	**E**	**C**	**D**	**E**	**C**
92904	04461	00518	85588	00079	01526	70138	04182	32526	50572
D	**E**	**C**	**D**	**E**	**C**	**D**	**E**	**C**	**D**
17372	31549	40736	28766	31052	30935	45763	30055	25054	55158
E	**C**	**D**	**E**	**C**	**D**	**E**	**C**	**D**	**E**
29568	20193	63357	28573	15371	58560	27071	10624	62164	28053
C	**D**	**E**	**C**	**D**	**E**	**F**			
88192	64957	28579	77182	28581	41314	51515			

No Permission Required

the one and tenths of a degree. If greater than 50, subtract 50 to obtain the value in whole degrees. Using this framework, the code **08863** corresponds to a temperature of 8.8°C and a dew point depression of 13°C (resulting in a dew point of −4.2°C). Similarly, the code **28766** represents a temperature of −28.7°C with a dew point depression of 16°C (giving a dew point of −44.7°C).

5. The superscript identifier **E** is used to indicate the wind speed and wind direction associated with the preceding pressure/height level. The middle digit in the group can be a **5** for direction resolution or a **1** if the speed is at least 100 kt. If this digit is expressing both direction and speed data, then 500 should be subtracted from the value. For example, the code **22645** corresponds to a wind speed of 145 kt from 225° (i.e., from the southwest). In the same way, the code **34016** corresponds to a wind speed of 16 knots from 340° (i.e., from the northwest).
6. The superscript identifier **F** marks the end of the repeatable three-code pattern.

3.3.2 Decoding significant level data

Let's now examine an example of a portion of **TTBB** data, as shown in Table 3.9.

Similar to the TTAA data, the letters above each code identify a repeatable two-code pattern, where **C** represents a significant level indicator and **D** represents the temperature and dew point reading at that level. The TTBB data is decoded as follows:

Table 3.8 Decoding of mandatory pressure/height levels

Mandatory level code	Level	Pressure or height of the level
99	Surface	Pressure in hPa
00	1000 hPa	Height in meters
92	925 hPa	Height in meters
85	850 hPa	Last 3 digits of height in meters
70	700 hPa	Last 3 digits of height in meters
50	500 hPa	First 3 digits of height in meters
40	400 hPa	First 3 digits of height in meters
30	300 hPa	First 3 digits of height in meters
25	250 hPa	Middle 3 digits of height in meters
20	200 hPa	Middle 3 digits of height in meters
15	150 hPa	Middle 3 digits of height in meters
10	100 hPa	Middle 3 digits of height in meters
88	Tropopause	Pressure in hPa
77	Maximum wind level	Pressure in hPa

No Permission Required

1. The superscript indicator **C** corresponds to the significant level indicator. The first two repeating digits indicate the start of the data at that level, while the final three digits represent the pressure in hPa. Thus, the code **66793** indicates a significant level of 793 hPa.
2. The superscript indicator **D** indicates the temperature and dew point reading at this level, using the same decoding rules as superscript D in TTAA. Thus, the code **01884** corresponds to a temperature of 1.8°C and a dew point depression of 34°C, resulting in a dew point temperature of −32.2°C.

Using this framework, the code **33856 00571** signifies a significant level at 856 hPa, with a temperature of −0.5°C and a dew point depression of 21°C (yielding a dew point temperature of −21.5°C). Likewise, the code **44831 01886** indicates a significant level at 831 hPa, with a temperature of −1.8°C and a dew point depression of 36°C (resulting in a dew point temperature of −34.2°C). A comparison of these two readings shows a slight increase in temperature with height (commonly referred to as a **temperature inversion) and a sharp decrease in moisture**.

Finally, let's examine a portion of **PPBB** data, as shown in Table 3.10. The letters above the code indicate a repeatable four-code pattern:

1. The superscript **C** marks the beginning of a new level. The first digit (**9**) signals the start of a new section. The second digit represents the base height (in ten thousands of feet) for the next three observations. The third, fourth, and fifth digits specify the heights relative to the base (in thousands of feet) for the first, second, and third observations, respectively. For example, the code **92057** indicates a base height of 20,000 feet, with the three observations at this level being 20,000 feet, 25,000 feet, and 27,000 feet, respectively.
2. The superscript **D** corresponds to the wind data for the first observation level, also adhering to the same format as the wind measurements in TTAA.
3. The superscript **E** corresponds to the wind data for the second observation level, also adhering to the same format as the wind measurements in TTAA.
4. The superscript **F** corresponds to the wind data of the third observation level, also adhering to the same format as the wind measurements in TTAA.

Using this framework, the code **91234 26044 23572 23078** indicates a 44-knot wind from 260° at an altitude of 12,000 feet, a 72-knot wind from 235° at 13,000 feet, and a 78-knot wind from 230° at 14,000 feet. If the wind data is unavailable for a given level, it will appear as "//" in the code. For example, the code **916//** means that wind data is available at the first observation level (16,000 ft), but not for the second or third levels.

3.4 Upper-air station plots

Just as surface station plots were created to simplify the processing of surface observations across large regions, upper-air station plots were developed for the

Table 3.9 An example of the TTBB code for a radiosonde launch in Albany, NY

		B	A	C	D	C	D	C	D
	TTBB	81000	72518	00090	08863	11012	11264	22925	04461
C	**D**	**C**	**D**	**C**	**D**	**C**	**D**	**C**	**D**
33856	00571	44831	01886	55823	03086	66793	01884	77677	05980
C	**D**	**C**	**D**	**C**	**D**	**C**	**D**	**C**	**D**
88662	05768	99572	12964	11554	14368	22540	14971	33527	15962

No Permission Required

Table 3.10 An example of the PPBB code for a radiosonde launch in Albany, NY

		B	**A**	**C**	**D**	**E**	**F**	**C**	**D**	**E**
	PPBB	80000	72518	90012	00000	01517	02521	90346	02004	32009
F	**C**	**D**	**E**	**F**	**C**	**D**	**E**	**F**	**C**	**D**
25026	90789	25531	25535	25041	91234	26044	23572	23078	916//	23095
C	**D**	**E**	**F**	**C**	**D**	**E**	**F**	**C**	**D**	
92057	24089	24594	24594	90327	23612	24126	25634	950//	25093=	

No Permission Required

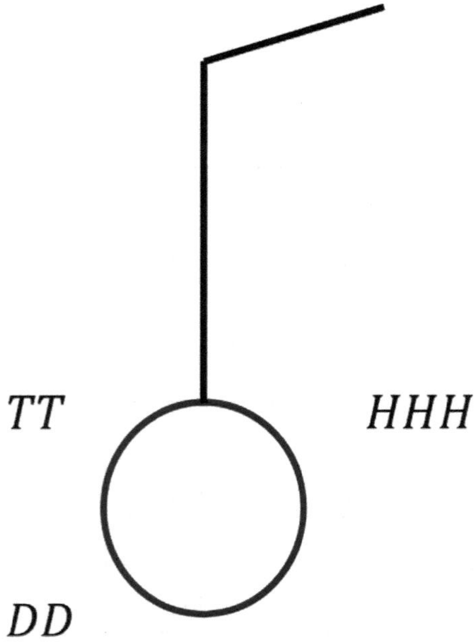

FIG. 3.7

Schematic of an upper-air station plot.

-No Permission Required

same purpose for radiosonde data.[4] Each weather station conducting rawinsonde observations uses a symbol to indicate the weather condition in the area, as shown in Figure 3.7.

Each symbol on the plot represents a measured variable from a rawinsonde launch. The temperature (**TT**) is in the upper left corner and is measured in degrees Celsius. The dew point or dew point depression (**DD**) is shown in the lower-left corner, also measured in degrees Celsius. If **DD** is greater than **TT**, then the dew point depression is plotted; otherwise, **DD** represents the dew point temperature. The name of the upper-air station is shown in the lower-right corner, while the three digits in the upper right corner represent the height (**HHH**) corresponding to a given pressure level.

As discussed in Ch. 5, since gravity slightly decreases with height, the radiosonde records the **geopotential height**, which is a modified height measurement that accounts for the variation in gravity. Decoding the geopotential height depends on the pressure level at which the reading is taken and follows the same decoding rules as indicated in Table 3.8. For example,

[4] For readers who are interested in visualizing upper-air observation data using Python, see the Jupyter notebook and other resources found on the textbook webpage.

- 850-hPa: The average geopotential height is about 1,500 m. The first digit is typically omitted, so a "1" should be added. Therefore, a reading of **478** at 850-hPa corresponds to 1,478 meters.
- 700-hPa: The average geopotential height is about 3,000 m. The first digit is typically omitted, so "2" or "3" should be added. Therefore, a reading of **150** at 700-hPa corresponds to 3,150 meters.
- 500-hPa or 300-hPa: The last digit is typically omitted, so a "0" should be added. Thus, a reading of **540** corresponds to 5,400 meters for a 500-hPa observation, while a reading of **930** corresponds to 9,300 meters for a 300-hPa observation.
- 200-hPa and 250-hPa: Both the leading "1" and trailing "0" are omitted. Therefore, a reading of **080** would correspond to 10,800 meters.

Box 3.2 provides three examples (shown in Figure 3.8) illustrating how these rules apply to upper-air station plots.

Now that we can decode surface weather observations (from METAR reports or surface station plots) and upper-air weather observations (from radiosonde codes or upper-air station plots), we are ready to analyze synoptic weather maps, which will be introduced in Ch. 4.

Questions

For Problems 1 to 3, decode the body of the METAR reports item-by-item. Convert the physical quantities into SI units and convert the time into local time. You may ignore the effects of daylight savings time.

Box 3.2 Decoding upper-air station plots

Figure 3.8 presents 3 examples of upper-air station plots.

In Figure 3.8a 700-hPa observation is shown with a temperature of –5°C and wind speed of 30 knots from the south. Since the dew point indicator is higher than the temperature indicator, we note the dew point depression. A dew point depression of 5°C indicates a dew point of –10°C for this observation. Given that this is a 700-hPa observation, the geopotential height should be about 3000 meters. Therefore, we add a "2" to the geopotential height indicator, resulting in a geopotential height observation of 2927 meters.

Figure 3.8b illustrates a 250-hPa observation where the temperature is –49°C, with a dew point depression of 12°C, leading to a dew point of –61°C. The wind is recorded at 70 knots from the northwest. As this is a 250-hPa observation, the geopotential height should be around 10,000 meters. Thus, adding a leading "1" and a trailing "0" to the geopotential height indicator provides a geopotential height observation of 10,690 meters.

Lastly, Figure 3.8c displays a 500-hPa observation with a temperature of –13°C and a dew point depression of 3°C, which corresponds to a dew point of –16°C. The wind speed is 25 knots from the southwest. For this 500-hPa observation, adding a trailing "0" to the geopotential height indicator gives a geopotential height observation of 5760 meters.

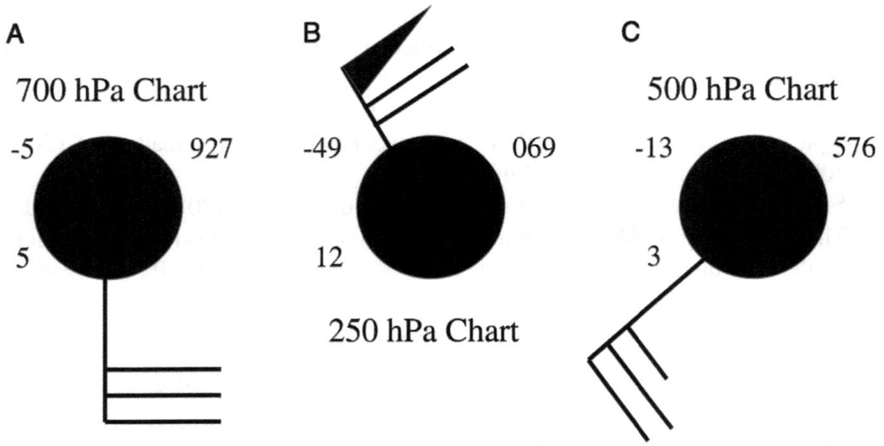

FIG. 3.8 Examples of upper-air station plots

(A) Corresponds to a 700-hPa station plot, (B) corresponds to a 250-hPa station plot, and (C) corresponds to a 500-hPa station plot.

-No Permission Required

1. KJFK 211951Z 21012KT 10SM BKN250 21/08 A3002 RMK AO2 SLP165 T02110083=
2. KMCB 022315Z AUTO 34010KT 1/4SM +TSRA FG BKN005 OVC035 24/23 A3009 RMK AO2 LTG DSNT ALQDS P0091 $
3. KDFW 022319Z 20003KT 10SM TS FEW037 SCT050CB BKN065 OVC130 27/20 A2998 RMK AO2 FRQ LTGICCG TS OHD MOV E-NE

For Problems 4 to 6, decode the remarks section of METAR reports item-by-item.

4. KSAT 022253Z 17010KT 10SM SCT034 BKN130 BKN250 28/23 A2998 RMK AO2 RAE42 SLP133 FEW CB DSNT NW-N P0001 T02830233.
5. KCON 131151Z AUTO 09009KT 1 3/4SM +RA BR OVC010 09/07 A3005 RMK AO2 CIG 007V013 SLP177 P0015 60056 70066 T00890072 10094 20089 53018
6. KABC 121755Z AUTO 21016G24KT 180V240 1SM R11/P6000FT -RA BR BKN015 OVC025 06/04 A2990 RMK AO2 PK WND 20032/25 WSHFT 1715 VIS 3/4V1 1/2 VIS 3/4 RWY11 RAB07 CIG 013V017 CIG 017 RWY11 PRESFR SLP125 P0003 60009 T00640036 10066 21012 58033 TSNO $

For Problems 7 to 9, draw the surface station plots associated with the following METAR reports using standard U.S. plotting conventions:

7. KJAN 021754Z 10005KT 10SM FEW046 31/20 A3008 RMK AO2 SLP182 T03110200 10311 20233 50004=

8. KMCI 022353Z 03004KT 10SM FEW025 25/16 A3002 RMK AO2 SLP159 T02500156 10261 20233 56008=

9. KDAB 251841Z 21019G34KT 1/2SM +TSRA FEW007 BKN022CB OVC30 23/21 A2983 RMK AO2 PK WND 27034/1837 TSB32RAB35 TS ALQDS FRQ LTGC ALQDS

For Problems 10 to 12, draw upper-air station plots associated with the upper-air data using U.S. conventions. Express your moisture in terms of dewpoint depression.

10. A temperature of 50°*F*, a dew point of 45°*F*, a wind velocity of 17 knots from the southeast, and a geopotential height of 1,560 m at 850-hPa.

11. A temperature of −7°*C*, a dew point of −28°*F*, a wind velocity of 33 knots from the north, and a geopotential height of 5,340 m at 500-hPa.

12. A temperature of −33°*C*, a dew point of −52°*C*, a wind velocity of 97 knots from the west, and a geopotential height of 10,570 m at 250-hPa.

13. Using the radiosonde data found in Table 3.11,

a. Calculate the geopotential height (in meters) at all mandatory levels. Present your data in tabular form, with one column for geopotential height (in meters) and another column for pressure (in hPa).

b. The relationship between pressure and geopotential height can be expressed as: $P = P_0 \quad e^{-z/H}$ where P_0 is the initial pressure (in hPa) and H is known as the scale height (in meters). After you have collected your data, create a graph plotting pressure (in hPa) vs geopotential height (in meters). Fit an exponential trendline to your data and use the equation of your trendline to calculate the scale height.

14. Using the radiosonde data found in Table 3.11,

a. Determine the wind speed and the wind directionat all significant levels from the surface up to 39,000 feet. Place your data in tabular form where one column corresponds to height (in ft), one column corresponds to wind speed (in knots), and one colum corresponds to wind direction (in degrees).

b. Use your data to plot wind speed (in knots) vs. height (in feet) and wind direction (in knots) vs. height (in feet). Place both curves on the same graph and use two different y-axes to plot your data. Give a brief description of how the wind speed and wind direction vary with height.

15. Vertical wind shear refers to the change in wind speed and/or direction with height. This can be expressed mathematically as: $\frac{\mathbf{V}_f - \mathbf{V}_i}{z_f - z_i}$ where $\mathbf{V}_f$is the horizontal wind vector at height $z_f$and $\mathbf{V}_i$ is the horizontal wind vector at height z_i, where $z_i < z_f$. Using the radiosonde data from Table 3.11, (a) Calculate the vertical wind shear between 925-hPa and 850-hPa and (b) Calculate the vertical wind shear between 850-hPa and 500-hPa. For part (a), express your answer in unit-vector notation. For part (b), express your answer in terms of magnitude and meteorological direction.

Table 3.11 Radiosonde data for Questions 13–16

	TTAA	61001	72208	99022	17029	05504	00196	15617	08017
92860	14023	12529	85569	09818	11524	70163	01817	16017	50582
12170	29009	40749	22781	32522	30953	39365	32525	25076	46566
33540	20220	57761	32533	15397	66962	30027	10640	72563	25529
88123	70163	28030	77999	31313	58208	82314	51515	10164	00004
10194	11024	13517							
	TTBB	61008	72208	00022	17029	11018	16427	22995	15414
33977	17425	44936	14614	55918	13831	66890	12029	77882	11850
88871	11018	99844	09424	11839	09013	22831	08828	33824	08412
44820	08423	55816	08411	66809	08214	77801	09441	88723	02401
99697	01820	11687	02650	22679	02658	33612	03543	44611	03550
55607	02761	66601	02771	77578	04174	88561	05968	99551	06383
11530	08770	22520	09575	33505	11573	44488	13363	55480	14170
66464	16171	77443	17182	88372	26976	99363	27576	11308	37963
22263	44963	33259	44964	44242	48366	55210	56360	66196	58562
77175	64356	88169	64360	99161	65761	11157	65561	22143	67762
33136	65964	44123	70163	55117	68764	66108	69164	31313	58208
82314	41414	856//							
	PPBB	61008	72208	90012	05504	10023	11528	90346	12523
26012	25514	9089/	26514	28516	91246	29523	29024	30528	92045
30535	31543	31043	93025	32037	32538	34061	939//	33540	945//
31034	9503/	31035	31021=						

No Permission Required

16. Using the radiosonde data found in Table 3.11,

a. Calculate the temperature and dew point at all significant levels from the surface to 175-hPa in Fahrenheit. Present your data in tabular form where one column corresponds to geopotential height (in meters), one colum corresponds to temperature (in degrees Celsius), and one colum refer to dewpoint (in Celsius). You will need to use your trendline equation from Problem 13 to estimate the geopotential height at each pressure level.

b. Use your data to graph dewpoint (in Celsius) vs. height (in km) and temperature (in Celsius) vs. height (in km). Place both curves on the same graph.

[illegible]

CHAPTER

Isopleth analysis

4

Chapter outline

Up to this point, we have shown that the state of the atmosphere is assessed by measuring temperature, pressure, moisture, and wind velocity at the surface (primarily using ASOS stations) and above the surface through radiosondes. Surface observations are converted into coded formats (such as METAR) and then plotted on surface charts. Similarly, upper-air observations are converted into radiosonde codes, which are then plotted on upper-air charts.

Because weather forecasters and analysts are typically interested in examining meteorological data across entire regions rather than at a single station, weather map analysis is an essential tool for understanding the state and evolution of the atmosphere. In this chapter, we will introduce the principles of synoptic weather map analysis and use them to identify common weather phenomena, such as fronts, troughs, ridges, and jet streams. We will begin by discussing the philosophy behind weather map analysis.

An Introduction to Synoptic-Dynamic Meteorology. https://doi.org/10.1016/B978-0-443-24858-0.00004-6

4.1 Philosophy of isopleth analysis

Weather forecasting and analysis require meteorologists to understand a wide range of processes and phenomena that operate on different spatial and temporal scales. For this reason, interpreting groups of meteorological observations is essential for determining atmospheric structure, understanding atmospheric phenomena, and forecasting the evolution of weather systems.

Meteorologists use a technique known as **contour analysis** (or **isopleth analysis**) to gain a clearer understanding of the information provided by the data. Contouring involves connecting points that represent equal values of a variable, forming lines known as **isopleths**. This technique allows meteorologists to create weather maps by linking data collected from surface observation stations and radiosondes. These measurement sites may be located just a few kilometers apart or separated by hundreds of kilometers. The scale of the analysis is determined by both the geographic area examined and the time interval between measurements. The most common isopleths plotted on weather maps are presented in Table 4.1.

Isopleth analysis enhances our understanding of synoptic-scale weather. For instance, consider a forecaster examining a map of surface temperature over a portion of the United States, as shown in Figure 4.1A. Even if data points are relatively sparse, it can take considerable time to discern which regions are warmer or colder. By drawing isotherms on the map, as shown in Figure 4.1B, the meteorologist can quickly identify patterns, such as the presence of cold fronts in northwestern Arkansas and in southern Nebraska.

Overall, isopleth analysis is crucial for identifying synoptic-scale features, including fronts, troughs, ridges, jet streams, and surface weather systems. When wind patterns are displayed alongside the isotherms, the forecaster can also determine where warm and cold air is being transported, which helps predict future temperature changes in different regions.

Isopleth analysis plays a crucial role in developing a mental model of the atmosphere. Research indicates that experienced forecasters often construct these models by synthesizing multiple types of data, then adjusting them to match the specific weather patterns present at a given time (Pliske et al., 2011). In essence, skilled forecasters use isopleth analysis, along with other tools, to assess the current state of the atmosphere

Table 4.1 Common isopleths for meteorological maps

Meteorological Variable	Isopleth Name
Pressure	Isobars
Temperature	Isotherm
Geopotential height	Isohypse
Dew point	Isodrosotherm
Wind speed	Isotach

No Permission Required

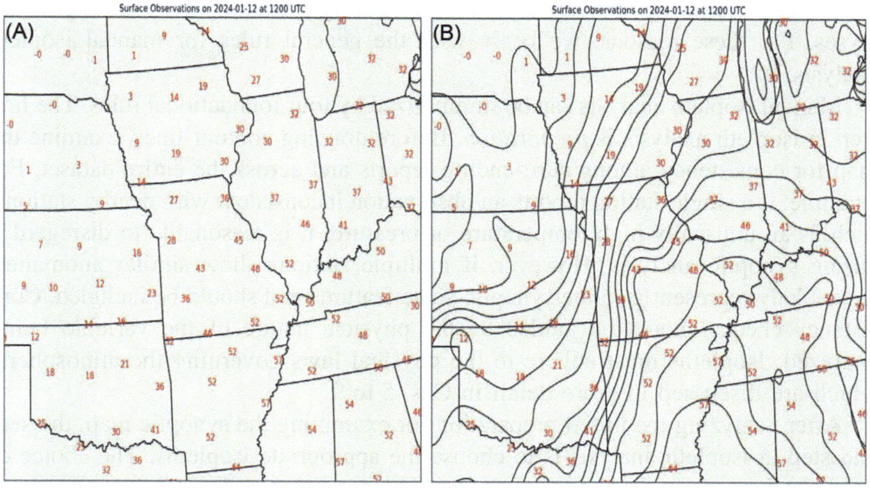

FIG. 4.1 Surface temperature map. (A) Uncontoured map of surface temperature on January 12, 2024, at 1200 UTC. (B) Contoured map of surface temperature on January 12, 2004, at 1200 UTC. Isotherms are given at 4°F intervals.

No Permission Required

and visualize the evolution of synoptic-scale weather systems.[1] Therefore, isopleth analysis is an essential part of the meteorologist's toolkit for analyzing weather patterns. In the next section, we will explore the basic principles of isopleth analysis.

4.2 Principles for manual isopleth analysis

Isopleth analysis can generally be divided into two types: manual and digital. Manual analysis is performed by hand, while digital analysis uses contouring algorithms on a computer. Despite advancements in computer-based methods, learning manual techniques remains valuable. One reason for this is the strong connection between our hands and minds- manual analysis fosters deeper learning. Research has shown that the ability to organize seemingly unrelated meteorological information (each requiring careful examination) is an important skill for any meteorologist (McNeal et al., 2019). Engaging in manual isopleth analysis teaches analysts how to process data and "see" the structure and order in the observations.[2] Another

[1] Pliske et al. (2011) also noted that "some forecasters reported visualizing the atmosphere as a fluid and picturing atmospheric dynamics within that fluid space." We will discuss atmospheric dynamics in Chs. 5–8 in the textbook and apply these insights in Chs. 9–12 in the textbook.

[2] The concept of harmonizing "cacophonous" data seems to map well to McNeal et al. (2019)'s concept of disembedding, who wrote "[D]isembedding skill positively affects meteorology skill with student (low knowledge) meteorologists as well as experienced (high knowledge) meteorologists. The advantage appears to lie in the ability to identify patterns, be they embedded in distracting background displays, repeated in weather scenarios, or present in observational data."

advantage of knowing manual methods is the ability to identify errors in digital analyses. For these reasons, we begin with the general rules for manual isopleth analysis.

Manual isopleth analysis can be summarized by four foundational rules. The first step in isopleth analysis is *preparation*. Before drawing contour lines, examine the map for consistency among surrounding reports and across the entire dataset. For example, if a single station reports an observation inconsistent with nearby stations, such as an unusually high temperature or pressure, it is reasonable to disregard it during synoptic analysis. However, if multiple stations show similar anomalies, these likely represent genuine synoptic-scale features and should be included. Consistency checks should be guided by the physical nature of the variable being analyzed. Isopleths must adhere to the physical laws governing the atmosphere, which are discussed in more detail in Chs. 5 to 9.

After analyzing the initial preparation for examining the synoptic map, the second step in isopleth analysis is to choose the appropriate isopleths. The choice of isopleths depends on the chart type and level being analyzed. Table 4.2 lists common conventions for surface and upper-air charts. The isopleths selected determine what the forecaster is emphasizing at a given level. For example, note that only isohypses and isotachs are typically plotted on 200-, 250-, and 300-hPa charts, indicating that wind features are the most significant at these altitudes. For 500-hPa charts, isohypses and isotherms are typically plotted, reflecting both dynamic and thermodynamic processes. For 700-hPa, 850-hPa, and surface charts, isohypses, isotherms, and isodrosotherms are typically plotted, emphasizing the role of moisture processes.

After selecting the isopleths, the third step is to create a "first estimate" by sketching the isopleths. Most station observations fall between two isopleth values, so interpolation is used to locate the contours. For example, if the geopotential height is 5,340 m at one station and 5,460 m at a nearby station, the 5400-m isopleth should be drawn approximately halfway between the stations. If the values are 5,420 m and 5,520 m, the 5460-m contour should be placed closer to the first station, reflecting proportional differences. Figure 4.2 illustrates this with the 9360-m isohypse positioned closer to Shreveport, LA (SHV) than to Little Rock, AR (LZK). This same principle applies to isotherms and isodrosotherms.

From this interpolation process follow two important implications. First, isopleths must be drawn smoothly, without unrealistic small-scale fluctuations unless clearly supported by data. Second, spacing between isopleths should be uniform, unless physical reasoning supports variations. For example, Figure 4.3 shows uneven isohypse spacing over Nebraska, Kansas, and Arkansas due to genuine geopotential height gradients, while isotherms exhibit greater small-scale variability. This pattern is common in many synoptic charts: isobars and isohypses are typically smooth, whereas isotherms and isodrosotherms contain more small-scale detail for synoptic-scale map analysis. At this stage, think of the map analysis as a first draft. Errors are expected, which is why pencil is preferred for manual analysis.

Table 4.2 Standard conventions for isopleth analysis on upper-air charts

Upper-Air Charts	Geopotential Height or Pressure	Temperature	Moisture	Other
200 hPa	120 m isohypses, centered at 12,000 m	N/A	N/A	Isotachs shaded at ≥ 75 kt
250 hPa	120 m isohypses, centered at 10,000 m	N/A	N/A	Isotachs shaded at ≥ 75 kt
300 hPa	120 m isohypses, centered at 9,000 m	N/A	N/A	Isotachs shaded at ≥ 75 kt
500 hPa	60 m isohypses, centered at 5,400 m	5°C isotherms, centered at 0°C	N/A	N/A
700 hPa	30 m isohypses, centered at 3,000 m	5°C isotherms, centered at 0°C	Dewpoint depression shaded at ≤ 3°C	N/A
850 hPa	30 m isohypses, centered at 1,500 m	5°C isotherms, centered at 0°C	5°C isodrosotherms centered at 0°C	N/A
Surface	4 hPa isobars, centered at 1000 hPa	4°F isotherms, centered at 50°F	5°F isodrosotherms centered at 50°F	N/A

No Permission Required

After creating the rough draft of the isopleth analysis, we proceed to the fourth and final step: refining and finalizing. In the final step, we refine the contours and check for accuracy. Two key rules apply:

- Isopleths cannot touch or cross each other, as this implies different values for a meteorological variable at a single point.
- Isopleths should not stop abruptly within the data region, nor extend beyond the limits of available data.

Because this is the final stage of analysis, the analyst should revise the contours so they reflect both the physical processes involved and the meteorological features observed. In Figures 4.2 and 4.3, faint erased pencil lines show where adjustments were made after review.[3]

[3] Even the most experienced analyst will revise their isopleth analysis at least once for some location on a synoptic map. Therefore, an analyst must be willing to make multiple revisions of a synoptic map to ensure accuracy.

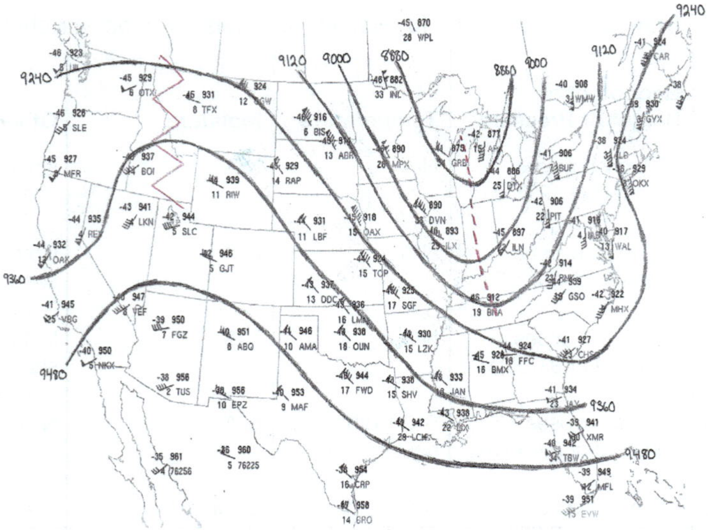

FIG. 4.2 Example of manual map analysis for 300-mb map. Manual isopleth analysis of 300-hPa map on December 18, 2023, at 12Z. The solid black contours are isohypses at 120-meter intervals, centered at 9000 m. Shaded contour fill are isotachs greater than 75 knots.

No Permission Required

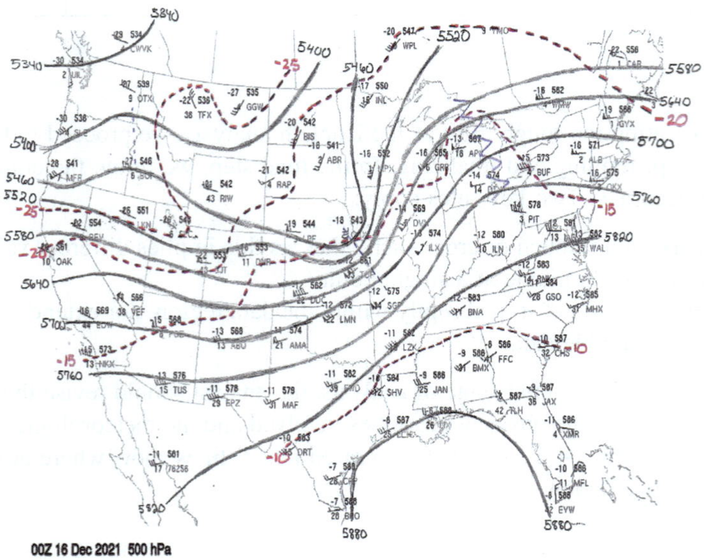

FIG. 4.3 Example of manual map analysis for 300-hPa map. Manual isopleth analysis of 500-hPa map on December 16, 2021, at 00Z. The solid black contours are isohypses at 60-meter intervals, centered at 5400 m. The dashed red contours are isotherms at 5°C intervals, centered at 0°C.

No Permission Required

Once the contours are finalized, they should be labeled clearly. As shown in Figures 4.2 and 4.3, labels are placed at both ends of each contour, and each contour is marked with a specific color. Generally, isohypses are represented by solid heavy black lines, isotherms by dashed red lines, isodrosotherms by dashed green contours, and isotachs by solid thin purple lines.

With contouring methods established, we are ready to identify synoptic systems on upper-air charts. Forecasters typically use a "top-down" approach, beginning synoptic analysis at the highest mandatory levels (such as 200, 250, and 300 hPa) and working downward toward the surface. Upper-tropospheric charts (above 500 hPa) are preferred starting points because they are less influenced by friction, heating, evaporation, and other surface effects. Analyzing these charts first reveals large-scale flow patterns that guide the placement of fronts and other weather systems on lower-level charts. In this way, upper-tropospheric charts serve as a "master key", simplifying complex weather situations into their most fundamental components. In the next sections, we will examine the major features commonly found on synoptic-scale charts.

4.3 Synoptic features on upper-tropospheric charts

In this section, we will examine the most used upper-tropospheric charts: the 200-, 250-, and 300-hPa charts, as well as the 500-hPa chart. At these altitudes, thermodynamic processes have minimal impact on the structure and evolution of synoptic-scale systems. However, features such as jet streams, jet streaks, troughs, ridges, short waves, lows, and highs are prominent at these levels. We will start by examining the 200-, 250-, and 300-hPa charts.

4.3.1 The 200-, 250-, and 300-hPa charts

Figure 4.4 shows a 300-hPa contoured chart of the continental United States from January 8, 2024, at 12 UTC. Many features of this chart are common to 200-, 250-, and 300-hPa charts. First, the geopotential heights are organized into a wave pattern that spans the continental United States. These patterns are known as **long waves** or **Rossby waves**. On a typical day, four or five long waves encircle the globe in the midlatitudes. Rossby waves are best observed between 300 and 200 hPa. As illustrated in Figure 4.4, a Rossby wave across the continental United States produces regions of lower geopotential height toward the equator and higher geopotential height toward the poles. The equatorward section, with lower geopotential height, is referred to as a **trough**, whereas the poleward section, with higher geopotential heights, is known as a **ridge**. In other words, ridges and troughs are large-scale undulations of Rossby waves. As we will discuss in Ch. 10, ridges are generally associated with warm, stable weathernear the surface, which is why regions downstream of ridges often experience fair weather. In contrast, troughs

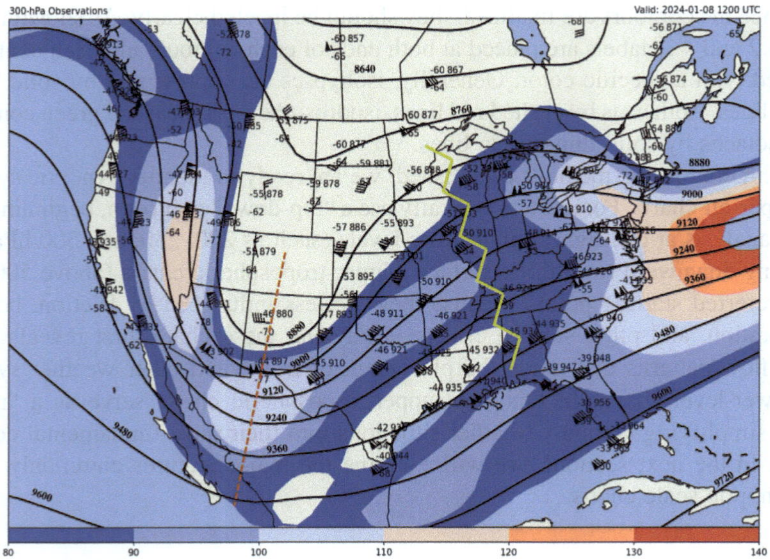

FIG. 4.4 300-hPa isobaric chart of the continental United States from January 8, 2024, at 12 UTC. The solid black contours correspond to isohypses at intervals of 120 meters. The contour fill colors correspond to isotachs greater than 80 knots. The polar jet is present across the continental United States. The dashed brown contour corresponds to the longwave trough axis, and the zigzag line corresponds to the longwave ridge axis.

No Permission Required

are typically linked to colder, more unstable conditions near the surface, leading to unsettled weather in downstream areas.

The center of the trough, referred to as the **trough axis** or the **trough line**, is typically denoted with a dashed line. In contrast, the center of the ridge, known as the **ridge axis** or the **ridge line**, is labeled with a zigzag symbol. Trough axes should be drawn through the base of the trough so that geopotential heights on either side are greater than those along the axis. Similarly, ridge axes should be drawn through the peak of the ridge so that the geopotential heights on either side are lower than those along the axis.

It is important to note that neither the trough axis nor the ridge axis is oriented strictly north-south. For instance, a trough axis that extends from the southwest to the northeast is an example of a **positively tilted trough**. Conversely, a ridge axis with a northwest-to-southeast orientation is known as a **negatively tilted ridge**.[4] As discussed in Ch. 11, the tilt of the trough and the ridge axes is dynamically linked to the evolution of the trough-ridge system. Generally, troughs start with a positive

[4] This convention is based upon the Northern Hemisphere. In the Southern Hemisphere, the positive tilt will be southeast to northwest and the negative one southwest to northeast.

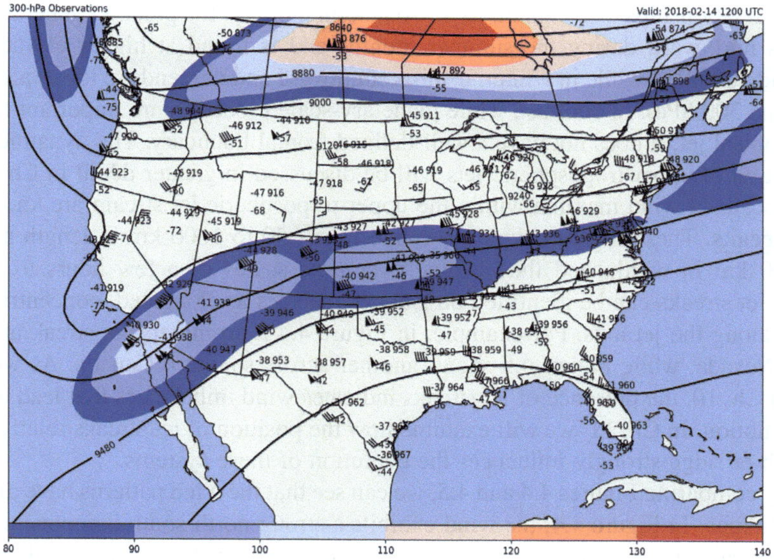

FIG. 4.5 300-hPa isobaric chart of the continental United States from February 14th, 2018, at 12 UTC. The solid black contours correspond to isohypses at intervals of 120 meters. The contour fill colors correspond to isotachs greater than 80 knots. The polar jet is present across southern Canada, and the subtropical jet is present across the Midwest portion of the United States.

No Permission Required

tilt as cold air is directed toward the equator. Over time, as cold air moves eastward, the trough evolves from positively tilted to negatively tilted. Therefore, a positively tilted trough often indicates the strengthening of a trough system, whereas a negatively tilted trough typically signals that a low-pressure system has reached its mature stage. The dynamics and evolution of Rossby waves, particularly their relation to trough and ridge dynamics, will be explored in more detail in Ch. 11.

The second major feature observed in Figure 4.4 is that wind patterns align with the wave patterns associated with Rossby waves. Specifically, Figure 4.4 shows a region of intense, narrow winds exceeding 80 knots, known as the **jet stream**. Jet streams range from 500 to 6000 km in length and 100 to 400 km in width, and they are significant features commonly seen on 200-, 250-, and 300-hPa maps.

In the Northern Hemisphere, the two jet streams that influence midlatitude weather are the **polar jet**, depicted in Figure 4.4, and the **subtropical jet**, shown across the continental United States in Figure 4.5. The polar jet is present year-round and can be identified on maps above 300 hPa, whereas the subtropical jet is primarily a winter phenomenon located between 20° and 35°N/S latitude.

As discussed in Ch. 10, the location and strength of the polar jet are associated with strong lower-tropospheric temperature gradients and strong vertical shear.

These relationships also explain its seasonal climatology: the polar jet generally migrates northward during the summer months and is found at higher elevations (around 200 hPa), while in winter, it shifts southward and descends to lower altitudes (around 300 hPa). In contrast, while planetary-scale processes are important for the subtropical jet, it does not have a well-defined frontal boundary. The dynamics and evolution of upper-tropospheric jets will be discussed in greater detail in Ch. 10.

Localized wind maxima within the upper-tropospheric jet stream are known as **jet streaks**. These features typically range from 250 to 1000 km in length and 50 to 200 km in width, and they evolve over time scales of a few hours to a few days. Jet streaks can be identified on isotach analyses as elongated, concentric patterns along the jet axis. For example, in Figure 4.4 a prominent jet streak appears over Nevada, while Figure 4.5 shows another across southern Canada. As we will see in Ch. 10, the presence of jet streaks indicates wind imbalances that lead to vertical motion. In Ch. 11, we will examine how the position of jet streaks relative to a trough or ridge strongly influences the evolution of these systems.

By comparing Figures 4.4 and 4.5, we can see that the wind patterns have distinct orientations. In Figure 4.4, the wind exhibits a strong north-south component, with equatorward flow at the base of the trough and poleward flow near the ridge. This type of flow is known as **meridional flow**, which is generally associated with strong troughs and ridges. Identifying meridional flow is important because it often signals the poleward transport of warm air or equatorward transport of cold air. As a result, the more meridional the flow is, the more extreme the associated weather. For example, heat waves and droughts often occur under strong ridges, whereas cold air outbreaks, cloudiness, and precipitation frequently occur downstream of strong troughs. In contrast, Figure 4.5 illustrates a predominantly west-east flow, known as **zonal flow**, which is typically linked to weak troughs and ridges Zonal flow is generally associated with relatively calm weather.

Finally, Figures 4.4 and 4.5 show that the wind field largely follows isohypses. In Figure 4.5, the zonal flow runs parallel to these isohypses. This type of upper-tropospheric wind pattern, characterized by straight paths on the synoptic scale, is called **geostrophic flow**. InFigure 4.4, the wind flows parallel to the curved Rossby wave pattern, producing a type of upper-tropospheric flow known as **gradient flow**. The physical mechanisms responsible for both geostrophic and gradient flow will be discussed in Ch. 7.

4.3.2 The 500-hPa chart

Figure 4.6 shows a 500-hPa contoured chart of the continental United States from January 8, 2024, at 12 UTC. The 500-hPa level is typically considered the middle of the troposphere. Many features observed in this chart are common for 500-hPa maps, and several large-scale features shown in Figure 4.4 also appear here. Jet stream patterns remain visible, and trough-ridge systems are superimposed on them. However, a comparison between Figures 4.4 and 4.6 reveals that the winds are generally weaker at 500 hPa than at 200-300 hPa. In Ch. 8, we will explain that this difference occurs throughout the atmosphere due to a fundamental synoptic-scale balance known as thermal wind balance. Another key difference is that the trough axis at 300 hPa is positioned west of the trough axis at 500 hPa. As we will show in Ch. 13, this westward tilt of the trough with increasing height often supports the development of surface low-pressure centers.

The second major feature observed in Figure 4.6 is the presence of small-scale disturbances superimposed on the large-scale structures, such as the trough over the Mountain West. These are called **short waves**. Their presence makes contouring at this level less smooth compared with the 200-300 hPa levels. Short waves can be thought of as smaller troughs and ridges embedded within the large-scale flow.

Rossby waves and short waves share similarities but also have important differences (discussed in Ch. 11). Two key differences are worth noting here. First, there are typically four to five Rossby waves across the hemisphere at any given time, whereas short waves span 15° to 40° longitude (roughly 1300 to 3400 km). Second, Rossby waves generally move eastward at a speed of 10 to 15 knots, while short waves tend to move faster and often have a meridional component. This allows short waves to propagate through Rossby waves, sometimes distorting their pattern northward or southward. As we will demonstrate in Ch. 11, short wave troughs can

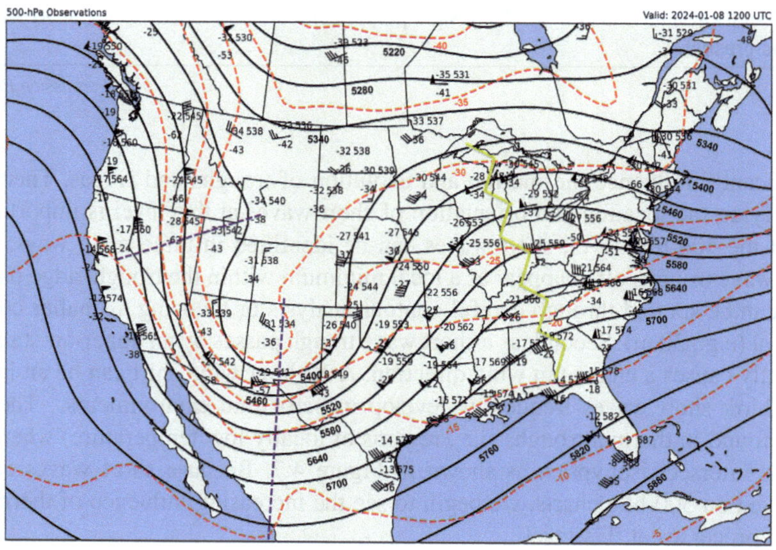

FIG. 4.6 500-hPa isobaric chart of the continental United States from January 8, 2024, at 12 UTC. The solid black contours correspond to isohypses at intervals of 60 meters, and the dashed red contours correspond to isotherms at intervals of 5°C. The dashed purple contour corresponds to the trough axes, and the zigzag contour correspond to the ridge axis.

No Permission Required

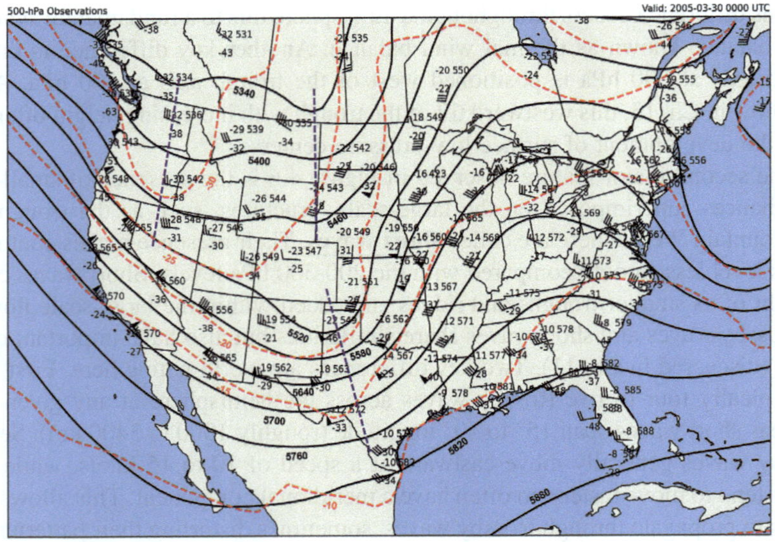

FIG. 4.7 500-hPa contoured map with many shortwaves. 500-hPa isobaric chart of the continental United States from March 30, 2005, at 00 UTC. The solid black contours correspond to isohypses at intervals of 60 meters, and the dashed red contours correspond to isotherms at intervals of 5°C. The dashed purple contours correspond to the short-wave troughs present.

No Permission Required

significantly influence the motion and evolution of troughs and ridges. Therefore, identifying the location and orientation of short waves at this level is important.

At the 500-hPa level, short waves can be identified in three main ways. First, short wave troughs often appear as a local minimum within the trough-ridge pattern. This can be spotted through careful contour analysis or by using a smaller contour interval (e.g., 30 m). Second, as a short wave trough passes by an upper-air station, it typically causes a change in wind direction. A 12-hour wind shift can be an indicator. Third, short waves frequently develop in baroclinic environments.[5] They are often found in thermal troughs (i.e., regions of locally low temperature) where isotherms intersect isohypses, as shown in Figure 4.7. Because short waves can be observed on 500-hPa charts, we begin to see the increasing influence of thermodynamic processes at this level.

[5] Within meteorology, baroclinic flow occurs when the air density depends on both temperature and pressure. On upper-air charts, this condition occurs when isotherms intersect isohypses (which is common at levels below 500 mb). In contrast, barotropic flow occurs when air density depends only upon the pressure. On upper-air charts, this condition occurs when isotherms remain parallel to the isohypses (which is common at levels above 500 mb).

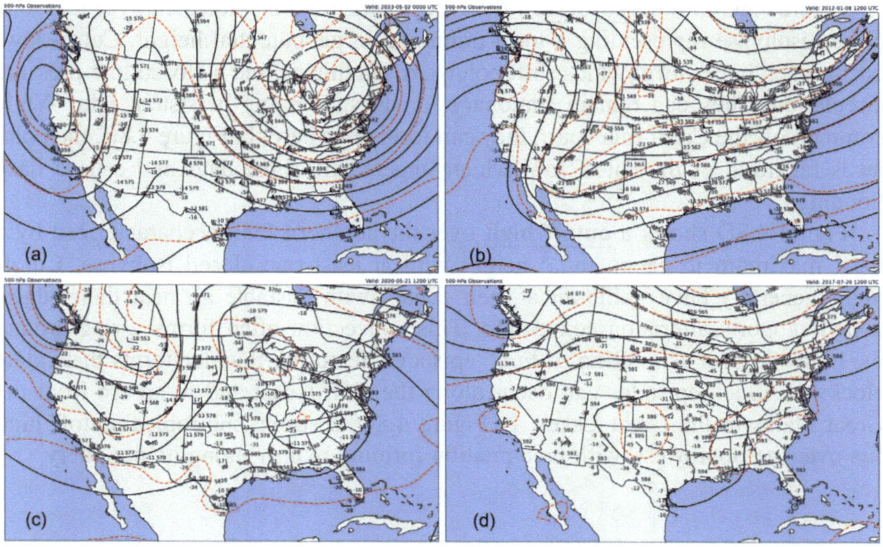

FIG. 4.8 Four examples of atmospheric blocking patterns at the 500-hPa level: (A) Omega block on May 2, 2023; (B) Rex block on January 8, 2012; (C) Cutoff low on May 21, 2020; and (D) Cutoff high on July 20, 2017. For each map, the solid black contours correspond to isohypses at intervals of 60 meters, and the dashed red contours correspond to isotherms at intervals of 5°C.

No Permission Required

The absence of short waves in the large-scale flow can cause trough-ridge systems to slow down or stall entirely, a phenomenon known as atmospheric blocking. Blocking results in persistent weather patterns that can last for days or even weeks, leading to flooding, drought, temperature extremes, and other impacts. Figure 4.8 illustrates four types of atmospheric blocking patterns.

Figure 4.8A shows an **omega block**, in which a high-pressure center is situated between two low-pressure centers. On an upper-air chart, this pattern resembles the Greek letter omega (Ω). In this case, air over the Southwest U.S. is forced north into Canada, then south into the Southeast U.S. by a deep ridge over the central U.S. The ridge's center experiences prolonged dry weather and light winds, while the adjacent troughs bring clouds and rain. Forecasting is simplified because conditions remain consistent for several days. Typically, the right side of the omega block has below-normal temperatures, while the left side is warmer than average.

Figure 4.8B shows a **rex block**, which resembles a "half figure 8" or "backwards S." It forms when a strong ridge is positioned directly north of a strong trough. In Figure 4.8B, a large ridge sits over the West Coast, north of a large trough in the Southwest. The flow from the North Pacific curves around the ridge, then around the trough, looping near the same longitude and forcing air far southward. As a result, the flow makes little eastward progress.

Figure 4.8C shows a **cutoff low** over the Southeast U.S., indicated by at least one closed isohypse representing a local minimum in geopotential height. Cutoff lows occur when upper-level winds shift poleward, leaving behind a closed low-pressure circulation. They can remain stationary for days, bringing persistent rain. While common off the California coast, they can form anywhere. In Figure 4.8C, the polar jet is far north of the cutoff low, which lingered for three days and caused flash flooding.

Figure 4.8D shows a **cutoff high** over the southern Plains, characterized by a local maximum in geopotential height with at least one closed isohypse. Cutoff highs generally have stable air at their center, with unstable air near their edges, which supports afternoon convection. This convection often forms a ring around the high - known as a "Ring of Fire" episode - featuring thunderstorms that rotate clockwise. Such episodes can occur along the Gulf Coast and in an arc from the Great Plains to the Great Lakes. In Figure 4.8D, a strong subtropical cutoff high sits over the Southwest, with precipitation forming mainly along its periphery.

4.4 Synoptic features on lower-tropospheric charts

In this section, we examine the most used lower-tropospheric charts: the 700-hPa map, the 850-hPa map, and the surface map. At these altitudes, thermodynamic processes and friction play a significant role in shaping isopleths and influencing the evolution of synoptic-scale systems. At these levels, we can identify low-level jets, fronts, surface cyclones, and surface anticyclones. We will discuss the characteristics of synoptic fronts in the following section.

4.4.1 The 700-hPa and 850-hPa charts

Figure 4.9 shows a 700-hPa contoured chart of the continental United States from January 8, 2024, at 12 UTC. The 700-hPa level serves as a bridge between the upper troposphere and the surface. The major upper-tropospheric features, such as troughs and ridges shown in Figure 4.6, are also evident in Figure 4.9. In this case, a low-pressure system is present at the 700-hPa level, located downstream of the trough shown at the 500-hPa level. This configuration, where the closed low is positioned relative to the trough, is connected to the westward tilt of the trough that occurs with increasing altitude, as mentioned previously.

Charts at 700 hPa are useful for identifying regions of saturated air that are favorable for cloud formation. Areas of significant moisture, where the dew point depression is less than 3°C (approximately corresponding to 70% relative humidity), are typically shaded in green. When these charts are combined with the wind field, they can identify scenarios where moist air is transported into a region (a process called **moisture advection**) and where warm air is transported into a region (a process called **warm advection**). Because of this, 700-hPa charts are frequently used to assess how thermodynamic processes affect the evolution of synoptic-scale

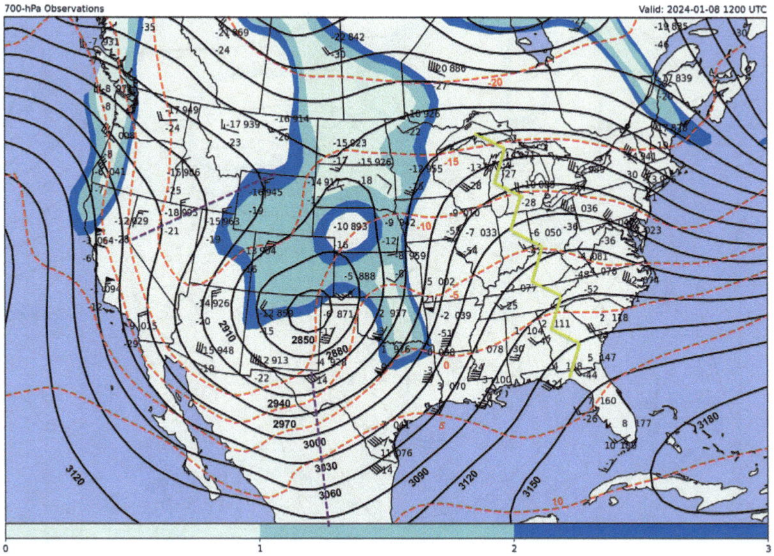

FIG. 4.9

700-hPa isobaric chart of the continental United States from January 8, 2024 at 12 UTC. The solid black contours correspond to isohypses at intervals of 30 meters, and the dashed red contours correspond to isotherms at intervals of 5°C. The dashed purple contours correspond to the trough axes, and the zig zag contour correspond to the ridge axis. The filled contours correspond to dew point depression below 3°C.

No Permission Required

systems, a topic discussed in more detail in Ch. 8. Additionally, the presence of small-scale features, such as short waves, can influence the development of synoptic-scale systems, similar to observations made from the 500-hPa chart. For this reason, the isopleths at this level tend to be less smooth compared with those at the 500-hPa level.

Figure 4.10 presents a contoured chart of the 850-hPa level over the continental United States from January 8, 2024, at 12 UTC. In addition to the thermodynamic processes discussed in the 700-hPa map, the 850-hPa chart is useful for understanding the structure of frontal systems at higher altitudes, which we will examine in the next section. As seen in Ch. 10, surface friction influences wind patterns and thermal fields at the 850-hPa level. This map is also instrumental in locating low-level jet streams, like the one found in the Gulf Coast states, as shown in Figure 4.10. These jet streams transport moisture and heat near the surface, making the 850-hPa map a useful tool for assessing warm advection in the lower troposphere.

4.4.2 The surface chart

Figure 4.11 shows a contoured surface weather chart of the continental United States from January 8, 2024, at 12 UTC. There are key differences between surface weather

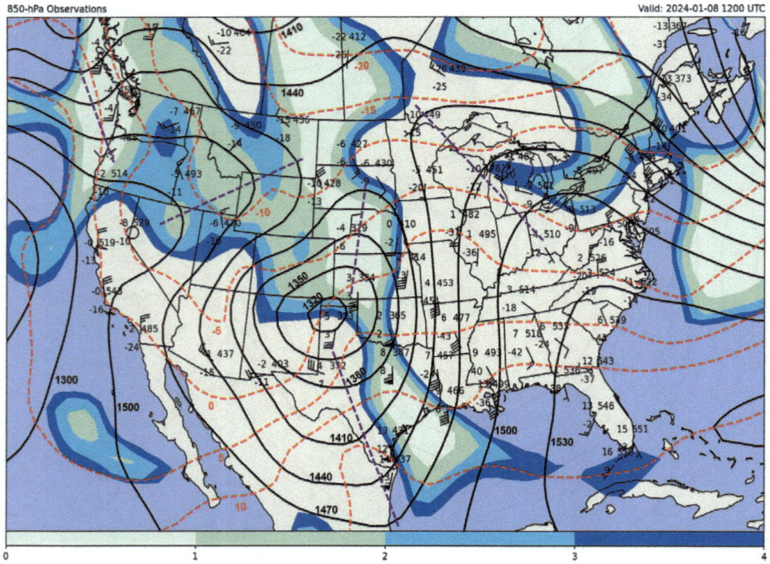

FIG. 4.10

850-hPa isobaric chart of the continental United States from January 8, 2024, at 12 UTC. The solid black contours correspond to isohypses at intervals of 30 meters, and the dashed red contours correspond to isotherms at intervals of 5°C. The dashed purple contours correspond to the trough axes. The filled contours correspond to dewpoint depression below 3°C.

No Permission Required

charts and other isobaric charts. Whereas isohypses are contoured on isobaric charts, surface charts use isobars. Importantly, surface charts generally do not plot isobars based directly on ASOS altimeter readings. As discussed in Ch. 3, surface pressure readings from METAR observations are based on **mean sea-level pressure (MSLP)**, which removes the effect of station elevation. Contouring MSLP allows analysts to examine horizontal variations in pressure without considering the impact of altitude.

In Figure 4.11, the closed isobars across Texas indicate a surface low-pressure system, with winds rotating counterclockwise around the low. More generally, in the Northern Hemisphere, a large air mass rotating counterclockwise around a low-pressure center is known as a **surface cyclone**, whereas one rotating clockwise around a high-pressure center is a **surface anticyclone**.[6]

[6] Surface cyclones and anticyclones rotate in the opposite direction within the Southern Hemisphere. Thus, surface cyclones are characterized by clockwise rotation in the Southern Hemisphere, whereas surface anticyclone are characterized by counterclockwise rotation in the Southern Hemisphere.

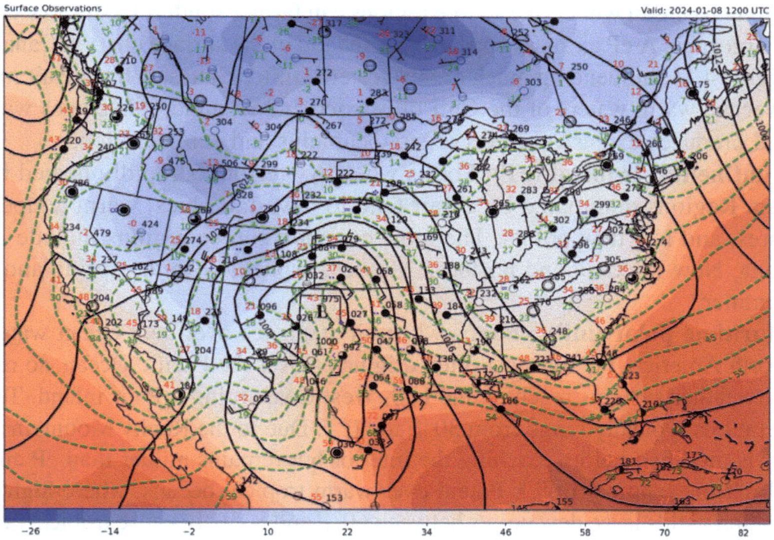

FIG. 4.11

Surface chart of the continental United States from January 8, 2024, at 12 UTC. The solid black contours correspond to isobars at intervals of 4 hPa, and the dashed green contours correspond to isodrosotherms at intervals of 5°F. The filled contours correspond to surface temperature.

No Permission Required

The effects of friction are most apparent on surface charts. In our discussion of upper-tropospheric charts, we noted that winds generally flow parallel to isohypses. In contrast, surface winds, as shown in Figure 4.11, blow *across* isobars. In particular, the surface cyclone in Figure 4.11 features winds spiraling toward the low-pressure center. As we will explain in Ch. 10, this pattern, known as **convergence**, results in vertical motion above the low-pressure area. Conversely, surface anticyclones are characterized by winds spiraling outward from high-pressure centers.

Whereas dew point depression is often shown on 700-hPa and 850-hPa maps, surface charts typically contour isodrosotherms (as shown in Figure 4.11). This helps analysts identify the air masses present at the surface. An **air mass** is a large body of air with relatively uniform temperature and humidity, whose characteristics depend on its source region. Air masses are classified based on moisture content and the temperature of the surface over which they form. The classification of air masses is summarized in Table 4.3. The four primary types of air masses that interact with the midlatitudes are as follows:

- **Continental polar (cP)**: Forms over cold or frozen terrain due to rapid cooling. In North America, cP air most often originates in northern Canada and the

Arctic basin. In winter, the source regions can farther south if fresh snow covers a large area. A cP air mass appears near the top of the surface chart in Canada in Figure 4.11, influencing much of the western United States. When cP air moves over the warm waters of the Gulf of Mexico and Atlantic, it becomes warmer and produces extensive streets of stratocumulus and cumulus clouds.

- **Continental tropical (cT)**: Forms when dry terrain is rapidly heated by the sun. Regions dominated by cT air tend to have fair skies, warm temperatures, and low dew points. The most common source regions include Arizona, New Mexico, and northern Mexico, especially in the spring and summer.
- **Maritime polar (mP)** : Forms when air stagnates over cool ocean waters. In North America, mP air most often originates over the Pacific Ocean, so nearly all weather systems reaching the U.S. West Coast bring mP air inland. Temperatures typically range from 30°F to 60°F. Due to the Rocky Mountains, mP air is modified before reaching the interior but remains milder than cP air.
- **Maritime tropical (mT)**: Found over warm tropical oceans, where high evaporation yields large dew points between 60°F and 80°F. Because tropical oceans cover a large portion of Earth, mT air is the most common air mass. In Figure 4.11, mT air occupies the Gulf of Mexico and influences southeastern Texas.

Many surface weather systems develop from the interaction of different air masses. For example, drylines in the Great Plains mark boundaries between cT and mT air. Similarly, surface fronts, discussed in the next section, often occur at the boundary between mT and cP air. In Figure 4.11, the cP air mass over the Southwest and the mT air mass across eastern Texas create a significant temperature and moisture contrast across central Texas, indicating a frontal boundary in this region. We will examine the formation and evolution of air masses in greater detail in Ch. 11 and take a closer look at surface fronts in the final section of this chapter.

4.5 Introduction to surface fronts

The *Glossary of Meteorology* defines a front as the "interface or transition zone between two air masses of different density." This means that a front marks the boundary between two distinct air masses. As mentioned earlier, the surface chart in Figure 4.11 suggests the presence of a front across central Texas. Figure 4.12 presents the surface chart alongside the 850-, 500-, and 300-hPa charts associated with the synoptic events depicted in Figure 4.11. By examining the surface chart together with the upper-air charts, we can identify patterns common to many surface fronts.

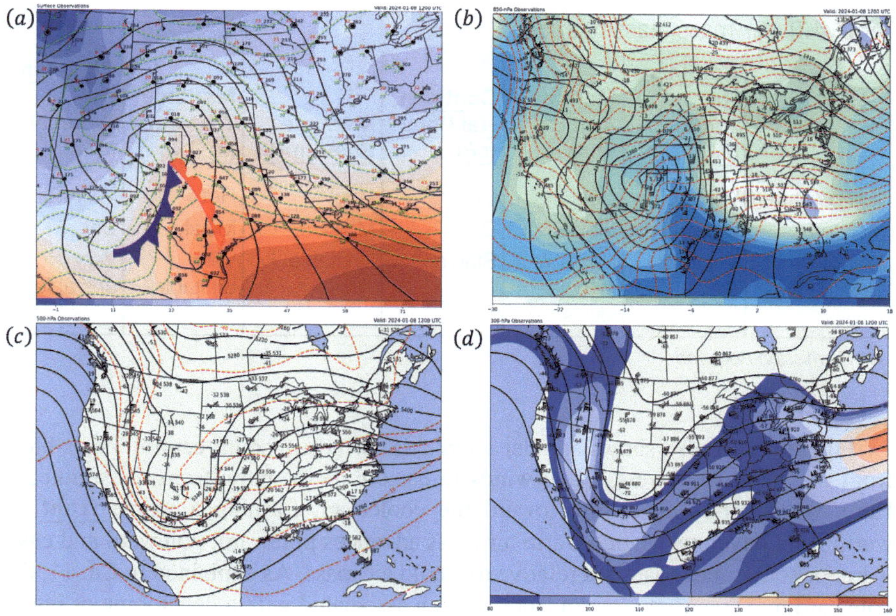

FIG. 4.12

Four-panel display of the continental United States from January 8, 2024, at 12 UTC. (A) Surface chart where solid black contours correspond to isobars at 4-mb intervals, the dashed green contours correspond to isodrosotherms at 5°F intervals , and the filled contours correspond to surface temperature. The surface cold front and warm front have been drawn. (B) 850-hPa chart where solid black contours correspond to isohypses at 30-meter intervals, the dashed red contours correspond to isotherms at 5°C intervals, and the filled contours correspond to dew point depression. (C) 500-hPa chart where solid black contours correspond to isohypses at 60-meter intervals and dashed red contours correspond to isotherms at 5°C. (D) 300-hPa chart where solid black contours correspond to isohypses at 120-meter intervals and the filled contours correspond to isotachs.

No Permission Required

4.5.1 Characteristics of surface fronts

In Figure 4.12A, there is a noticeable wind shift across southern Texas: winds blow from the west in western Texas, while in south-central Texas they original from the south. The strongest winds occur in the cooler air to the west, suggesting that cold air is advancing toward the warmer air in south-central Texas. This pattern indicates the presence of a **cold front**. A cold front develops when a cold air mass advances equatorward, equatorward and eastward, or eastward relative to a warm air mass. As discussed in Ch. 11, cold fronts are usually oriented equatorward from a surface cyclone and followed by a cold surface anticyclone to the west.

Table 4.3 Air mass classification

	Continen-tal Arctic (cA)	**Maritime Arctic (mA)**	**Continen-tal Polar (cP)**	**Maritime Polar (mP)**	**Maritime Tropical (mT)**	**Continen-tal Tropical (cT)**
Temperature	Very cold	Cold	Cold	Cool	Hot	Very hot
Moisture	Dry	Moist	Dry	Moist	Moist	Dry
Stability	Very stable	Unstable in lower levels	Stable	Unstable	Very unstable	Very unstable

No Permission Required

Figure 4.12A also shows another wind shift: winds in south-central Texas blow from the south-southwest, while winds in eastern Texas blow from the southeast. Here, warm air is advancing toward the cooler air, forming a **warm front**. A warm front develops when a warm air mass advances poleward, poleward and eastward, or sometimes eastward relative to a cold air mass. As explained in Ch. 11, the poleward movement of a warm front is often associated with a developing or intensifying surface cyclone. However, not all surface cyclones have warm fronts or well-defined wind-shift lines. Warm fronts are generally weaker than cold fronts, with gentler slopes and smaller temperature contrasts.

Wind shifts along warm fronts are usually less pronounced than those along cold fronts, reflecting the weaker temperature gradients typical of warm fronts. The patterns observed in Figure 4.12A are representative of cold and warm fronts across the United States, where changes in both wind speed and direction mark the frontal boundary. The wind shift in a frontal zone is usually accompanied by a pressure minimum (i.e., a low-pressure center), another hallmark of surface fronts. In Figure 4.12A-C, the temperature and moisture contrasts across central Texas are strongest near the surface and weaken with height - a pattern typical of most surface fronts. For this reason, frontal boundaries are most often observed below 850 hPa and rarely above 500 hPa.[7]

Wind speed increases sharply with height near a front. For example, Figure 4.12B shows 850-hPa exceeding 50 knots, compared with surface winds near 15 knots in Figure 4.12A. This is a common feature of surface fronts, which exhibit strong vertical wind shear along with the horizontal wind shift.

Comparing Figure 4.12A with Figure 4.12C shows the surface front downstream of the upper-level trough - a typical feature for intensifying fronts. A comparison between Figure 4.12A and Figure 4.12D reveals that the surface front is also near an

[7] Frontal zones are not only found at the surface. Observational studies have documented the formation of strong fronts near the tropopause as well. A discussion of upper-tropospheric fronts is beyond the scope of this text.

upper-tropospheric jet streak, consistent with the tendency for strong cold fronts to form beneath such features.

When neither the cold air mass nor the warm air mass moves significantly, the front is called a **stationary front**. Rising air above a stationary front is typically caused by warm air overrunning cooler air, producing light, stratiform precipitation due to high stability in the frontal zone. Cold fronts that stall can transition into stationary fronts. Stationary fronts may later move poleward and become warm fronts, especially when a new cyclone deepens and tracks poleward and eastward. Persistent precipitation along and north of a stationary front can lead to flooding.

In Ch. 11, we will examine the vertical structure and evolution of surface fronts in greater detail. In the final part of this section, we will apply our isopleth analysis to surface fronts.

4.5.2 Identifying surface fronts

Although there is no universally approved method for analyzing frontal boundaries on surface charts, several general guidelines can help in drawing them accurately. When identifying a frontal boundary, it is important to look for consistent signals across multiple meteorological variables. This consistency can be evaluated by examining five key features that typically occur near surface fronts.

1. **Temperature contrast:** Surface fronts are marked by strong horizontal temperature differences. Look for a sharp temperature contrast, which can be identified by examining isotherms on surface charts. Cold and warm fronts are generally located along the leading edge of a distinct temperature change.
2. **Moisture gradient:** Surface fronts also exhibit horizontal moisture contrasts. A strong dew point gradient, visible through isodrosotherms, can indicate the location of a front. Cold fronts often feature a dramatic change in moisture content because they typically separate mT and cP air masses. The moisture contrast along warm fronts is usually weaker than that along cold fronts.
3. **Wind shift patterns:** Surface fronts are associated with distinct changes in both wind direction and speed, which can be observed in surface station plots. For example, ahead of a cold front, winds commonly come from the south-southeast, while behind the front, they often shift to the west. A general rule of thumb for warm fronts is that easterly winds are frequently found ahead of them; however, exceptions occur so this guideline should be applied with caution.
4. **Pressure gradient:** Wind shifts along a frontal boundary are often accompanied by a local minimum in surface pressure. Look for a strong pressure gradient, which can be identified by examining isobars on surface charts. Cold fronts are commonly found within a pressure trough. Reviewing the 3-hour pressure tendency can also be useful: pressure generally falls steadily ahead of a cold front and rises rapidly behind it. For warm fronts, pressure typically decreases ahead of the front and rises more gradually behind it.

5. **Cloud and precipitation patterns:** These can be assessed using surface station plots. Cold fronts are often accompanied by heavy precipitation and cumulonimbus cloud development as the front passes. In contrast, warm fronts tend to produce light to moderate rain when they pass.

After completing a surface map analysis, it is helpful to verify the position of the front using upper-air maps. This ensures consistency across different analyzed fields. The 850-hPa map is often used to identify areas of temperature and moisture advection, which can be aligned with the surface frontal boundary. As mentioned in the previous section, jet streaks are typically located near surface fronts, so their positions on the 200-, 250-, and 300-hPa maps can assist in accurately locating the fronts. Because surface fronts are often guided by upper-level winds, the 500-hPa map often can be used to determine the speed and direction of frontal motion.

Once all frontal systems have been identified on the surface chart and confirmed with additional upper-air charts, the frontal boundaries can be sketched. The cold front should be drawn at the leading edge of the cold air, while the warm front should be positioned at the leading edge of the warm air. The final step is to adjust the pressure fields to match the pressure tendencies. As noted earlier, the abrupt change in pressure tendency across a frontal boundary implies that pressure tendency is discontinuous there. This discontinuity is represented on surface charts by "kinking" the isobars appropriately around the frontal system, as shown in Figure 4.13.

So far in this textbook, we have discussed how meteorological data is collected (Ch. 2) and how it is processed and decoded (Ch. 3). We have also examined how to analyze meteorological data across entire regions using isopleth analysis.Now that we have described the major synoptic-scale features of the atmosphere and developed the tools necessary to examine them, we can turn to explaining these systems in detail. In Chs. 5 to 8, we will investigate the physical laws that govern the evolution of the atmosphere, and in Chs. 9 to 12, we will apply these insights to specific synoptic weather systems.

Questions

1. Answer the following questions regarding air masses:
   a. Explain why the central United States is typically not a good air-mass source region.
   b. Explain why maritime polar (mP) air masses along the east coast of the United States are usually colder than those along the nation's west coast? Why are they also *less* prevalent?
2. Consider the 300-hPa uncontoured map given in Figure 4.14. Perform manual isopleth analysis using the standard conventions. Mark the location of jet streaks, troughs, ridges, and short waves.
3. Consider the 300-hPa contoured map in Figure 4.13. Shade the jet stream using light-blue colored pencil. Place an "X" at the location of jet streaks within the

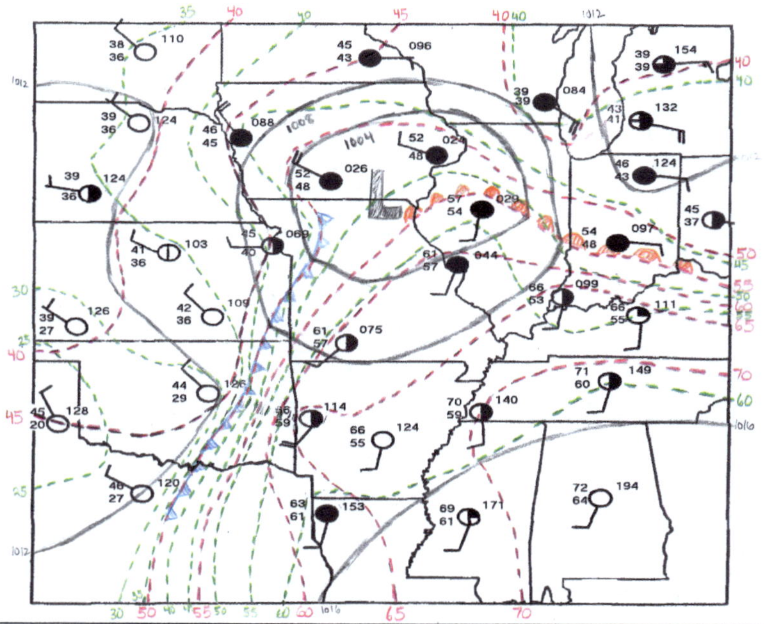

FIG. 4.13

Surface front analysis on a surface chart. Red dashed contours correspond to isotherms. Green dashed contours correspond to isodrosoterms. Solid black contours correspond to isobars where the isobars are "kinked" as it passes through the frontal zone. The cold front and the warm front symbols have been drawn on the chart.

No Permission Required

jet stream using a purple-colored pencil. Is this the polar jet stream or subtropical jet stream? Defend your answer.

4. Consider the 500-hPa uncontoured map in Figure 4.15.
   - **a.** Perform manual isopleth analysis using standard conventions. Mark the location of jet streaks, troughs, ridges, and short waves.
   - **b.** Are there any blocking patterns present on your chart? Defend your answer.

5. Consider the 500-hPa contoured map in Figure 4.16. Indicate the location of long-wave troughs with a black colored pencil. Indicate the location of long-wave troughs using a yellow-colored pencil. Indicate the location of short-wave troughs with a purple-colored pencil.

6. Consider the 700-hPa uncontoured map in Figure 4.17.
   - **a.** Perform manual isopleth analysis using standard conventions. Mark the location of low-level jets, troughs, ridges, lows, and highs and short waves.
   - **b.** Use a green pencil to shade the regions of dewpoint depression in which $\leq 3°C$. Based on your isopleth analysis, where is the moist air located?

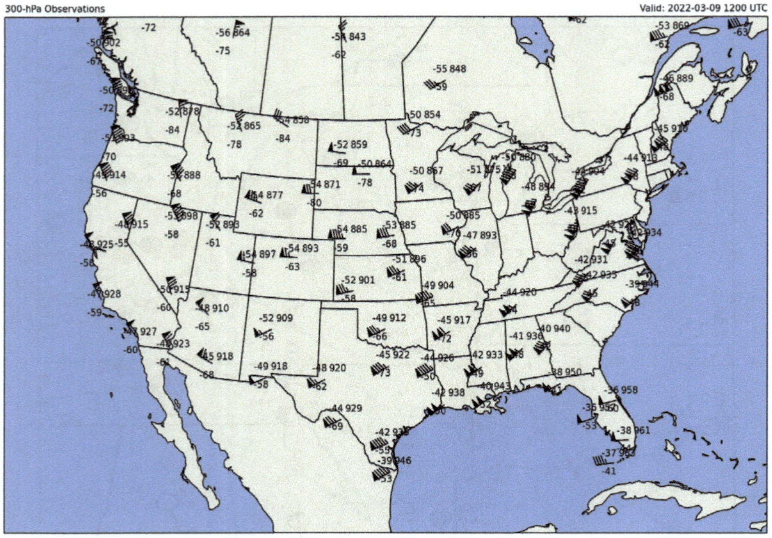

FIG. 4.14

300-hPa observations from March 9, 2022, at 12 Z.

No Permission Required

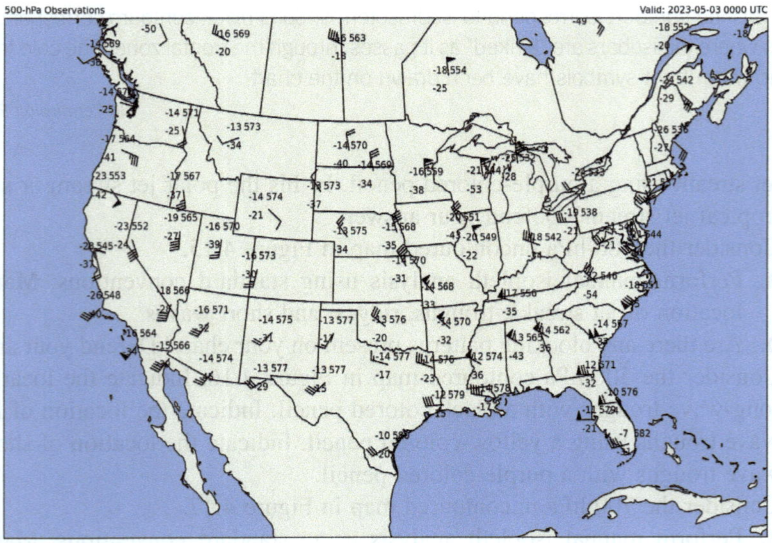

FIG. 4.15

500-hPa observations from May 3, 2023, at 00Z.

No Permission Required

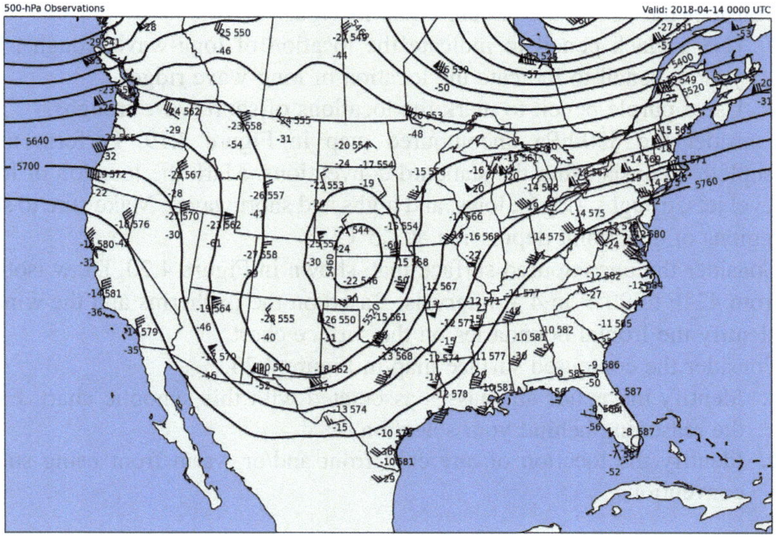

FIG. 4.16

500-hPa observations from April 14, 2018 at 00Z. Solid black contours correspond to isohypses at 60-meter intervals.

No Permission Required

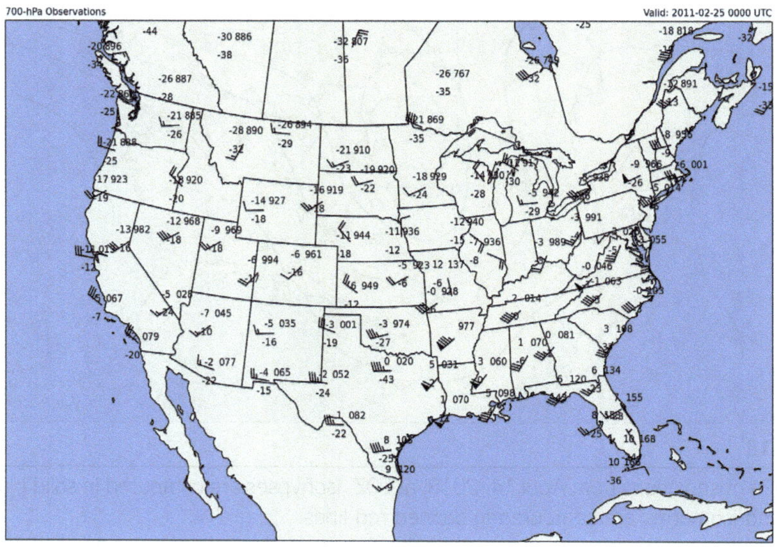

FIG. 4.17

700-hPa observations from February 25, 2011 at 00Z.

No Permission Required

7. Consider the 700-hPa contoured map in Figure 4.18.
   a. Use a black pencil to indicate the location of long-wave troughs. Use a yellow pencil to indicate the location of long-wave ridges.
   b. Use a purple pencil to mark the locations of short-wave troughs.
8. Consider the 850-hPa uncontoured map in Figure 4.19. Perform manual isopleth analysis using the standard conventions. Mark the location of low-level jets, troughs, ridges, lows, and highs and short waves. Make sure to shade regions of dew point depression at ≤3°C.
9. Consider the uncontoured surface map shown in Figure 4.20. Draw isotherms from 42°F to 90°F in 4°F intervals. Based on the isotherms and the wind, identify the frontal boundaries on the surface chart.
10. Consider the contoured surface map in Figure 4.23.
    a. Identify the major air masses associated with this synoptic chart. Explain the reasoning behind your selection.
    b. Identify the location of any cold front and/or warm front using standard conventions.

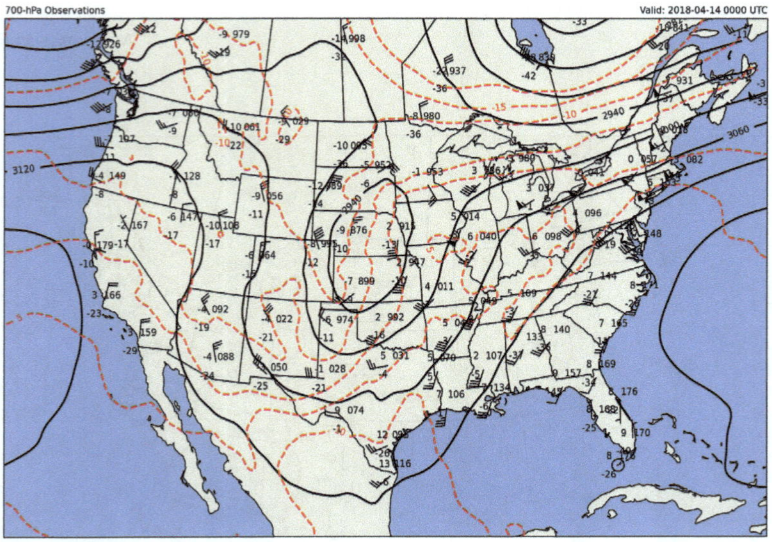

FIG. 4.18

700-hPa observations from April 14, 2018, at 00Z. Isohypses are contoured in solid black lines and isotherms are contoured in dashed red lines.

No Permission Required

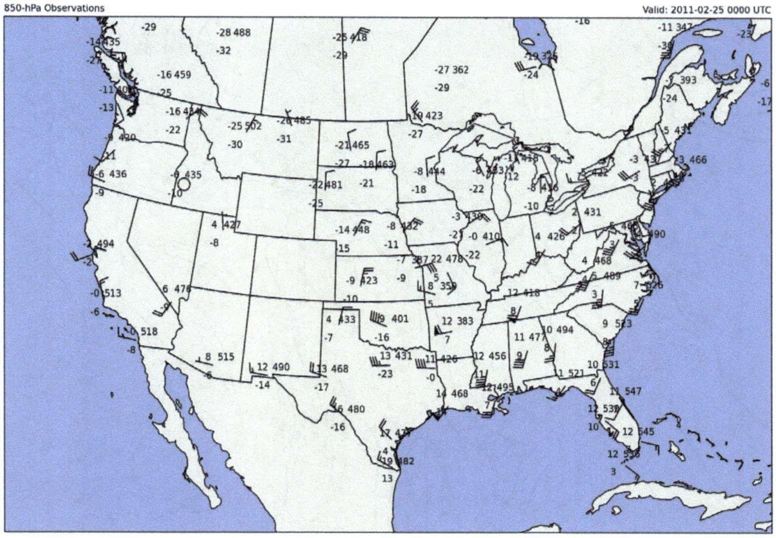

FIG. 4.19

850-hPa observations from February 25, 2011, at 00Z.

No Permission Required

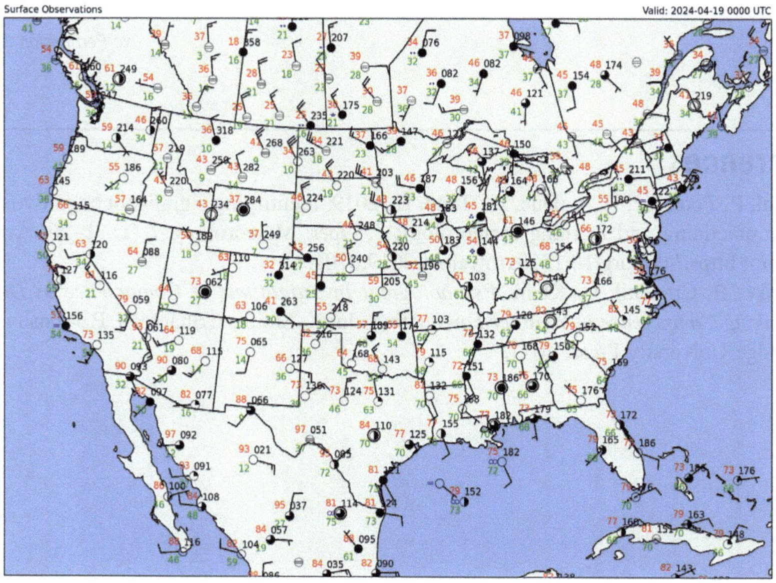

FIG. 4.20

Uncontoured surface observations from April 19, 2024, at 00Z.

No Permission Required

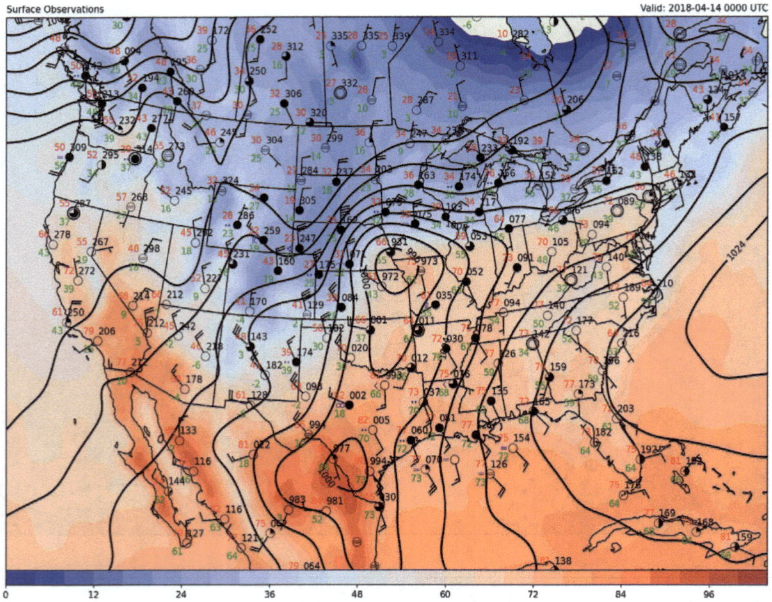

FIG. 4.21

Contoured surface map observations from April 14, 2018 at 00Z. Isohypses are contoured in solid black lines and isotherms are contour filled.

No Permission Required

References

McNeal, P., Petcovic, H., LaDue, N., Todd, E., 2019. Identifying significant cognitive factors for practicing and learning meteorology. J. Oper. Meteorology 7 (1), 1–26. Available from: https://doi.org/10.15191/nwajom.2019.0701.

R. Pliske, B. Crandall, G. Klein, *Psychological Investigations of Competence in Decision Making Competence in Weather Forecasting* . In: K. Smith, J. Shanteau, P. Johnson. Cambridge University Press, 2011.

CHAPTER

The fundamental forces

5

Chapter outline

In the first unit of this textbook, covering Chs. 1 to 4, we provided an overview of the tools used to measure and analyze meteorological data. In Ch. 1, we discussed the four variables necessary to specify the state of the atmosphere: pressure, temperature, water vapor, and wind. As discussed in Ch. 2, these variables are continuously measured and monitored at the surface -primarily using the instruments associated with ASOS weather stations- and within the troposphere using radiosondes. In Ch. 3, we explained how surface observations are transmitted in surface observation codes, such as METAR, and the upper-air observations are transmitted into radiosonde code. To analyze meteorological data over large regions, these observations are typically plotted on synoptic charts using station plot models. Ch. 4 concludes the unit

An Introduction to Synoptic-Dynamic Meteorology. https://doi.org/10.1016/B978-0-443-24858-0.00005-8

by discussing how synoptic analysis begins with the examination of weather charts through isopleth analysis.

In the second unit of this textbook, covering Chs. 5 to 8, we will examine the physical laws and mechanisms that describe atmospheric evolution. This chapter focuses on the forces that drive atmospheric motion, beginning with an examination of the properties of moving fluids.

5.1 Properties of an air parcel

As mentioned in Ch. 1, Earth's atmosphere can be described as a continuously evolving fluid system, characterized by constantly changing temperature, pressure, water vapor, and wind. To understand the basic mechanisms of atmospheric flow, it is helpful to think of the atmosphere as being composed of numerous fluid elements known as air parcels. An **air parcel** is a small, three-dimensional volume of air that is large enough to contain many molecules (i.e., more than one mole) but small enough for its temperature, mixing ratio, and pressure to remain uniform throughout. For many contexts, a cubic foot of air serves as a reasonable size for an air parcel.

Two key assumptions are commonly made about air parcels:

1. **Insulation from surroundings**. An air parcel does not exchange mass with the surrounding environment - there is no mixing between the air parcel and the outside air. Because of its size, an air parcel can move through the atmosphere without significantly altering the surrounding air. This point will become especially relevant when we discuss thermodynamic processes in Ch. 8.
2. **Adjustment to environmental pressure**. An air parcel's "walls" are flexible, so its internal pressure always matches the external environmental pressure. As the surrounding pressure changes, the parcel's volume changes accordingly.

With this understanding of an air parcel, we can now examine how Newton's laws of motion govern its movement, illustrating how air responds to various forces.

5.1.1 Center of mass and linear momentum

Based on the discussion above, we can treat an air parcel as a system composed of numerous particles in continuous, random motion. Rather than tracking each individual molecule, introductory physics teaches us that the motion of a system with many particles can be represented by the motion of its **center of mass**. Conceptually, the center of mass is the average position of the system's total mass (as illustrated in Figure 5.1). Mathematically it is defined as:

$$\boldsymbol{R} = \frac{1}{M}\sum_{i=1}^{N} m_i \boldsymbol{r}_i \tag{5.1}$$

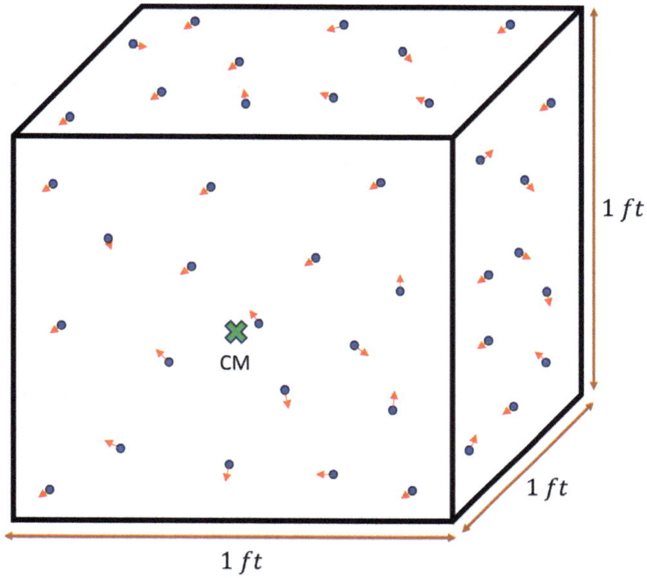

FIG. 5.1 The center of mass of an air parcel. An air parcel in which the molecules are in random, linear motion. The center of mass (CM) is marked with the green "X."

No Permission Required

where M is the total mass of the system, m_i is the mass of the ith particle, and $\boldsymbol{r}_i$ is the position of the ith particle. If the density of an air parcel is constant, then its mass is uniformly distributed, and the center of mass coincides with the geometric center of the parcel.

In introductory physics, the motion of any object is described by its **linear momentum**, defined as $\boldsymbol{p} = m\boldsymbol{v}$, where m is the mass of the object and $\boldsymbol{v}$ is its velocity. We can show that the total linear momentum of the air parcel is equivalent to the linear momentum of its center of mass. Rearranging Eq. (5.1) gives:

$$M\boldsymbol{R} = \sum_{i=1}^{N} m_i \boldsymbol{r}_i$$

Differentiating this expression with respect to time gives:

$$M\frac{d\boldsymbol{R}}{dt} = \sum_{i=1}^{N} m_i \frac{d\boldsymbol{r}_i}{dt}$$

$$\Rightarrow M\mathbf{v}_{\mathrm{CM}} = \sum_{i=1}^{N} m_i \boldsymbol{v}_i \tag{5.2}$$

where $\boldsymbol{v}_{cm}$ is the velocity of the center of mass and $\boldsymbol{v}_i$ is the velocity of the ith particle. In words, Eq. (5.2) states that the linear momentum of a system of particles (such as an air parcel) is equal to the product of the parcel's total mass and the velocity of its center of mass. The center of mass thus provides a convenient way to describe the motion of the entire air parcel. Put differently,the center of mass of the air parcel is the point that moves *as if* all the system's mass were concentrated there.

5.1.2 Newton's 1st and 2nd laws

Because an air parcel can be viewed as a collection of particles, its evolution over time is governed by Newton's laws of motion. For our purposes, Newton's 1st and 2nd laws are particularly relevant for understanding the motion of an air parcel. In terms of linear momentum, Newton's 1st law states:

If no net external force acts on an object, then the object's linear momentum cannot change.

To apply this law, we must distinguish between external and internal forces. *Internal forces* arise from within a system. For an air parcel, the internal forces are the contact forces between the molecules as they collide. *External forces* act on the system from its surroundings, originating from interactions between the air parcel and its environment. The net external force is the vector sum of all the external forces acting on an object:

$$\boldsymbol{F}_{net} = \boldsymbol{F}_1 + \boldsymbol{F}_2 + \ldots + \boldsymbol{F}_N = \sum_{i=1}^{N} \boldsymbol{F}_i$$

When external forces are balanced, the net force ($\boldsymbol{F}_{net}$) is equal to zero. According to Newton's 1st law, an air parcel's linear momentumremains constant if external forces are balanced. Because the parcel's mass is constant, its velocity must also remain unchanged if the external forces remain balanced. In atmospheric terms, we can state Newton's 1st law as follows:

If the wind speed and direction remain constant, the external forces acting on the air must be balanced.

Thus, observing constant wind velocity on synoptic charts indicates balanced forces - a concept explored further in Ch. 7.

Whereas Newton's 1st law describes the motion of an object under balanced forces, Newton's 2nd law describes the motion of an object under unbalanced forces. In terms of linear momentum, Newton's 2nd law states the following:

When the net external force upon an object is non-zero, its linear momentum changes according to

$$\boldsymbol{F}_{net} = \frac{d\boldsymbol{P}}{dt}$$

where $\boldsymbol{P}$ is the total momentum of the object. Newton's 2nd law states that*any* change in linear momentum is caused by unbalanced forces. Since an air parcel's mass remains constant, we can rewrite Newton's 2nd law as:

$$\boldsymbol{F}_{net} = \frac{d}{dt}(M\boldsymbol{v}_{CM}) = M\frac{d\boldsymbol{v}_{CM}}{dt}$$

$$\Rightarrow \boldsymbol{F}_{net} = M\boldsymbol{a}_{CM} \tag{5.3}$$

where $\boldsymbol{a}_{CM}$ is the acceleration of the parcel's center of mass. Thus, an air parcel *must* accelerate if the external forces acting upon it are unbalanced according to Eq. (5.3). This leads to two key points:

1. Acceleration is **ALWAYS** in the same direction as the net force. To determine acceleration, we must identify the net external forces acting on the object.
2. Because velocity is a vector, acceleration can result from a change in magnitude (i.e., changing its speed) or a change in direction (e.g., motion along a curved path).

In atmospheric terms, we can state Newton's 2nd law as follows:

If the wind speed and/or direction changes, the forces acting on the air are unbalanced.

On synoptic charts, changing wind velocity indicates unbalanced flow. As shown in Chs. 10 and 11, such flow is closely tied to vertical motion in the atmosphere, which is central to the dynamics of fronts, jets, and cyclones.

Mathematically, Newton's 2nd law can be expressed in unit-vector notation as:

$$a_x\boldsymbol{i} + a_y\boldsymbol{j} + a_z\boldsymbol{k} = \frac{1}{M}\left(F_{net,x}\boldsymbol{i} + F_{net,y}\boldsymbol{j} + F_{net,z}\boldsymbol{k}\right)$$

This expression is equivalent to three scalar component equations for the acceleration:

$$a_x = \frac{F_{net,x}}{M}, \qquad a_y = \frac{F_{net,y}}{M}, \qquad a_z = \frac{F_{net,z}}{M} \tag{5.4}$$

Based on Newton's 2nd law, atmospheric motion can be described by understanding the nature of the forces acting on the air at any given time. Box 5.1 provides an example of how these equations can be applied to analyze the motion of an air parcel.

The external forces acting on an air parcel on the rotating Earth can be grouped into two categories: fundamental forces and inertial forces. Fundamental forces act on an air parcel even in the absence of Earth's rotation. The most significant fundamental forces are the gravitational force, the pressure gradient force, and the turbulent drag force. Inertial forces arise solely due to Earth's rotation, with the two most important being the Coriolis force and the centrifugal force. We will examine fundamental forces in this chapter and inertial forces in Ch. 6.

5.2 Newtonian gravitation

Consider two objects of masses M and m that are separated by a distance r. The force of gravitational attraction between these two objects is described by Newton's law of universal gravitation. In vector form, this law is written as:

Box 5.1 Motion of an air parcel

Suppose that the air parcel that is initially at the surface with temperature $T = 20°C$ is acted on by a constant net force of $\boldsymbol{F}_{net} = (1.70 \times 10^{-5}\ \mathrm{N})\boldsymbol{j}$. If the air parcel is initially at rest, determine its speed after 30 minutes.

First, determine the mass of the air parcel by using the ideal gas law:

$$\rho = \frac{P}{R_d T} \Rightarrow m = \frac{PV}{R_d T} = \frac{(101325\ \mathrm{Pa})(0.0283\ \mathrm{m}^3)}{\left(287\ \frac{\mathrm{J}}{\mathrm{kg{\cdot}K}}\right)(293.15\ \mathrm{K})} = 0.0341\ \mathrm{kg}$$

where $1ft^3 = 0.0283\ \mathrm{m}^3$. Since the net force is only in the $\boldsymbol{j}$ direction, then $a_x = 0 = a_z$ and by writing Eq. (5.4) in terms of velocity, we have

$$\frac{dv_y}{dt} = \frac{F_{net,y}}{m}$$

Since the net force is constant, then we can integrate this expression to obtain

$$\int_0^{1800\ s} \frac{dv_y}{dt} dt = \int_0^{1800\ s} \frac{F_{net,y}}{m} dt$$

$$\Rightarrow v_y(1800\ \mathrm{s}) - v_y(0) = \frac{F_{net,y}}{m}(1800\ \mathrm{s})$$

where 30 minutes = 1800 s. Since the air parcel is initially at rest, then $v_y(0) = 0$. Therefore, we have

$$v_y = \frac{(1.70 \times 10^{-5}\ \mathrm{N})}{(0.0341\ \mathrm{kg})}(1800\ \mathrm{s}) = 0.897 \frac{\mathrm{m}}{\mathrm{s}}$$

Notice that the wind speed changes as the temperature and pressure change (since density is a function of pressure and temperature). This emphasizes an important point. Although winds are driven by forces acting on the air, the motion of air parcels can be altered by thermodynamic changes in the atmosphere.

$$\boldsymbol{F}_G = -\frac{GMm}{r^2}\boldsymbol{r} \tag{5.5}$$

where $G = 6.674 \times 10^{-11}\ \mathrm{N{\cdot}m^2\ kg^{-2}}$ is the universal gravitational constant, and $\boldsymbol{r}$ is the unit vector directed from one mass toward the other, as shown in Figure 5.2.

We can apply Eq. (5.5) to the gravitational interaction between an air parcel and Earth. In this case, M represents Earth's mass, m is the air parcel's mass, r is the distance between the air parcel's center of mass and Earth's center, and $\boldsymbol{r}$ is the unit vector pointing from the air parcel to Earth's center, as illustrated in Figure 5.3. According to Newton's law of universal gravitation, the gravitational attraction between Earth and an air parcel is equivalent toassuming that all of Earth's mass is concentrated at its center.[1]

[1] This statement is a consequence of the shell theorem, which states that a spherically symmetric object affects other objects gravitationally as if all of its mass were concentrated at its center.

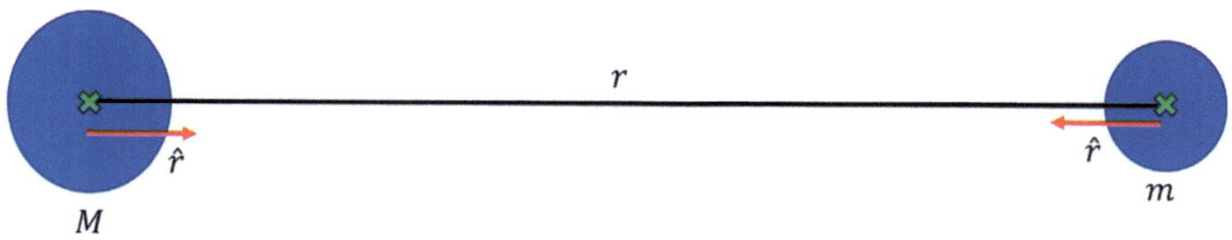

FIG. 5.2 Schematic of Newton's law of gravitation

Schematic of Newton's law of universal gravitation between two masses *M* and *m*. The unit vector r is directed between the centers of mass of both objects.

No Permission Required

In synoptic meteorology, Newton's law of universal gravitation is applied primarily in two ways: (1) to describe how gravitational acceleration varies with height and (2) to define geopotential height.

5.2.1 Gravitational acceleration

Applying Newton's 2nd law to an air parcel, the gravitational acceleration can be written as:

$$\boldsymbol{a}_g = \frac{\boldsymbol{F}_g}{m} = -\frac{GM}{r^2}\boldsymbol{r} \tag{5.6}$$

According to Eq. (5.6), when an air parcel is released near Earth's surface with no other external forces acting on it, it will accelerate toward Earth's center due to gravity, as shown in Figure 5.3. Relative to Earth's surface, this acceleration is vertically downward so $\boldsymbol{a}_g$ points in the $-\boldsymbol{k}$ direction.

At sea level, the magnitude of the gravitational acceleration is given by:

$$g = \frac{GM_E}{R_E^2} = \frac{\left(6.674 \times 10^{-11}\,\frac{\mathrm{N{\cdot}m^2}}{\mathrm{kg^2}}\right)\left(5.9722 \times 10^{24}\ \mathrm{kg}\right)}{\left(6.3781 \times 10^{6}\ \mathrm{m}\right)^2} = 9.80\,\frac{\mathrm{m}}{\mathrm{s^2}}$$

where M_E is the mass of Earth and R_E is the average radius of Earth[2]. This value matches the familiar free-fall acceleration discussed in introductory physics.

However, Eq. (5.6) shows that gravitational acceleration varies with altitude. For an air parcel at height z above sea level, the magnitude of the gravitational acceleration is:

$$g = \frac{GM}{(R_E + z)^2} \tag{5.7}$$

---

[2] It should also be noted that Earth is not a perfect sphere. Earth is approximately an ellipsoid, which is flattened at the poles and bulging at the equator. Its equatorial radius (from its center point to the equator) is greater than its polar radius (from its center point out to either north or south pole) by 21 km. Therefore, a point at the poles is closer to Earth's center than is a point on the equator. The reason for this will be explained in Ch. 6.

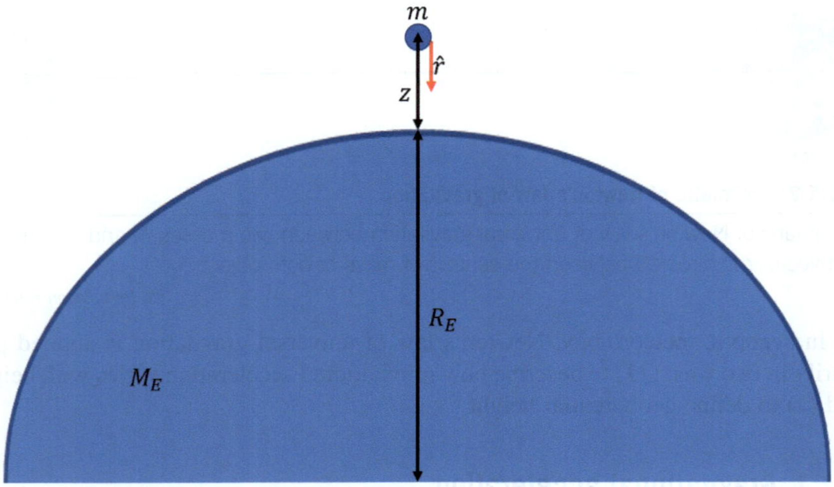

FIG. 5.3 Gravitational attraction between air parcel and the Earth

Schematic of the gravitational attraction between an air parcel (of mass m) and the Earth (of mass M_E).

No Permission Required

Table 5.1 lists g for various altitudes within the lowest 20 km of the atmosphere, where weather phenomena occur. Although g decreases with altitude, the reduction is less than 1% in this layer. Consequently, it is standard practice to treat g as constant near Earth's surface. An alternative approach is to define a modified height that accounts for this variation - a concept introduced in the next section.

5.2.2 Geopotential height

Consider an object of mass m that we wish to raise vertically to a height z. According to introductory physics, the work required is given by:

$$W = \int_0^h F_z dz$$

where F_z is the z-component of the applied force and dz is the vertical displacement. If the object is raised slowly at a constant speed, the applied force must balance the weight of the object. In this case, the work can be expressed as:

$$W = \int_0^h mg(z)dz$$

This shows that raising an object upward requires work against the downward-directed gravitational force. To focus on Earth's gravitational influence rather than

Table 5.1 Gravitational acceleration as a function of altitude

z (km)	g (m/s^2)	Example
0	9.80	Mean sea level on Earth
1.50	9.79	Average elevation at 850 hPa
5.50	9.78	Average elevation at 500 hPa
12.0	9.76	Near the tropopause
20.0	9.74	Near the stratosphere

No Permission Required

the object's mass, we define a new variable known as **geopotential** Φ as the work done per unit mass against gravity:

$$\Phi(z) = \int_0^z g(z')dz' \tag{5.8}$$

The geopotential Φ can also be related to the average gravitational acceleration at sea level g_0 by noting that:

$$g_0 = \frac{GM}{R_E^2}$$

$$\Rightarrow GM = g_0 R_E^2$$

Hence, the magnitude of the gravitational acceleration can be written as:

$$g = \frac{GM}{(R_E + z)^2} = \frac{g_0 R_E^2}{(R_E + z)^2} = g_0 \left(\frac{R_E}{R_E + z}\right)^2$$

Substituting this expression into Eq. (5.8) yields:

$$\Phi(z) = \int_0^z g(z')dz' = g_0 R_E^2 \int_0^z \frac{dz'}{(R_E + z')^2}$$

$$\Rightarrow \Phi(z) = g_0 R_E^2 \left[-\frac{1}{R_E + z}\right]_0^z = g_0 R_E^2 \left(\frac{1}{R_E} - \frac{1}{R_E + z}\right)$$

$$\Rightarrow \Phi(z) = g_0 \left(\frac{R_E z}{R_E + z}\right) \tag{5.9}$$

In words, Eq. (5.9) represents the work (per unit mass) required to lift a 1-kg mass from sea level to a height z.

Geopotential height is defined as

$$Z = \frac{\Phi}{g_0}$$

Using Eq. (5.9), this can be expressed as:

$$Z = \frac{R_E z}{R_E + z} \tag{5.10}$$

Geopotential height compensates for the decrease in gravitational acceleration with altitude. It is, therefore, a modified measure of height that accounts for the variation of gravity above Earth's surface. From the definition of geopotential height, we have

$$dZ = \frac{d\Phi}{g_0} = \frac{g(z)\,dz}{g_0}$$
$$\Rightarrow g_0\,dZ = g(z)\,dz$$

This relation shows that when using Z rather than z, we can treat $g = g_0$ as a constant - even though gravitational acceleration actually decreases with altitude. The difference between geometric height and geopotential height increases from 0 meters at sea level to about 16 meters at an altitude of 10 km.

5.3 The pressure gradient force

A quick look at any surface map reveals significant variations in pressure. This spatial variability indicates the presence of pressure differences throughout the atmosphere. The fundamental force that describes how air responds to these differences is called the **pressure gradient force**. Before examining its nature, we must first introduce the concept of the gradient vector.

5.3.1 The gradient vector

To motivate this concept, consider a function of one variable $f(x)$. This is known as a *scalar function* because it outputs a single scalar value. According to introductory calculus, the derivative df/dx describes how rapidly $f(x)$ changes when x increases by dx. The formal definition of a derivative is given by:

$$\frac{df}{dx} = \lim_{\Delta x \to 0} \frac{f(x + \Delta x) - f(x)}{\Delta x} \tag{5.11}$$

Geometrically, Eq. (5.11) states that if we draw a secant line through the points $(x,\ f(x))$ and $(x + \Delta x,\ f(x + \Delta x))$ with Δx is very small, we obtain the slope of the tangent line at x, as shown in Figure 5.4A. Thus, the derivative corresponds to the slope of the tangent line at a given point.

Now, suppose we have a scalar function of two variables, denoted as $f(x, y)$. As shown in Figure 5.4B, the rate at which this function changes depends on the direction of movement. For such a function, variation can occur in both the x-direction and the y-direction. The rate of change in a specific direction is called the **directional derivative** or, in many cases, the **partial derivative**. We write the directional derivative in the x-direction as $\partial f/\partial x$ and in the y-direction as $\partial f/\partial y$, as shown in Figure 5.4B.

A partial derivative measures how the function changes in a single specified direction while keeping all other variables constant. For example, $\partial f/\partial x$ examines how $f(x,y)$ changes *only* in the x-direction, treating any y-dependence as a constant. Likewise, $\partial f/\partial y$ examines changes *only* in the y-direction, treating any x-dependence as a constant. In short, evaluating a partial derivative means measuring the rate of change along one direction while holding the others fixed. Box 5.2 provides examples of partial derivative calculations.

With the concept of the directional derivative established, we can extend it to functions of three variables. For a function $f(x,y,z)$, which can vary in the x-, y-, and z-directions, we generalize the derivative into a vector form known as the **gradient vector**. In Cartesian coordinates, the gradient vector is expressed as:

$$\nabla f = \frac{\partial f}{\partial x}\boldsymbol{i} + \frac{\partial f}{\partial y}\boldsymbol{j} + \frac{\partial f}{\partial z}\boldsymbol{k}$$

Whereas the function $f(x,y,z)$ is a scalar function, the gradient vector ∇f is a **vector function** because, at every point in space (x,y,z), it produces a vector whose components correspond to directional derivatives. Box 5.3 provides an example of how to calculate the gradient vector.

The gradient is important because it describes the rate of change of a scalar function that depends on multiple variables. In the atmosphere, for example, temperature, pressure, and moisture can vary zonally, meridionally, and vertically. The rate of change of these scalar meteorological variables can therefore be expressed using the gradient vector. With an understanding of the gradient vector, we can now examine the pressure gradient force.

5.3.2 Derivation of the pressure gradient force

To examine the pressure gradient force, we will consider the pressure exerted by the atmosphere on sides A and B of an air parcel, as illustrated in Figure 5.5. The

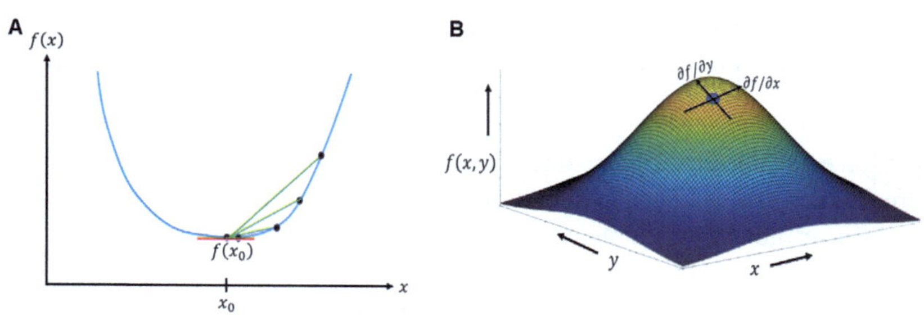

FIG. 5.4 Schematic of partial derivatives. (A) A schematic that illustrates the definition of the derivative for a scalar function of one variable. (B) A schematic that illustrates the definition of the directional derivative for a scalar function of two variables.

No Permission Required

Box 5.2 Evaluating partial derivatives

Calculate the partial derivatives of $f(x,y) = x^2 + y^2$ and evaluate them at $(x,y) = (1,2)$.

Answer: For $\partial f/\partial x$, we treat the y^2 term as a constant, and thus, we have

$$\frac{\partial f}{\partial x} = \frac{\partial}{\partial x}(x^2 + y^2) = \frac{\partial}{\partial x}(x^2) + \frac{\partial}{\partial y}(y^2) = 2x$$

Thus, at $(x,y) = (1,2)$, we have

$$\left[\frac{\partial f}{\partial x}\right]_{(1,2)} = 2(1) = 2$$

For $\partial f/\partial y$, we treat the x^2 term as a constant, and thus, we have:

$$\frac{\partial f}{\partial y} = \frac{\partial}{\partial y}(x^2 + y^2) = \frac{\partial}{\partial y}(x^2) + \frac{\partial}{\partial y}(y^2) = 2y$$

Thus, at $(x,y) = (1,2)$, we have

$$\left[\frac{\partial f}{\partial y}\right]_{(1,2)} = 2(2) = 4$$

This result indicates that the rate of change in the y-direction is twice the rate of change in the x-direction at $(x,y) = (1,2)$.

Box 5.3 Calculating the gradient vector

Calculate the gradient of the function $f(x,y,z) = x^2y^3z^4$ at $(x,y,z) = (1,1,1)$.

Answer: First, evaluate the directional derivatives

$$\frac{\partial f}{\partial x} = \frac{\partial}{\partial x}(x^2y^3z^4) = y^3z^4\frac{\partial}{\partial x}(x^2) = 2xy^3z^4$$
$$\frac{\partial f}{\partial y} = \frac{\partial}{\partial y}(x^2y^3z^4) = x^2z^4\frac{\partial}{\partial y}(y^3) = 3x^2y^2z^4$$
$$\frac{\partial f}{\partial z} = \frac{\partial}{\partial z}(x^2y^3z^4) = x^2y^3\frac{\partial}{\partial z}(z^4) = 4x^2y^3z^3$$

The gradient is given by

$$\nabla f = \left(\frac{\partial f}{\partial x}\right)\boldsymbol{i} + \left(\frac{\partial f}{\partial y}\right)\boldsymbol{j} + \left(\frac{\partial f}{\partial z}\right)\boldsymbol{k} = (2xy^3z^4)\boldsymbol{i} + (3x^2y^2z^4)\boldsymbol{j} + (4x^2y^3z^3)\boldsymbol{k}$$

Evaluating the gradient at $(x,y,z) = (1,1,1)$ gives

$$[\nabla f]_{(1,1,1)} = [2(1)(1)^3(1)^4]\boldsymbol{i} + [3(1)^2(1)^2(1)^4]\boldsymbol{j} + [4(1)^2(1)^3(1)^3]\boldsymbol{k} = 2\boldsymbol{i} + 3\boldsymbol{j} + 4\boldsymbol{k}$$

pressure on each side arises from air molecules continuously striking the walls of the air parcel, transferring linear momentum to them. According to Newton's 2nd law, the total linear momentum transferred per second produces the force that the atmosphere exerts on the parcel's walls.

Based on this argument, any change in pressure across the parcel will lead to a net force. In other words, the magnitude of the pressure gradient force is proportional to the pressure gradient. Because pressure is defined as force per unit area, the expression for the pressure gradient force between two points can be written as:

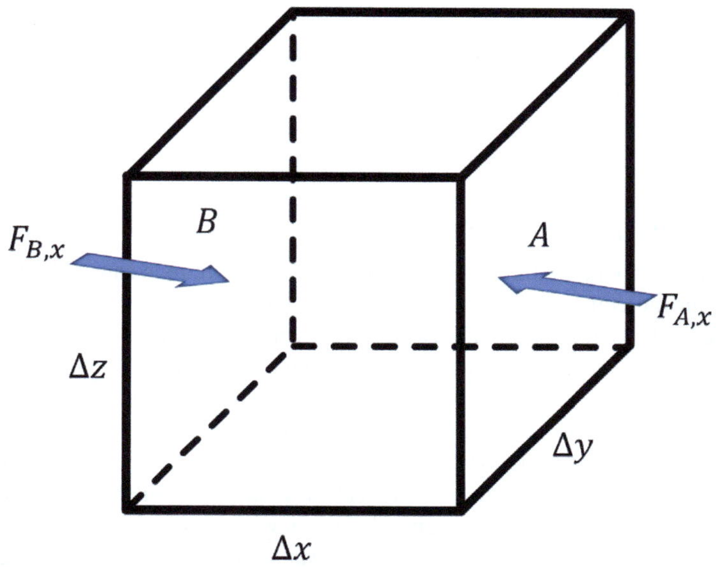

FIG. 5.5 Schematic of pressure gradient force on an air parcel

Schematic of pressure gradient force. In this figure, we are examining the x-component of the forces acting on an air parcel.

No Permission Required

$$F_{PG,x} = F_{B,x} - F_{A,x} = -\Delta P \cdot A = -\frac{\Delta P}{\Delta x} V$$

where ΔP is the pressure difference between the two points, Δx is the distance between them, and V is the volume of the air parcel. The minus sign indicates that air accelerates from regions of higher pressure to regions of lower pressure. Taking the limit as $\Delta x \rightarrow 0$, this expression becomes:

$$F_{PG,x} = -\frac{\partial P}{\partial x} V$$

Repeating this process in the *y*- and *z*-directions gives us:

$$\boldsymbol{F}_{PG} = -\left(\frac{\partial P}{\partial x}\boldsymbol{i} + \frac{\partial P}{\partial y}\boldsymbol{j} + \frac{\partial P}{\partial z}\boldsymbol{k}\right) V = -(\nabla P)V$$

Physically, the pressure gradient force shows that *any* spatial variability in atmospheric pressure produces acceleration of an air parcel. According to Newton's 2nd law, the acceleration due to the pressure gradient is :

$$\boldsymbol{a}_{PG} = \frac{\boldsymbol{F}_{PG}}{m} = -\nabla P\left(\frac{V}{m}\right)$$

$$\Rightarrow \mathbf{a}_{PG} = -\frac{1}{\rho}\nabla P \tag{5.12}$$

where ρ is the local air density. Box 5.4 provides an example of how the acceleration from the pressure gradient force depends upon the thermodynamic state of the environment.

5.4 Hydrostatic balance

As mentioned in Ch. 1, atmospheric pressure at any altitude is caused by the weight of all the air molecules above that altitude. Consequently, as you travel higher in the atmosphere, there are fewer molecules above you than at the surface, leading to a decrease in pressure with height. This creates an upward-directed pressure gradient force throughout the atmosphere, as established in the previous section.

The vertical component of the pressure gradient force must be significant, given that 90% of Earth's atmosphere is contained within the bottom 20 km. However, observations of the synoptic-scale atmosphere do not reveal large vertical accelerations. This suggests that another force of comparable magnitude - the downward-directed gravitational force discussed in Ch. 5.2 - opposes the pressure gradient force. In other words, the large-scale atmosphere tends to exhibit a balance between the vertical component of the pressure gradient force and the gravitational force.

If these forces are balanced, then, according to Newton's 2nd law, we have:

$$0 = -g - \frac{1}{\rho}\frac{\partial P}{\partial z}$$

Box 5.4 Calculate the acceleration due to the pressure gradient force

Consider an air parcel with a temperature of 12.3°C and a pressure 1004 hPa that is located at the surface. If the pressure near the air parcel changes 8.00 hPa over 500 km and if the air parcel is initially at rest, determine its speed after 30 minutes if only the pressure gradient force acts on the air parcel.

Answer: First, determine the air density via the ideal gas law:

$$\rho = \frac{P}{R_d T} = \frac{100{,}400\ \text{Pa}}{\left(287\ \frac{\text{J}}{\text{kg}\cdot K}\right)(285.45\ \text{K})} = 1.226\ \frac{\text{kg}}{\text{m}^3}$$

The magnitude of the acceleration produced by pressure gradient force is:

$$a_{PG} = \frac{1}{\rho}\nabla P \approx \frac{1}{\rho}\frac{\Delta P}{\Delta x} = \frac{8.00\ \text{hPa}}{\left(1.226\ \frac{\text{kg}}{\text{m}^3}\right)(500.0\ \text{km})} = \frac{800\ \text{Pa}}{\left(1.226\ \frac{\text{kg}}{\text{m}^3}\right)(5.00\times 10^5\ \text{m})}$$

$$= 1.31\times 10^{-3}\ \frac{\text{m}}{\text{s}^2}$$

If the acceleration remains constant, then the speed after 30.0 minutes is given by:

$$V_y = a_{PG}t = \left(1.31\times 10^{-3}\ \frac{\text{m}}{\text{s}^2}\right)(1800\ \text{s}) = 2.36\ \frac{\text{m}}{\text{s}}$$

Since the pressure gradient force is proportional to the spacing between isobars, tightly packed isobars imply a strong pressure gradient force, whereas loosely packed isobars imply a weak pressure gradient force. Thus, the acceleration of air parcels will vary based on the spacing of the isobars. Moreover, notice that the pressure gradient force changes as the temperature and pressure change (since density is a function of pressure and temperature).

$$\Rightarrow \frac{dP}{dz} = -\rho g \tag{5.13}$$

Eq. (5.13) is known as the **hydrostatic equation**, and we call this balance of forces **hydrostatic balance**. Since the vertical velocity is zero at Earth's surface, hydrostatic balance is exact only if the entire atmosphere is at rest. Nevertheless, observations indicate that hydrostatic balance is maintained with high accuracy for nearly all synoptic-scale motions within Earth's atmosphere.

To illustrate this, we can examine the vertical velocities typically found on synoptic weather charts. Vertical velocity is generally understood as the rate at which geometric height changes: rising motion indicates an increase in height, whereas sinking motion indicates a decrease in height. However, because synoptic weather charts and radiosonde observations are presented as functions of pressure rather than altitude, it is useful to express vertical velocity in terms of the rate of change of pressure with time. In meteorology, this pressure-based vertical velocity is known as **omega velocity** ω (or simply **omega**). For a rising air parcel, the pressure decreases over time, which means that $\omega < 0$ (rising motion). Conversely, for a sinking air parcel, the pressure increases over time, so $\omega > 0$ (sinking motion). ω is measured in decipascals per second $dPa\ s^{-1}$. For reference, a vertical velocity of 1 dPa/s in the lower troposphere corresponds to approximately 1 cm/s of motion.

Figure 5.6 provides 500-hPa observations on January 8, 2024, at 12Z. As air parcels move through the trough located over the south-central United States, horizontal winds exceed 70 knots. The strongest rising motion is associated with ω values around −20 dPa/s (or approximately 20 cm/s of rising motion), whereas the strongest sinking motion corresponds to ω values around 30 dPa/s (or approximately −30 cm/s of sinking motion). This demonstrates that horizontal wind speeds are roughly two orders of magnitude greater than vertical velocities.

Table 5.2 outlines the typical ω values for various weather conditions. Across most of the continental United States, as shown in Figure 5.6, omega velocities remain close to zero, indicating very weak synoptic-scale motion.

Based on these observations and arguments, hydrostatic balance is a fundamental characteristic of Earth's atmosphere. We will now examine two important implications associated with hydrostatic balance: (1) the relationship between pressure and height in the atmosphere, and (2) the use of pressure as a vertical coordinate.

5.4.1 The barometric equation

In Ch. 2, it was asserted that the relationship between pressure and height is nonlinear function of virtual temperature. We can justify the pressure–height relationship using the hydrostatic equation. Applying the ideal gas law to Eq. (5.13) gives

$$\frac{dP}{dz} = -\rho g = -\frac{gP}{R_d T_v}$$

$$\Rightarrow \frac{dP}{P} = -\frac{g\,dz}{R_d T_v}$$

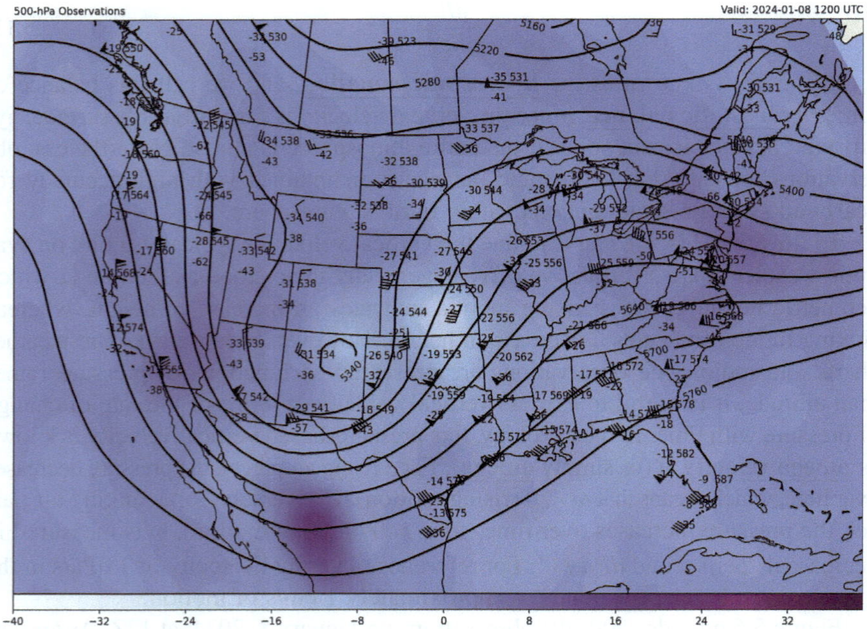

FIG. 5.6 500-hPa observations of vertical wind

500-hPa observations from January 8, 2024, at 12 Z. Solid black contours correspond to isohypses, and contour color fill corresponds to omega velocity in units of dPa/s.

No Permission Required

Table 5.2 Representative values for the omega velocity

ω (dPa/s)	Typical weather conditions
–0.5	Stratus clouds
–5	Light rain
–50	Heavy rain; cumulus clouds
–500	Thunderstorms

No Permission Required

We can rewrite this expression in terms of geopotential height by recalling that $g_0\ dZ = g\ dz$ where $g_0 = 9.80\ \text{m/s}^2$ is the average gravitational acceleration at sea level. Therefore, the hydrostatic equation becomes:

$$\frac{dP}{P} = -\frac{g_0\ dZ}{R_d T_v}$$

Integrating both sides of this expression gives us:

$$\int_{P_0}^{P} \frac{dP}{P} = -\frac{g_0}{R_d} \int_{Z_0}^{Z} \frac{dZ'}{T_v}$$

$$\Rightarrow \ln\left(\frac{P}{P_0}\right) = -\frac{g_0}{R_d} \int_{Z_0}^{Z} \frac{dZ'}{T_v}$$

$$P = P_0 \exp\left(-\frac{g_0}{R_d} \int_{Z_0}^{Z} \frac{dZ'}{T_v}\right) \tag{5.14}$$

Eq. (5.14) is known as the **barometric equation**, which describes how atmospheric pressure varies with height. As expected, the relationship between pressure and height is a nonlinear function of virtual temperature.

To obtain a closed-form expression for Eq. (5.14), we need to know the vertical profile of temperature and moisture as a function of height. Fortunately, this information can be obtained from radiosonde observations. A simple approximation of the barometric equation assumes that the virtual temperature at a given height $T_v(z)$ is approximately equal to the layer-averaged virtual temperature, denoted as $\overline{T}_v$. In this case, Eq. (5.14) simplifies to:

$$P \approx P_0 \exp\left(-\frac{g_0}{R_d \overline{T}_v} \int_{Z_0}^{Z} dZ'\right)$$

$$\Rightarrow P \approx P_0 \exp\left(-\frac{Z - Z_0}{H_P}\right), \qquad H_P = \frac{R_d \overline{T}_v}{g_0} \tag{5.15}$$

where H_P is known as the **scale height**, which indicates how rapidly pressure decreases with height. When $Z - Z_0 = H_P$, Eq. (5.15) predicts that the pressure will be $P \approx P_0 e^{-1} \approx 0.368 P_0$, or about 37% of its surface value. This equation is useful for estimating atmospheric pressure at various geometric heights, as illustrated in Box 5.5.

According to Eq. (5.15), pressure decreases nearly exponentially with height. If the virtual temperature remains approximately constant within an atmospheric layer, then density is proportional to pressure according to the ideal gas law. Therefore, Eq. (5.15) also predicts that density decreases nearly exponentially with height in the atmosphere, following the same mathematical form:

$$\rho \approx \rho_0 \exp\left(-\frac{Z - Z_0}{H_P}\right), \qquad H_P = \frac{R_d T_v}{g_0} \tag{5.16}$$

where ρ_0 is the density at Z_0. At sea level, where $Z_0 = 0$ m, the surface density is $\rho_0 = 1.225$ kg/m^3.

5.4.2 The thickness equation

As discussed in the previous section, the barometric equation defines the relationship between pressure and geopotential height, which depends on virtual temperature. However, as noted in Ch. 2, radiosondes measure geopotential height, temperature,

Box 5.5 Estimating the atmospheric pressure at various geometric heights using the barometric equation

In Ch. 4, it was asserted that the 250-hPa roughly corresponds to a height of 10,000 meters. Let's justify this statement using the barometric equation. Let's suppose that the average temperature between sea level and 10,000 meters is 245 K and that the average mixing ratio between these two elevations is 3 g kg^{-1}. To determine the atmospheric pressure associated with a geometric height of 10,000 meters, we first must determine the scale height, which is given by:

$$H_P = \frac{R_d T_v}{g_0} \approx \frac{R_d T(1+0.608\,\mathrm{r})}{g_0} = \frac{\left(287\,\frac{\mathrm{J}}{\mathrm{kg{\cdot}K}}\right)(245\ \mathrm{K})\left[1+0.608\left(0.003\,\frac{\mathrm{kg}}{\mathrm{kg}}\right)\right]}{\left(9.80\,\frac{\mathrm{m}}{\mathrm{s}^2}\right)} = 7.188\ \mathrm{km}$$

where the virtual temperature approximation from Ch. 1 was used. Because we are examining the portion of the atmosphere between sea level and 500 hPa, then $Z_O = 0$ m, $P_O = 1013.25$ hPa, and the geopotential height associated with a height of 10,000 m is given by:

$$Z = \frac{R_E z}{R_E + z} = \frac{(6378.1\ \mathrm{km})(10.0\ \mathrm{km})}{(6378.1\ \mathrm{km} - 10.0\ \mathrm{km})} = 10.016\ \mathrm{km}$$

Therefore, by Eq. (5.15), we have

$$P \approx P_0 \exp\left(-\frac{Z}{H_P}\right) = (1013.25\ \mathrm{hPa})\exp\left(-\frac{10.016\ \mathrm{km}}{7.188\ \mathrm{km}}\right) = 252\ \mathrm{hPa}$$

Therefore, 10,000 m elevation roughly corresponds to 250-hPa within Earth's atmosphere.

and moisture as functions of pressure. Therefore, it is useful to derive an expression for geopotential height as a function of pressure.

By combining the ideal gas law with the hydrostatic equation, we obtain:

$$\frac{dP}{dz} = -\rho g = -\frac{gP}{R_d T_v}$$

Rearranging this equation to solve for dz yields:

$$dz = -\frac{R_d T_v}{g}\frac{dP}{P}$$

Since $g\,dz = g_0\,dZ$, then this equation can be rewritten as:

$$dZ = -\frac{R_d T_v}{g_0}\frac{dP}{P}$$

Integrating both sides gives:

$$\int_{Z_0}^{Z} dZ = -\int_{P_0}^{P}\frac{R_d T_v}{g_0}\frac{dP'}{P'}$$

$$\Rightarrow \Delta Z = \frac{R_d}{g_0}\int_{P}^{P_0} T_v \frac{dP'}{P'}$$

Typically, the virtual temperature is not constant with respect to pressure. However, if we approximate it using the layer-averaged virtual temperature $\overline{T}_v$ between P_O and P, we can approximate the above expression:

$$\Delta Z \approx \frac{R_d \overline{T}_v}{g_0} \int_P^{P_0} \frac{dP'}{P'} = \frac{R_d \overline{T}_v}{g_0} \ln\left(\frac{P_0}{P}\right) \tag{5.17}$$

ΔZ represents the **thickness** of the layer between two isobaric surfaces. Consequently, Eq. (5.17) is known as the **thickness equation** or the **hypsometric equation**. According to this equation, the thickness between two isobaric surfaces is directly proportional to the layer-averaged virtual temperature between the two isobaric surfaces. In other words, warm, moist air produces a greater thickness than dry, cool air between the same two isobaric surfaces.

The thickness equation has several important applications that are essential for analyzing and understanding midlatitude weather systems, which we will explore throughout the rest of the textbook. Boxes 5.6 and 5.7 provide examples of how the thickness equation can be applied in various meteorological settings.

5.4.3 Isobaric vertical coordinates

Based on the arguments presented in Chs. 5.4.1 and 5.4.2, radiosonde measurements can be used to determine the geopotential height as a function of pressure. Because of the relationship between pressure and geopotential height, meteorologists construct upper-air maps at constant pressure rather than at constant height. Since

Box 5.6 Calculating geopotential height from radiosonde measurements

Since radiosondes measure the vertical profiles of temperature, moisture, and pressure, the thickness equation can be used to determine the geopotential heights at which radiosonde data are found. As an example of this procedure, suppose that a radiosonde makes the following measurements as shown in Table 5.3.

Let's suppose that the geopotential height at 1012.3 hPa is 0 meters. Within the layer between 1012.3 hPa and 965 hPa, the column-averaged virtual temperature is given by:

$$\overline{T}_v = \frac{1}{2}(23.9°C + 21.1°C) = 22.5°C = 295.65\ \text{K}$$

Therefore, the geopotential height at 965.0 hPa is:

$$Z = Z_0 + \frac{R_d \overline{T}_v}{g_0} \ln\left(\frac{P_0}{P}\right) = 0\ \text{m} + \frac{\left(287\ \frac{\text{J}}{\text{kg}\cdot K}\right)(295.65\ \text{K})}{\left(9.80\ \frac{\text{m}}{\text{s}^2}\right)} \ln\left(\frac{1012.3\ \text{hPa}}{965\ \text{hPa}}\right) = 414.3\ \text{m}$$

Within the layer between 965 hPa and 960 hPa, the column-averaged virtual temperature is given by:

$$\overline{T}_v = \frac{1}{2}(21.1°C + 21.6°C) = 22.35°C = 294.50\ \text{K}$$

Therefore, the geopotential height at 960 hPa is:

$$Z = Z_0 + \frac{R_d \overline{T}_v}{g_0} \ln\left(\frac{P_0}{P}\right) = 414.3\ \text{m} + \frac{\left(287\ \frac{\text{J}}{\text{kg}\cdot\text{K}}\right)(294.5\ \text{K})}{\left(9.80\ \frac{\text{m}}{\text{s}^2}\right)} \ln\left(\frac{965\ \text{hPa}}{960\ \text{hPa}}\right) = 459.1\ \text{m}$$

Repeating the same procedure, the geopotential height at 953 hPa is 522.5 m.

Box 5.7 Reduced sea-level pressure and the altimeter equation

As discussed in Ch. 4, isopleths of constant sea-level pressure are plotted on surface maps, allowing for the identification of areas with high and low pressure. However, because most surface observation stations are not situated at sea level, the differences in pressure between these stations can be attributed to both horizontal pressure gradients and differences in elevation. To account for the effects of elevation, meteorologists calculate what the sea-level pressure at a station would be if it were at an elevation of 0 meters. This calculated value is known as **reduced sea-level pressure (RSLP)**, and is typically reported in the remarks section of a METAR report as discussed in Ch. 3.

To calculate the RSLP, we can use the thickness equation. For example, let's consider a scenario where the station pressure in Denver, CO, is 828 hPa and the station elevation is 1609 meters.

To determine the RSLP, we will define the following variable:

- Let Z represent the elevation at Denver (Z_{DEN}) and Z_0 equals 0 meters, which is the geopotential height at sea level.
- Let P be the observed station pressure (P_{STA}) and P_0 denote the value we want to calculate, which is the sea-level pressure (P_{SLP}) for Denver.
- Finally, $\overline{T}_V$ represents the average virtual temperature between the sea-level pressure at Denver and the station's elevation.

Now we can rearrange the thickness equation using these variables:

$$\frac{g_0 Z_{DEN}}{R_d \overline{T}_v} = \ln\left(\frac{P_{SLP}}{P_{STA}}\right)$$

$$\Rightarrow P_{SLP} = P_{STA}\exp\left(\frac{g_0 Z_{STA}}{R_d T_v}\right) \tag{5.18}$$

Eq. (5.18) is known as the **altimeter equation**, and it is the standard expression for obtaining the RSLP. We can determine the geopotential height to be:

$$Z_{STA} = \frac{R_E z_{STA}}{R_E + z_{STA}} = \frac{(6371.8\text{ km})(1.609\text{ km})}{(6371.8\text{ km} + 1.609\text{ km})} = 1.608\text{ km}$$

Thus, by Eq. (5.18), the reduced sea-level pressure is given by:

$$P_{SLP} = P_{STA}\exp\left(\frac{g_0 Z_{STA}}{R_d T_v}\right) = (828\text{ hPa})\exp\left[\frac{\left(9.80\frac{\text{m}}{\text{s}^2}\right)(1608\text{ m})}{\left(287\frac{\text{J}}{\text{kg}\cdot\text{K}}\right)(293.15\text{ K})}\right] = 998.6\text{ hPa}$$

A reduced sea-level pressure calculation allows us to determine how much of the observed pressure difference between two locations is due to horizontal pressure difference rather than vertical variations of pressure due to elevation.

Table 5.3 Sample radiosonde data at an upper-air station

Pressure (hPa)	Virtual Temperature (°C)
1012.3	23.9
965	21.1
960	21.6
953	21.2

No Permission Required

pressure is fixed on isobaric charts, we can apply the hydrostatic equation to examine how geopotential height varies across that surface. Consequently, instead of plotting isobars on a geometric height surface (as on surface weather charts), we plot isohypses on upper-air isobaric surfaces.

This approach suggests that the pressure gradient force should be reformulated in terms of geopotential height. In Cartesian coordinates, the pressure gradient force on a surface of constant geopotential height is expressed as:

$$-\frac{1}{\rho}\nabla_Z P = -\frac{1}{\rho}\left(\frac{\partial P}{\partial x}\boldsymbol{i} + \frac{\partial P}{\partial y}\boldsymbol{j}\right)$$

where the subscript "Z" indicates that the pressure gradient force is evaluated on a surface of constant geopotential height. Given that $|dP| = \rho g\ dz = \rho g_0\ dZ$ by the hydrostatic equation, we can rewrite the pressure gradient force (per unit mass) associated with the horizontal wind as:

$$-\frac{1}{\rho}\left(\frac{\partial P}{\partial x}\boldsymbol{i} + \frac{\partial P}{\partial y}\boldsymbol{j}\right) = -\frac{1}{\rho}(\rho g_0)\left(\frac{\partial Z}{\partial x}\boldsymbol{i} + \frac{\partial Z}{\partial y}\boldsymbol{j}\right)$$

$$\Rightarrow -\frac{1}{\rho}\nabla_Z P = -g_0 \nabla_P Z \tag{5.19}$$

In this expression, the subscript "P" indicates that the pressure gradient force is evaluated on a surface of constant pressure (i.e., an isobaric surface). Eq. (5.19) provides a more straightforward expression for calculating the pressure gradient force because it does not explicitly depend on air density. This simplification is advantageous because density varies with temperature, pressure, and water vapor according to the ideal gas law. Instead, the dependence on density is implicit through the calculation of geopotential height via the hypsometric equation. We will use Eq. (5.19) when we examine atmospheric flow above the surface in more detail in Ch. 7.

5.5 Turbulent drag force

So far, we have examined atmospheric flow influenced by the pressure gradient force and gravity. However, as air moves near Earth's surface, various obstacles (such as grass, trees, and buildings) partially block the wind. The net effect of these obstacles is to decelerate the wind close to the surface, similar to how surface friction slows an object moving on a table. In this section, we will explore the nature of this frictional interaction within the atmosphere. To understand it properly, we must first introduce the concept of viscosity.

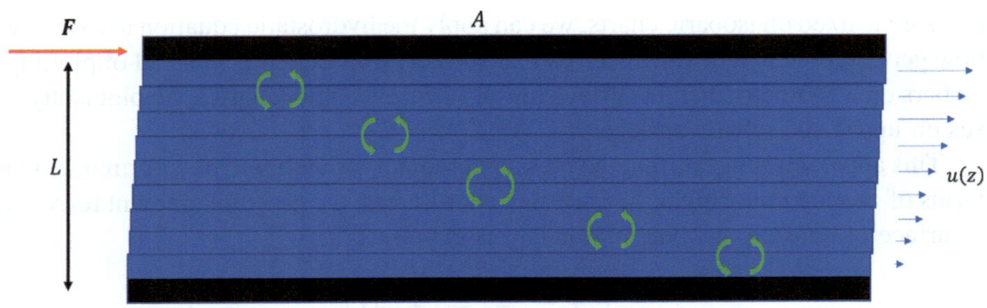

FIG. 5.7 Viscous fluid flow

Schematic of fluid flow between two plates of area A where the bottom plate is fixed. When the top plate is pushed to the right by a force F, it drags the fluid along with it. The green arrows indicate that momentum is transferred between each layer.

No Permission Required

5.5.1 Introduction to viscosity

As discussed in Ch. 1, the behavior of gases can be understood through kinetic theory. A modified version of this theory can also describe the molecular behavior of fluids in general.[3] Based on kinetic theory, the molecules of a fluid are in constant motion, exchanging kinetic energy and momentym through collisions. This process causes energy and momentum to spread throughout the fluid at the molecular level.

To illustrate this, consider two parallel surfaces with a small gap filled with fluid, as shown in Figure 5.7. Suppose the bottom surface is at rest while the top surface moves in the $\boldsymbol{i}$ direction, generating a shearing stress on the fluid. At the molecular level, a thin layer of fluid "sticks" to the top surface so that the average molecular motion in this layer matches the top plate. Similarly, a thin layer of fluid adheres to the bottom surface, where the average molecular motion is approximately zero. Consequently, vertical shear, represented as $u(z)$, develops in the gap between the plates. In summary, when a shear stress is applied to a fluid, one part of the fluid flows past another part.

Although fluids cannot sustain shear stress without flowing, they will resist this shearing flow. This resistance is known as **viscosity**. In the context of Figure 5.7, viscosity can be thought of as an internal shear stress opposing the applied external stress. On a molecular scale, the top layer transfers some of its forward momentum to the layer below, and this process continues down to the bottom layer, which exerts a forward force on the bottom surface. Simultaneously, according to Newton's 3rd

[3] The properties of liquids can be explained using a modified version of the kinetic theory of gases. Liquids exhibit short-range order because the strong intermolecular attractive forces cause the molecules to group together. For this reason, liquids are much denser than gases, and the molecules within a liquid are not completely random.

law, the loss of momentum in the top layer causes it to exert a backward force on the top surface. The efficiency of this mechanism depends on the fluid's viscosity.

Based on the viscosity of the fluid, a force is required to keep the top plate moving at constant velocity. This force is referred to as the **viscous force**. Using the schematic in Figure 5.7, we can identify the factors that determine its magnitude. We expect this force to be proportional to the cross-sectional area (since larger plates will transfer more momentum), proportional to the velocity of the plate (because more work is required to maintain this velocity), and inversely proportional to the plate separation (as molecules near the bottom plate can transfer momentum to molecules above).

Experimental data indicates that the magnitude of this force within a layer is expressed as:

$$F = \mu \frac{Au}{L} \tag{5.20}$$

where μ represents the **dynamic viscosity coefficient**. Dimensional analysis shows that the units of viscosity are $N{\cdot}s/m^2$ or $Pa{\cdot}s$.

Table 5.4 presents the dynamic viscosity coefficients for common fluids. Fluids with high viscosity, such as glycerin and oil, require significant force to maintain shearing stress. In contrast, air has very low viscosity, which leads to two important effects regarding the viscous force in the atmosphere.

First, due to air's low viscosity, the mechanism for momentum transfer is relatively inefficient, meaning that the effects of viscosity are felt primarily near the surface. This region, where shear stress is significant, is known as the **planetary boundary layer (PBL)**, which we will examine in more detail in Ch. 5.5.3.

Second, when shear stress is applied to air, it produces relative acceleration between adjacent layers because air's low viscosity generates weak internal stress. This enables us to derive an expression for the viscous force within the atmosphere, which will be examined in the next section.

Table 5.4 The dynamic viscosity coefficient for various liquids and gases

Fluids	$\mu(Pa{\cdot}s)$
Glycerin	1.49
Heavy oil	0.66
Liquid mercury	0.0016
Liquid water at 20°C	0.0010
Air	0.000018

No Permission Required

5.5.2 Derivation of the viscous force

Let's return to the experimental setup where a viscous fluid is situated between two plates, as shown in Figure 5.7. If the top plate is moving at a constant speed, then every horizontal layer of fluid must exert the same force on the layer beneath it. Therefore, if a layer of fluid has a depth Δz, then the viscous force can be expressed as:

$$F = \mu A \frac{\Delta u}{\Delta z}$$

where Δu represents the velocity shear across the layer with depth Δz. If we take the limit as $\Delta z \to 0$, the viscous force per unit area (equivalent to the shearing stress applied to the fluid) becomes:

$$\tau = \frac{F}{A} = \mu \frac{\partial u}{\partial z} \tag{5.21}$$

In summary, Eq. (5.21) describes the shearing stress applied to a fluid layer due to molecular viscosity. As illustrated in Figure 5.7, the speed of the fluid increases with height. This means that a fluid layer near the top has more linear momentum compared with a fluid layer near the bottom. Consequently, the shearing stress results from a net downward transport of momentum due to molecular motion.

This discussion assumes that the fluid is moving at constant speed, leading to a constant vertical shear. Now, let's explore what happens when there is a relative acceleration between adjacent layers of air in the atmosphere. In this case, the shearing stress originates from the Earth's surface, which decelerates the air. Because of the relative acceleration between adjacent layers of air, we do not expect each horizontal layer to exert the same force on the layer above it. Instead, the shearing stress should

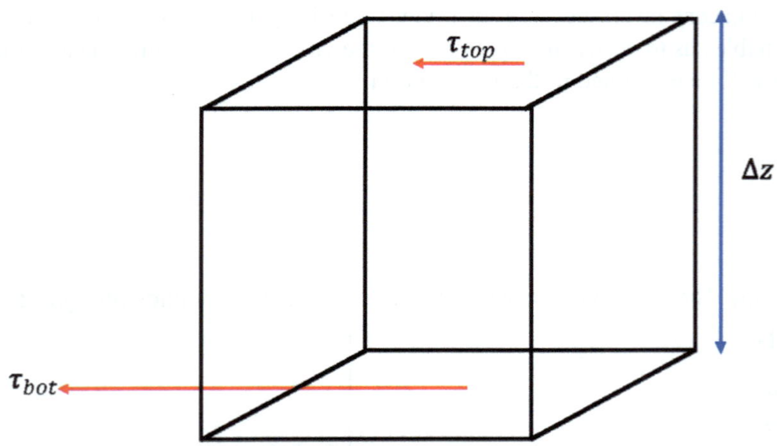

FIG. 5.8 Schematic of air parcel under vertical shearing stress

Schematic of air parcel under the influence of a vertical shearing stress. Here, we focus on the x-component of this vertical shearing stress, and it is assumed that the air parcel is moving in the *i* direction.

No Permission Required

be greater for layers closer to Earth's surface compared with those further away. Applied to an air parcel, this means that the shear stress on the bottom face is greater than on the top face, as shown in Figure 5.8.

On the bottom face of the air parcel, Earth's surface exerts a stress of magnitude τ_{bot}. Since the surface has less momentum than the moving air parcel, the parcel experiences a stress in the $-\boldsymbol{i}$ direction. Similarly, on the top face of the air parcel, the fluid layer above exerts a stress of magnitude τ_{top}. Because the air parcel moves slower than the layer above, the top face exerts a shear stress in the $-\boldsymbol{i}$ direction on the fluid above. By Newton's 3rd law, this means that the fluid layer above exerts a shear stress in the $+\boldsymbol{i}$ direction. The combination of the shear stress from Earth's surface and the shear stress from the air layer above results in a net deceleration, where $\tau_{top} < \tau_{bot}$.

The net force generated by this vertical shearing stress can be written as:

$$F_{net,x} = -A\Delta\tau = -V\frac{\Delta\tau}{\Delta z}$$

where A is the cross-sectional area of the air parcel, V is its volume, and the minus sign indicates that the turbulent drag force slows down the air parcel. Taking the limit as $\Delta z \rightarrow 0$, this expression becomes:

$$F_{drag,x} = -V\frac{\partial\tau_x}{\partial z}$$

Repeating this process for shear stress in the y- and z-directions gives:

$$\boldsymbol{F}_{drag} = -\left(\frac{\partial\tau_x}{\partial z}\boldsymbol{i} + \frac{\partial\tau_y}{\partial z}\boldsymbol{j} + \frac{\partial\tau_z}{\partial z}\boldsymbol{k}\right)V = \frac{\partial\boldsymbol{\tau}}{\partial z}V$$

where τ_y and τ_z correspond to the shear stress acting in the y-direction and in the z-direction, respectively. Thus, the acceleration caused by the viscous drag force is given by:

$$\boldsymbol{a}_{drag} = \frac{\boldsymbol{F}_{drag}}{m} = -\frac{1}{\rho}\frac{\partial\boldsymbol{\tau}}{\partial z} \quad (5.22)$$

To summarize, Eq. (5.22) describes the viscous drag force per unit mass acting on an air parcel due to molecular viscosity, where $\boldsymbol{\tau}$ represents the shear stress experienced by the air due to vertical transport of momentum. Viscosity functions as an internal shear stress that opposes the relative acceleration of the air, and vertical momentum transport acts to reduce vertical wind gradients. As shown in Figure 5.8, the shear stress at Earth's surface acts opposite to the surface wind. For this reason, surface shear stress can be visualized as a drag force that acts on the surface wind.

According to Eq. (5.22), the strength of the viscous force depends on the rate of momentum transport throughout the PBL. However, in the PBL, turbulence, rather than molecular viscosity, dominates this momentum transfer. For this reason, the viscous force in the PBL is often referred to as the **turbulent drag force**. In the

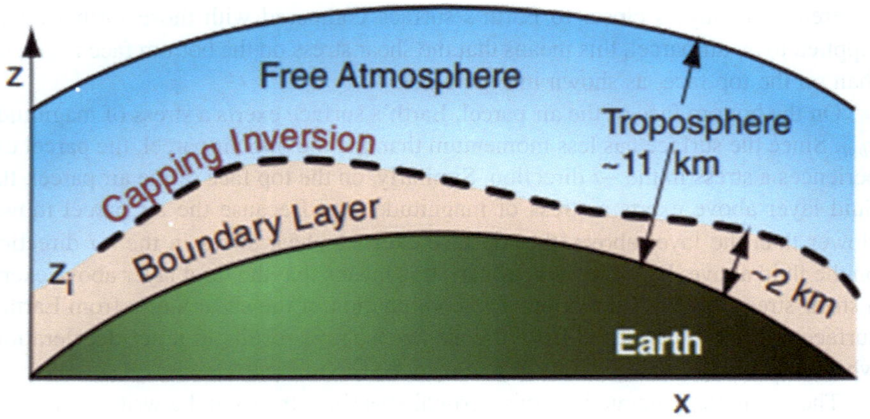

FIG. 5.9

A schematic of the planetary boundary layer (PBL). Notice that the PBL ranges in thickness across the globe, and that a capping inversion separates the PBL from the free atmosphere within the troposphere.

next section, we provide an overview of the PBL and the mechanisms underlying turbulent drag force.

5.5.3 Overview of the planetary boundary layer

As shown in Figure 5.9, the vertical structure of the troposphere can be broadly divided into two regions: the PBL and the free atmosphere. The PBL typically extends from 1 to 2 km above the surface, occupying roughly the lowest 20% of the troposphere. However, this depth can vary significantly, from tens of meters to more than 4 km, depending on location[4]. Much of this variability is associated with turbulence, which is the primary mechanism that causes PBL air to adopt the characteristics of the underlying surface.

Turbulence consists of numerous complex swirling features called **eddies** that interact with one another. When analyzed statistically, identifiable patterns emerge that can be described and measured. For our purposes, we can classify turbulence into two types: mechanical turbulence and thermal turbulence.

Mechanical turbulence arises from wind shear, which can result from frictional drag (slower winds near the ground compared with those aloft) or from **wake turbulence** created by obstacles such as trees, buildings, and islands. **Thermal turbulence** is characterized by rising plumes, or thermals, of warm air, accompanied by sinking cold air, driven by buoyant forces.

[4] The structure and depth of the PBL also depends upon the time of day since the PBL undergoes a well-known diurnal cycle. This will be discussed in more detail in Ch. 8.

Turbulence develops within the PBL as a natural response to flow instabilities and acts to reduce these instabilities. For example, on a sunny day, the warm ground heats the lower layers of air, making them unstable. The flow responds by generating thermal circulations that transport warm air upward and cold air downward until a new equilibrium is reached. Turbulence subsides once these thermal circulations stabilize, indicating that thermal turbulence persists only if there is a continuous destabilizing mechanism, such as solar heating.

A similar principle applies to mechanical turbulence. Vertical shear in the horizontal wind creates dynamic instabilities that generates turbulence. This turbulence mixes faster- and slower-moving air, leading to more uniform wind velocities with increasing height. Mechanical turbulence ceases when the vertical shear is removed. As with thermal turbulence, persistent mechanical turbulence in the atmosphere requires ongoing destabilizing mechanisms, such as synoptic-scale weather patterns.

Because air has low molecular viscosity, the viscous force is generally much weaker than the turbulent drag force. For this reason, we focus our attention on the turbulent drag force within the PBL, as this forms the basis for approximating its effects. As discussed previously, the magnitude of shear stress in the PBL is proportional to the vertical wind shear. Therefore, we can approximate the turbulent drag term by expressing it in terms of the vertical wind shear within the PBL as follows:

$$\frac{1}{\rho}\frac{\partial \boldsymbol{\tau}}{\partial z} = w_T \frac{\boldsymbol{V}_{\mathrm{H}}}{z_i} \tag{5.23}$$

where z_i is the PBL depth, $\boldsymbol{V}_H$ is the horizontal velocity vector at the top of the PBL, and w_T is known as the **turbulent transport velocity, which provides a measure** of the speed at which turbulent eddies move within the PBL. Based on Eq. (5.23), the strength of turbulent drag decreases as the PBL depth increases, as a thicker layer reduces the efficiency of momentum transport by turbulence.

Our estimate of the turbulent drag force also depends on the turbulent transport velocity. When thermal turbulence dominates, such as during daytime light-wind conditions with a warm surface, the turbulent transport velocity can be estimated using the equation:

$$w_T = b_D w_B$$

where $b_D = 1.83 \times 10^{-3}$ is a dimensionless constant, and w_B is known as the **buoyancy velocity scale**, which measures the vertical velocity of rising thermals. Observations show that the buoyancy velocity scale varies depending on the PBL depth and environmental stability.[5] This formulation indicates that stronger thermals, characterized by a higher buoyancy velocity, create more intense thermal turbulence, promoting greater momentum transport within the PBL.

[5] The buoyancy velocity scale is ranges from 10 to 50 m/s. According to Stull (2017), typical updraft speeds in thermals are about 2% of the buoyancy velocity scale.

When mechanical turbulence dominates, such as on windy, overcast days, the turbulent transport velocity is expressed as:

$$w_T = C_D V_H$$

where V_H is the horizontal wind speed and C_D is a dimensionless drag coefficient that quantifies flow resistance over a surface. The drag coefficient is typically small (around 10^{-3}) over smooth surfaces but larger (approximately 10^{-2}) over rougher surfaces. This expression of the turbulent transport velocity indicates that stronger winds interacting with rough surfaces generate more rigorous mechanical turbulence, which promotes greater momentum transport within the PBL.

Questions

1. Suppose that an air parcel is heated at constant pressure. According to the ideal gas law, the increase in temperature should lead to an increase in volume. As the parcel expands in volume, what happens to its center of mass?
2. Based on the description of an air parcel, would a weather balloon be a good description of an air parcel? Defend your answer.
3. Use the kinetic theory of gases to explain why the viscosity of a low-density gas should increase as the temperature increases.
4. Suppose an aircraft flies along a 500-mb pressure surface from standard temperature into anomalous warm air. If there are no temperature corrections, how would the measured geopotential height of the aircraft compare with the true geopotential height of the aircraft? Defend your answer.
5. In Ch. 5.1, it was noted that a cubic foot of air is a reasonable size for an air parcel because it contains a significant number of molecules. Estimate the number of molecules in a cubic foot of air near the surface. You may assume that the density of air is 1.15 kg/m^3 and that the molar mass of dry air is 28.96 g/mol.
6. In our discussion of Newtonian gravitation, it was assumed that Earth can be modeled as a perfect sphere. However, the true shape of the Earth is an ellipsoid, which is flattened at the poles and bulging at the equator. Its equatorial radius (from its center point to the equator) is greater than its polar radius (from its center point out to either north or south pole) by 21 km. Therefore, a point at the poles is closer to Earth's center than is a point on the equator.
   a. Assuming that Earth's equatorial radius is 6,378 km, calculate the difference in gravitational acceleration at Earth's equator vs. Earth's poles.
   b. Based on your answer in (a), how does the variation in gravitational acceleration with latitude compare with the variation in gravitational acceleration with height?
7. Suppose that Earth's atmosphere can be modeled such that pressure and density decrease exponentially with height with a scale height of 7.50 km.
   a. Determine the average virtual temperature of the atmosphere.

Table 5.5 Radiosonde measurements for Problems #11 and #12

Pressure (hPa)	Geometric Height (m)	Temperature (°C)	Wind Direction	Wind Speed (kts)
1000	185	16.0	250	4
925		9.8	354	20
850		10.4	0	25

No Permission Required

b. Estimate the geopotential height in the atmosphere where $\rho = 0.960\ \mathrm{kg/m^3}$.

c. Estimate the geometric height in the atmosphere where the pressure is 100 hPa.

8. A meteorological station is located 50 meters below sea level. If the surface pressure at this station is 1020 mb, the temperature at the surface is 15°C, and the mean temperature for the 1000- to 500-mb layer is 0°C, compute the height of the 500-mb pressure level above sea level at this station.

9. Calculate the partial derivatives of $f(x, y) = \cos(2x) + \sin(2y)$ and evaluate them at $(x, y) = (2, 3)$.

10. Calculate the gradient of the function $f(x, y, z) = x\cos(xy) + z^2y^4/48$ at $(x, y, z) = (1, 1, 1)$

11. Table 5.5 provides radiosonde data near Earth's surface.

a. Use the thickness equation to provide an estimate of the geometric height at 925 hPa and 850 hPa. You may ignore the effects of moisture.

b. Using the geometric heights determined in part (a), estimate the vertical component of the pressure gradient force (per unit mass) of the layer between 1000 hPa and 850 hPa. Based on your answer, is the atmosphere in hydrostatic balance?

12. As discussed in the text, the turbulent stress due to molecular viscosity is typically much smaller than the turbulent stress due to turbulence. Use the data in Table 5.5 to answer the following questions related to the turbulent drag force.

a. Determine the shear stress in the layer between (i) 1000 hPa and 925 hPa and (ii) 925 hPa and 850 hPa. You will need to use the geometric heights from Problem 11(a) to evaluate the shear stress.

b. Use your answers to part (a) to estimate the turbulent drag force (per unit mass) between 1000 hPa and 850 hPa due to molecular viscosity.

The following problems are meant to demonstrate that turbulent stress due to molecular viscosity is many orders of magnitude less than other forces.

13. Suppose that the turbulence near the surface is driven primarily by thermals. Typically, the buoyancy velocity scale is of the order of 10 m/s. Using this buoyancy velocity scale, determine the turbulent drag force due to thermal turbulence. Compare your answer to the turbulent stress produced by molecular viscosity as shown in Problem #12(b). Assume that the top of the PBL is at the 850.0-hPa pressure level.

14. Suppose that the turbulence near the surface is driven primarily by mechanical turbulence. Use your answer in Problem #13 to provide an estimate of the drag coefficient. Based on your answer, is the shear stress caused by turbulent motions in the atmosphere likely caused by molecular viscosity? Defend your answer.

Reference

Stull, R., 2017. Practical Meteorology: An Algebra-Based Survey of Atmospheric Science -version 1.02b. Univ. of British Columbia. 940 pages, 978-0-88865-283-6. https://www.eoas.ubc.ca/books/Practical_Meteorology/, 978-0-88865-283-6.

CHAPTER

The inertial forces

6

Chapter Outline

In the previous chapter, we began our discussion on the laws governing atmospheric motion by examining the impact of the fundamental forces (i.e., the gravitational force, pressure gradient force, and turbulent drag force) on an air parcel. In that discussion, we implicitly assumed that the motion of an air parcel was being analyzed within a non-accelerating reference frame, known as an **inertial reference frame**. However, when analyzing a synoptic map, we are observing wind patterns relative to a reference frame fixed to the rotating Earth, which is a non-inertial reference frame. In this chapter, we will apply Newton's laws of motion within such a rotating reference frame. This shift in perspective has important consequences for understanding atmospheric flow. To explore the nature of rotating reference frames and the resulting inertial forces, we must first review the concepts of the scalar product and the vector product.

An Introduction to Synoptic-Dynamic Meteorology. https://doi.org/10.1016/B978-0-443-24858-0.00006-X

6.1 Scalar Product

The **scalar product** of any two vectors $\boldsymbol{A}$ and $\boldsymbol{B}$ is a scalar quantity that is defined as:

$$\boldsymbol{A} \cdot \boldsymbol{B} \equiv AB \cos\theta \tag{6.1}$$

where θ is the angle between the vectors.[1] Geometrically, $\boldsymbol{A}\cdot\boldsymbol{B}$ can be seen as the product of $\boldsymbol{B}$ and the component of $\boldsymbol{A}$ that is parallel to $\boldsymbol{B}$, as shown in Figure 6.1.

6.1.1 Mathematical Properties of the Scalar Product

The scalar product has the following mathematical properties:

1. $\boldsymbol{A}\cdot\boldsymbol{B}$ is a maximum when $\theta = 0°$. Thus, the scalar product produces its maximum value when **A** is parallel to $\boldsymbol{B}$.
2. If $\boldsymbol{A}$ is perpendicular to $\boldsymbol{B}$, then $\boldsymbol{A}\cdot\boldsymbol{B} = 0$.
3. If $\boldsymbol{A}$ is anti-parallel to $\boldsymbol{B}$ such that $\theta = 180°$, then $\boldsymbol{A}\cdot\boldsymbol{B}$ is a minimum.

These three properties demonstrate that the scalar product provides a measure of how well two vectors $\boldsymbol{A}$ and $\boldsymbol{B}$ align with each other.

Eq. (6.1) can be used to evaluate the scalar product when we know the magnitude of each vector and the angle between them. Additionally, we can express the scalar product in terms of vector components as follows:

$$\boldsymbol{A} \cdot \boldsymbol{B} = (A_x\boldsymbol{i} + A_y\boldsymbol{j} + A_z\boldsymbol{k}) \cdot (B_x\boldsymbol{i} + B_y\boldsymbol{j} + B_z\boldsymbol{k})$$

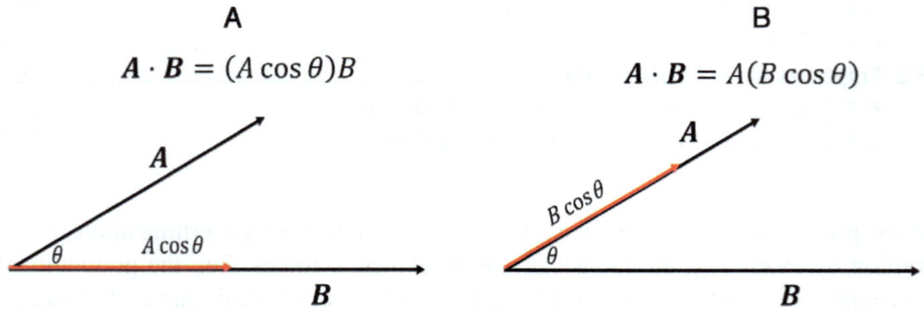

FIG. 6.1

Schematic of the geometric interpretation of the scalar product. As shown in (A), the scalar product of **A** and **B** is the magnitude of **A** times the component of **B** that is parallel to **A**. Notice that this is exactly equivalent to the magnitude of B times the component of **A** that is parallel to **B**, as shown in (B).

-No Permission Required

[1] Because of the notation, the scalar product between two vectors A and B is often read "A dot B." For this reason, the scalar product is also known as the dot product.

$$= A_xB_x(\boldsymbol{i}\cdot\boldsymbol{i}) + A_xB_y(\boldsymbol{i}\cdot\boldsymbol{j}) + A_xB_z(\boldsymbol{i}\cdot\boldsymbol{k}) + A_yB_x(\boldsymbol{j}\cdot\boldsymbol{i}) + A_yB_y(\boldsymbol{j}\cdot\boldsymbol{j}) + A_yB_z(\boldsymbol{j}\cdot\boldsymbol{k}) + A_zB_x(\boldsymbol{k}\cdot\boldsymbol{i})$$
$$+ A_zB_y(\boldsymbol{k}\cdot\boldsymbol{j}) + A_zB_z(\boldsymbol{k}\cdot\boldsymbol{k})$$

Because the unit vectors are mutually perpendicular and each unit vector has a magnitude of 1, we have:

$$\boldsymbol{i}\cdot\boldsymbol{i} = 1 = \boldsymbol{j}\cdot\boldsymbol{j} = \boldsymbol{k}\cdot\boldsymbol{k}, \boldsymbol{i}\cdot\boldsymbol{j} = 0 = \boldsymbol{j}\cdot\boldsymbol{i}, \boldsymbol{j}\cdot\boldsymbol{k} = 0 = \boldsymbol{k}\cdot\boldsymbol{j}, \boldsymbol{i}\cdot\boldsymbol{k} = 0 = \boldsymbol{k}\cdot\boldsymbol{i}$$

Therefore, the scalar product in component form is given by:

$$\boldsymbol{A}\cdot\boldsymbol{B} = A_xB_x + A_yB_y + A_zB_z \tag{6.2}$$

Based on Eq. (6.2), we can see that the scalar product is commutative, meaning $\boldsymbol{A}\cdot\boldsymbol{B} = \boldsymbol{B}\cdot\boldsymbol{A}$. Thus, the order of multiplication does not matter for the scalar product.

Finally, by applying the product rule of differentiation, we can express the derivative of the scalar product as follows:

$$\frac{d}{dx}(\boldsymbol{A}\cdot\boldsymbol{B}) = \frac{d\boldsymbol{A}}{dx}\cdot\boldsymbol{B} + \boldsymbol{A}\cdot\frac{d\boldsymbol{B}}{dx}$$

Box 6.1 provides several examples that illustrate how the scalar product is used.

How is the scalar product used in synoptic meteorology? In the following section, we will show that the scalar product is connected to the concept of advection.

Box 6.1 Examples using the scalar product

Consider the following three vectors:

$$\boldsymbol{A} = -3.00\boldsymbol{i} + 3.00\boldsymbol{j} + 2.00\boldsymbol{k}, \boldsymbol{B} = -2.00\boldsymbol{i} - 4.00\boldsymbol{j} + 2.00\boldsymbol{k}, \boldsymbol{C} = 2.00\boldsymbol{i} + 3.00\boldsymbol{j}$$

Let's perform the vector operation $\boldsymbol{C}\cdot(\boldsymbol{A} - \boldsymbol{B})$. First, we perform the operation $\boldsymbol{A} - \boldsymbol{B}$

$$\boldsymbol{A} - \boldsymbol{B} = (A_x - B_x)\boldsymbol{i} + (A_y - B_y)\boldsymbol{j} + (A_z - B_z)\boldsymbol{k}$$
$$= [-3.00 - (-2.00)]\boldsymbol{i} + [3.00 - (-4.00)]\boldsymbol{j} + [2.00 - 2.00]\boldsymbol{k} = -1.00\boldsymbol{i} + 7.00\boldsymbol{j}$$

Next, we perform the scalar product. Let's call $\boldsymbol{D} = \boldsymbol{A} - \boldsymbol{B}$, and therefore, we have

$$\boldsymbol{C}\cdot\boldsymbol{D} = C_xD_x + C_yD_y = (2.00)(-1.00) + (3.00)(7.00) = 19.0$$

The large positive value indicates that $\boldsymbol{C}$ and $\boldsymbol{D}$ are well-aligned. We can determine how well-aligned these vectors are by determining the angle between $\boldsymbol{C}$ and $\boldsymbol{D}$. Using Eq. (6.1), we have

$$\boldsymbol{C}\cdot\boldsymbol{D} = CD\cos\theta$$
$$\Rightarrow \cos\theta = \frac{\boldsymbol{C}\cdot\boldsymbol{D}}{CD}$$

Since the magnitude of each vector can be written as $C = \sqrt{C_x^2 + C_y^2}$ and $D = \sqrt{D_x^2 + D_y^2}$, where $C_x = 2.00$, $C_y = 3.00$, $D_x = -1.00$, $D_y = 7.00$. Therefore, we have

$$\cos\theta = \frac{\boldsymbol{C}\cdot\boldsymbol{D}}{\left(\sqrt{C_x^2 + C_y^2}\right)\left(\sqrt{D_x^2 + D_y^2}\right)} = \frac{19.0}{\left(\sqrt{(2.00)^2 + (3.00)^2}\right)\left(\sqrt{(-1.00)^2 + (7.00)^2}\right)} = 0.745$$

Therefore, the angle between these vectors is

$$\theta = \cos^{-1}(0.745) = 41.8°$$

6.1.2 Eulerian and Lagrangian Reference Frames

In Ch. 5, we applied Newton's 2nd law to analyze the motion of an air parcel. There are two general approaches for doing this. One natural approach is to use a reference frame that moves with the air parcel, known as a **Lagrangian reference frame**. However, when analyzing synoptic maps, we observe changes in wind at fixed locations. This approach uses an **Eulerian reference frame**.

Based on these definitions, we can describe the evolution of atmospheric flow in two ways. When we track changes relative to a moving air parcel (Lagrangian frame), we refer to this as the **Lagrangian rate of change**. In contrast, when we track changes at a fixed position (Eulerian frame), we refer to this as the **Eulerian rate of change**.

How are these two reference frames related? To answer this, consider acceleration at a fixed point in space. The change in velocity at that point (Eulerian rate of change) must equal the velocity change experienced by a moving air parcel (Lagrangian rate of change) plus the change due to air flowing past the point. This latter contribution is known as **advection.** Therefore, to apply Newton's 2nd law at a fixed location on Earth, we must account for advection.

To illustrate this, consider an air mass with slow westerly winds in the north and fast westerly winds in the south, as shown in Figure 6.2A. This implies that the zonal wind decreases with latitude such that $\partial u/\partial y < 0$. Now suppose this air mass is advected northward over a fixed location by a southerly wind ($v > 0$), also shown in Figure 6.2A. Over time, the westerly wind at point O increases. This means that the rate of change of the zonal wind u at point O increases when v increases or when $|\partial u/\partial y|$ increases (i.e., $\partial u/\partial y$ becomes more negative).

Therefore, at point O, the zonal wind speed changes according to:

$$\frac{\partial u}{\partial t} = -v\frac{\partial u}{\partial y}$$

Here, $\partial u/\partial t$ represents the Eulerian rate of change at point O due to advection. This equation shows that advection can produce changes in wind at a fixed location (i.e., within the Eulerian reference frame). Although advection is not a true force, the wind flowing past a fixed point on Earth can carry linear momentum (per unit mass) through that point. Since linear momentum can only change in the presence of a force, we may treat advection as a force per unit mass in this context. Returning to our earlier discussion of the linear momentum of an air parcel, advection can be understood as the transport of linear momentum by the wind itself,. This explains how different wind conditions can be advected to a fixed location by the flow.

However, this thought experiment also reveals that the velocity vector and the gradient vector must be properly aligned for advection to result in acceleration. To understand why this alignment matters, consider a northward-moving air mass encountering slow westerly winds to the west and fast westerly winds to the east, as shown in Figure 6.2B. In this case, the zonal wind increases toward the east,

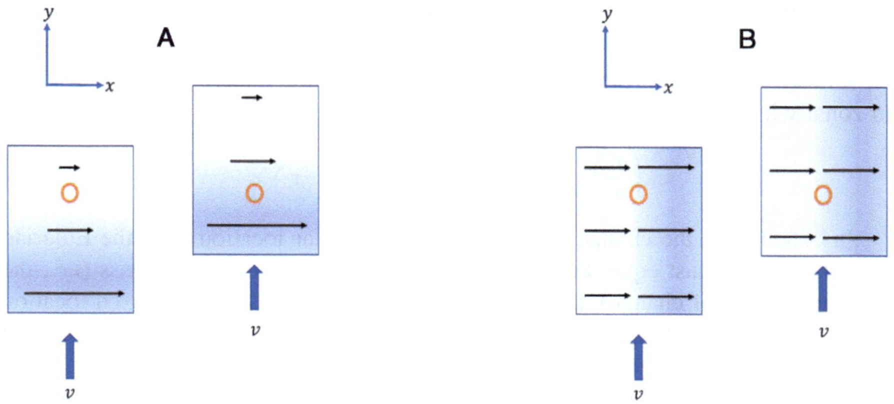

FIG. 6.2

Schematic illustration of northward advection of zonal wind. (A) The air mass has a meridional gradient in zonal wind. (B) The air mass has a zonal gradient in meridional wind. For both sections of the figure, the change in contour shading indicates a change in wind speed.

-No Permission Required

meaning $\partial u/\partial x > 0$. As the air mass is advected northward, we observe no change in wind speed at the point O, indicating that $\partial u/\partial t = 0$.

Likewise, if the air mass had a vertical gradient such that $\partial u/\partial z > 0$, the wind speed at point O would remain unchanged. Therefore, Figure 6.2 illustrates that for a southerly wind where $\boldsymbol{v} = v\boldsymbol{j}$, acceleration occurs *only* when there is a wind gradient in the $\boldsymbol{j}$ direction. In particular, the change in zonal wind speed in Figure 6.2 can be expressed as:

$$\frac{\partial u}{\partial t} = -v\frac{\partial u}{\partial y} = (-v\boldsymbol{j}) \cdot \left(\frac{\partial u}{\partial y}\boldsymbol{j}\right) = -\boldsymbol{v} \cdot \nabla u$$

Based on the definition of the scalar product, the acceleration generated by advection depends on the magnitude of the wind velocity (i.e., the wind speed), the gradient of the wind, and the angle between the two vectors. The acceleration due to advection reaches its maximum when the wind vector ($\boldsymbol{v}$) is parallel to the wind gradient vector (∇u), as shown in Figure 6.2A. Conversely, when $\boldsymbol{v}$ is perpendicular to ∇u, the acceleration is zero, as illustrated in Figure 6.2B.

Generalizing this approach for the full three-dimensional wind field, we can express changes in zonal wind due to advection as a force (per unit mass) using the following equation:

$$\left(\frac{\partial u}{\partial t}\right)_{adv} = -\boldsymbol{v} \cdot \nabla u = -u\frac{\partial u}{\partial x} - v\frac{\partial u}{\partial y} - w\frac{\partial u}{\partial z} \tag{6.3}$$

Based on Eq. (6.3), zonal, meridional, and vertical winds can each contribute to zonal acceleration through the wind's spatial gradient. For example, the term $-u\,\partial u/\partial x$ indicates that zonal wind can generate acceleration when there is a zonal gradient in zonal wind. Similarly, we can write:

$$\left(\frac{\partial v}{\partial t}\right)_{adv} = -\boldsymbol{v}\cdot\nabla v,\ \left(\frac{\partial w}{\partial t}\right)_{adv} = -\boldsymbol{v}\cdot\nabla w$$

In conclusion, the change in velocity at a particular location (i.e., the Eulerian rate of change) must equal the change in velocity following the air mass (i.e., the Lagrangian rate of change), plus the rate at which air flows past the fixed location due to advection. Mathematically, the relationship between the Eulerian and Lagrangian rates of change can be expressed as follows:

$$\frac{\partial u}{\partial t} = \frac{Du}{Dt} - \boldsymbol{v}\cdot\nabla u,\ \frac{\partial v}{\partial t} = \frac{Dv}{Dt} - \boldsymbol{v}\cdot\nabla v,\ \frac{\partial w}{\partial t} = \frac{Dw}{Dt} - \boldsymbol{v}\cdot\nabla w$$

where $\partial/\partial t$ signifies the Eulerian rate of change, while D/Dt represents the Lagrangian rate of change. Box 6.2 provides a practical example illustrating how advection influences wind acceleration.

The concept of advection applies not only to the wind but also to any scalar quantity carried by the wind. For example, the rapid transport of cold or warm air into a location can change the temperature at a given observation station. This is known as **thermal advection** and can be written mathematically as follows:

$$\left(\frac{\partial T}{\partial t}\right)_{adv} = -\boldsymbol{v}\cdot\nabla T$$

Thermal advection depends on three factors: (1) the thermal gradient (represented by ∇T), (2) the wind speed across that gradient (denoted by $\boldsymbol{v}$), and (3) the angle at which the wind intersects the gradient. To maximize the magnitude of temperature advection, strong winds should blow perpendicular to tightly packed isotherms, as shown in Figure 6.3A. Conversely, if the winds are parallel to isotherms, no thermal advection occurs. Based on the above definitions, warm advection takes place when $-\boldsymbol{v}\cdot\nabla T > 0$ (meaning the velocity vector points towards lower temperatures, whereas cold advection occurs when $-\boldsymbol{v}\cdot\nabla T < 0$ (indicating that the velocity vector points towards higher temperatures).

The rapid transport of moist or dry air can change the moisture at a given observation station. This is referred to as **moisture advection**, which can be expressed as:

$$\left(\frac{\partial r}{\partial t}\right)_{adv} = -\boldsymbol{v}\cdot\nabla r$$

where r is the water vapor mixing ratio. Since dew point is a function of the water mixing ratio, moisture advection is typically assessed by examining dew point changes. Following the reasoning given above, moisture advection is maximized when a strong wind blows perpendicular to tightly packed isodrosotherms, as shown

Box 6.2 Calculating the acceleration of the wind due to advection

Minneapolis, MN, is about 400 km north of Des Moines, IA. In Minneapolis, the wind is 7.21 m/s from 236°, whereas at Des Moines, IA, the wind is 10.2 m/s from 191°. Let's calculate the acceleration generated by southerly advection at a location halfway between Minneapolis, MN and Des Moines, IA.

As shown in Ch. 2, the wind components can be determined if the wind speed and the meteorological direction are known by:

$$u = -V\sin\alpha, v = -V\cos\alpha$$

where V is the horizontal wind speed and α is the meteorological direction. For Minneapolis, we have

$$u_{MN} = -\left(7.21\frac{m}{s}\right)\sin(236°) = 5.98\frac{m}{s}, v_{MN} = -\left(7.21\frac{m}{s}\right)\cos(236°) = 4.03\frac{m}{s}$$

For Des Moines, we have

$$u_{DM} = -\left(10.2\frac{m}{s}\right)\sin(191°) = 1.94\frac{m}{s}, v_{DM} = -\left(10.2\frac{m}{s}\right)\cos(191°) = 10.0\frac{m}{s}$$

Based on the definition of the gradient, we have

$$\frac{\partial u}{\partial y} \approx \frac{u_{MN} - u_{DM}}{400.0km} = \frac{\left(5.98\frac{m}{s}\right) - \left(1.94\frac{m}{s}\right)}{4.00 \times 10^5 m} = 1.01 \times 10^{-5} s^{-1}$$

$$\frac{\partial v}{\partial y} \approx \frac{v_{MN} - v_{DM}}{400.0km} = \frac{\left(4.03\frac{m}{s}\right) - \left(10.0\frac{m}{s}\right)}{4.00 \times 10^5 m} = -1.49 \times 10^{-5} s^{-1}$$

The x-component of the acceleration due to advection is

$$-v\frac{\partial u}{\partial y} \approx -\bar{v}\frac{\partial u}{\partial y} = -\left(\frac{4.03\frac{m}{s} + 10.0\frac{m}{s}}{2}\right)(1.01 \times 10^{-5} s^{-1}) = -7.08 \times 10^{-5}\frac{m}{s^2}$$

Notice that the slower zonal winds from Des Moines are being blown by northward winds toward Minneapolis, causing the zonal wind speed to decrease at Minneapolis. The y-component of the acceleration due to advection is

$$-v\frac{\partial v}{\partial y} \approx -\bar{v}\frac{\partial v}{\partial y} = -\left(\frac{4.03\frac{m}{s} + 10.0\frac{m}{s}}{2}\right)(-1.49 \times 10^{-5} s^{-1}) = 1.05 \times 10^{-4}\frac{m}{s^2}$$

Here, the faster meridional winds from Des Moines are being blown northward toward Minneapolis, causing the meridional winds to increase at Minneapolis. Thus, the total acceleration of the wind due to advection is

$$\mathbf{a} = \left(-7.08 \times 10^{-5}\frac{m}{s^2}\right)\mathbf{i} + \left(1.05 \times 10^{-4}\frac{m}{s^2}\right)\mathbf{j}$$

Although the acceleration in this problem appears to be small, advection can have a noticeable impact on the wind at a given location. For example, if the acceleration by advection is applied over a 12-hour span, this will lead to a velocity change of:

$$\Delta\mathbf{v} = \mathbf{a}\Delta t = \left[\left(-7.08 \times 10^{-5}\frac{m}{s^2}\right)\mathbf{i} + \left(1.05 \times 10^{-4}\frac{m}{s^2}\right)\mathbf{j}\right](12 \quad hr \times 3600 \quad s)$$

$$= \left[\left(-3.06\frac{m}{s}\right)\mathbf{i} + \left(4.54\frac{m}{s}\right)\mathbf{j}\right]$$

in Figure 6.3B. Conversely, if the winds are parallel to isodrosotherms, no moisture advection occurs. We will explore the significance of thermal and moisture advection in greater detail in Ch. 8.

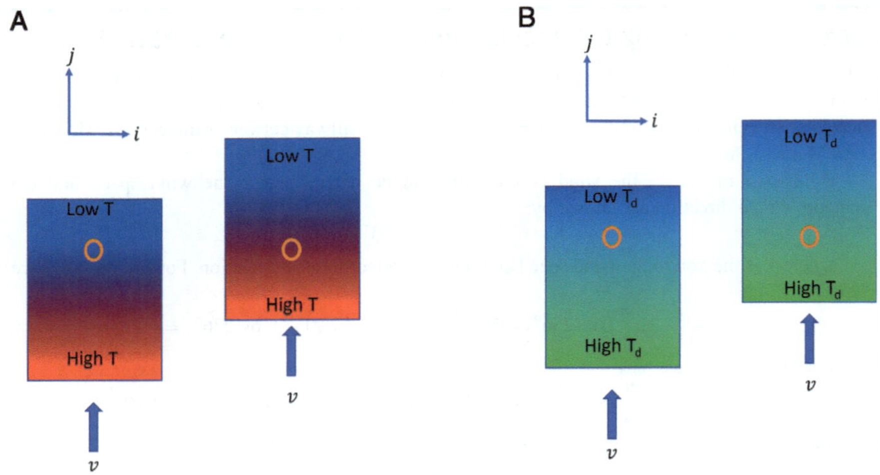

FIG. 6.3 Schematic of thermal advection and moisture advection. (A) Warm advection by a southerly wind. The air mass has a meridional gradient in temperature as shown by the contour shading. The temperature gradient points to the south and the velocity vector points to the north, which produces warm advection. (B) Moist advection by a southerly wind. The air mass has meridional gradient in dew point as shown by the contour shading. The dew point gradient points to the south and the velocity vector points to the north, which produces moist advection.

-No Permission Required

6.2 Vector Product

In Ch. 6.1, we showed that two vectors can be multiplied to produce a scalar using the scalar product. In this section, we demonstrate that two vectors can also be multiplied to produce another vector, known as the **vector product** (or cross product). For any two vectors $\boldsymbol{A}$ and $\boldsymbol{B}$, the **vector product** $\boldsymbol{A} \times \boldsymbol{B}$ is defined as

$$\boldsymbol{A} \times \boldsymbol{B} = |AB \sin\theta| \boldsymbol{n} \tag{6.4}$$

where θ is the angle between the vectors, which ranges from 0° to 90°.[2] Geometrically, the quantity $AB \sin \theta$ corresponds to the area of the parallelogram formed by $\boldsymbol{A}$ and $\boldsymbol{B}$, as shown in Figure 6.4.

6.2.1 Mathematical Properties of the Vector Product

Based on Eq. (6.4), the vector product has the following properties:

[2] Because of the notation, the vector product of two vectors $\boldsymbol{A}$ and $\boldsymbol{B}$ is often read "A cross B." For this reason, the vector product is also called the cross product.

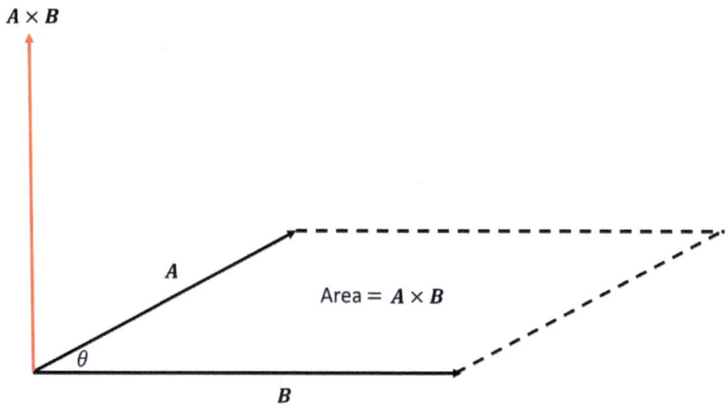

FIG. 6.4

Schematic of the geometric interpretation of the vector product. In contrast to the scalar product, notice that the area of this parallelogram (which corresponds to the magnitude of the vector product) is maximized when **A** and **B** are perpendicular.

-No Permission Required

1. The magnitude of $\boldsymbol{A} \times \boldsymbol{B}$ is maximized when $\theta = 90°$ (i.e., when $\boldsymbol{A}$ is perpendicular to $\boldsymbol{B}$). This is the opposite of the scalar product, which produces a value of zero when $\boldsymbol{A}$ is perpendicular to $\boldsymbol{B}$.
2. The magnitude of $\boldsymbol{A} \times \boldsymbol{B}$ is zero when $\theta = 0°$ (i.e., when $\boldsymbol{A}$ is parallel to $\boldsymbol{B}$). This is also the opposite of the scalar product, which produces its maximum value when $\boldsymbol{A}$ is parallel to $\boldsymbol{B}$.

The unit vector $\boldsymbol{n}$, as given in Eq. (6.4), is perpendicular to the plane formed by vectors $\boldsymbol{A}$ and $\boldsymbol{B}$. For this reason, $\boldsymbol{n}$ is called the **unit normal vector**. The direction of $\boldsymbol{n}$ can be determined using the following procedure:

1. Point the four fingers of your right hand in the direction of $\boldsymbol{A}$.
2. Wrap your fingers in the direction that would rotate $\boldsymbol{A}$ into $\boldsymbol{B}$ through the angle θ.
3. The direction of your upright thumb indicates the direction of $\boldsymbol{n}$.

This procedure is known as the **right-hand rule**, and it illustrates an important property of the cross-product: $\boldsymbol{A} \times \boldsymbol{B}$ points in the opposite direction of $\boldsymbol{B} \times \boldsymbol{A}$. Therefore, unlike the scalar product, the vector product is not commutative - the order of multiplication affects the result.

Eq. (6.4) can be used to calculate the vector product when the magnitudes and directions of the vectors are known. To compute the vector product in component form, we first need to determine the products of the unit vectors, which are summarized in the cyclic diagram shown in Figure 6.5.

To evaluate the vector product in component form, we calculate it term by term as follows:

$$\boldsymbol{A} \times \boldsymbol{B} = \left(A_x\boldsymbol{i} + A_y\boldsymbol{j} + A_z\boldsymbol{k}\right) \times \left(B_x\boldsymbol{i} + B_y\boldsymbol{j} + B_z\boldsymbol{k}\right)$$

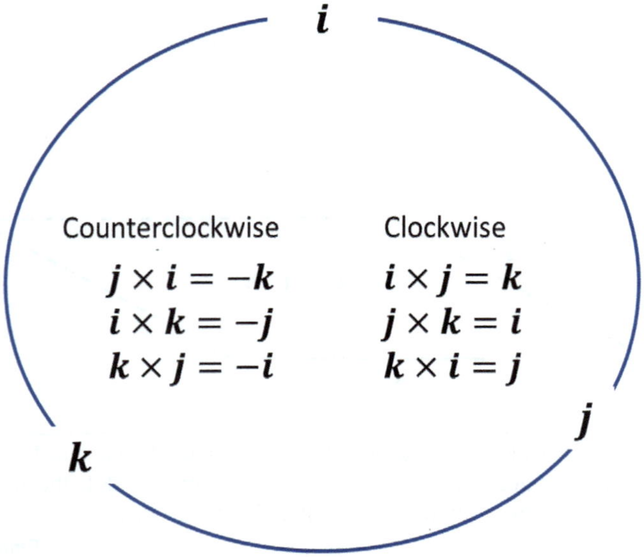

FIG. 6.5 Cyclic permutation of vector products. Summary of the vector products associated with the unit vectors. Notice that the vector product is not commutative (i.e., changing the order of the vector product reverses the direction of the vector).

-No Permission Required

$$= \left[A_xB_y(\boldsymbol{i}\times\boldsymbol{j}) + A_xB_z(\boldsymbol{i}\times\boldsymbol{k})\right] + \left[A_yB_x(\boldsymbol{j}\times\boldsymbol{i}) + A_yB_z(\boldsymbol{j}\times\boldsymbol{k})\right] + \left[A_zB_x(\boldsymbol{k}\times\boldsymbol{i}) + A_zB_y(\boldsymbol{k}\times\boldsymbol{j})\right]$$

$$= \left[A_xB_y(\boldsymbol{k}) + A_xB_z(-\boldsymbol{j})\right] + \left[A_yB_x(-\boldsymbol{k}) + A_yB_z(\boldsymbol{i})\right] + \left[A_zB_x(\boldsymbol{j}) + A_zB_y(-\boldsymbol{i})\right]$$

$$= \boldsymbol{i}(A_yB_z - A_zB_y) - \boldsymbol{j}(A_xB_z - A_zB_x) + \boldsymbol{k}(A_xB_y - A_yB_x)$$

The above expression can also be represented in determinant form as follows:

$$\boldsymbol{A}\times\boldsymbol{B} = \begin{vmatrix} \boldsymbol{i} & \boldsymbol{j} & \boldsymbol{k} \\ A_x & A_y & A_z \\ B_x & B_y & B_z \end{vmatrix} \tag{6.5}$$

Finally, using the product rule of differentiation, the derivative of the vector product can be expressed as:

$$\frac{d}{dx}(\boldsymbol{A}\times\boldsymbol{B}) = \frac{d\boldsymbol{A}}{dx}\times\boldsymbol{B} + \boldsymbol{A}\times\frac{d\boldsymbol{B}}{dx} \tag{6.6}$$

Box 6.3 illustrates how the vector product is used.

In the following section, we will demonstrate how the vector product is used to describe the rotation of any object, including air parcels on the rotating Earth.

Box 6.3 Example using the vector product

Vector $\boldsymbol{A}$ has a magnitude of 2.00 with a direction of 45°. Vector $\boldsymbol{B}$ has a magnitude of 2.50 with a direction of 300°. Assume that the direction of each vector is measured from the positive x-axis. Calculate $\boldsymbol{A} \times (\boldsymbol{A} \times \boldsymbol{B})$ and express your answer in unit vector notation.

Answer: The components of our vectors are given by

$$A_x = 2.00\cos(45.0°) = 1.41, A_y = 2.00\sin(45.0°) = 1.41$$
$$B_x = 2.50\cos(300.0°) = 1.25, B_y = 2.50\sin(300.0°) = -2.17$$

First, we calculate $\boldsymbol{A} \times \boldsymbol{B}$ using Eq. (6.5)

$$\begin{aligned}\boldsymbol{A} \times \boldsymbol{B} &= \boldsymbol{i}(A_yB_z - A_zB_y) - \boldsymbol{j}(A_xB_z - A_zB_x) + \boldsymbol{k}(A_xB_y - A_yB_x) \\ &= 0\boldsymbol{i} + 0\boldsymbol{j} + \boldsymbol{k}[(1.41)(-2.17) - (1.41)(1.25)] = -4.82\boldsymbol{k}\end{aligned}$$

If we define $\boldsymbol{E} = \boldsymbol{A} \times \boldsymbol{B}$, we compute the cross product $\boldsymbol{A} \times \boldsymbol{E}$ using Eq. (6.5)

$$\begin{aligned}\boldsymbol{A} \times \boldsymbol{E} &= \boldsymbol{i}(A_yE_z - A_zE_y) - \boldsymbol{j}(A_xE_z - A_zE_x) + \boldsymbol{k}(A_xE_y - A_yE_x) \\ &= \boldsymbol{i}[(1.41)(-4.82) - 0] - \boldsymbol{j}[(1.41)(-4.82) - 0] + 0\boldsymbol{k} \\ &= -6.80\boldsymbol{i} + 6.80\boldsymbol{j}\end{aligned}$$

6.2.2 General Properties of Rotation

With the formal definition of the vector product, we can introduce some important results related to rotational motion. Although a comprehensive study of rotations is beyond the scope of this textbook, we will highlight two properties of rotation that are relevant for Earth's rotating reference frame.

The first general result associated with rotation is known as **Euler's theorem**, which states

The most general motion of any rotating object relative to a fixed point is a rotation about some axis through the fixed point.

According to Euler's theorem, to uniquely specify the rotation about a given point, we must know two things: (1) the direction of the axis about which the rotation occurs, and (2) the angle through which the object is rotated. This theorem allows us to specify Earth's rotation. In particular, Earth's rotation can be uniquely determined when we know the direction of its rotation axis and the magnitude of Earth's angular velocity.

Observations indicate that Earth's angular velocity vector points toward the North Pole, which means that Earth rotates toward the east. The magnitude of Earth's angular velocity can be determined by noting that Earth completes one full rotation on its axis every 23 hours and 56 minutes. This implies that:

$$\Omega = \frac{\Delta\theta}{\Delta t} = \frac{2\pi rad}{23\,hr, 56\,min} = \frac{2\pi}{86,160s} = 7.29 \times 10^{-5}\frac{rad}{s}$$

The second result involves the relationship between vectors in a rotating frame and those in a stationary reference frame. Let frame S_0 be a stationary reference frame, and let frame S be Earth's rotating reference frame with angular velocity $\boldsymbol{\Omega}$. We assume both frames share a common origin O, so the only motion of S relative to S_0 is a rotation with angular velocity $\boldsymbol{\Omega}$.

The rate of change of any vector in the stationary frame S_0 can be related to its rate of change in the rotating frame S by the following theorem:

For any arbitrary vector $\boldsymbol{A}$, its time derivative in the stationary frame S_0 is related to its time derivative in the rotating frame S by the following relationship:

$$\left(\frac{d\boldsymbol{A}}{dt}\right)_{S_0} = \left(\frac{d\boldsymbol{A}}{dt}\right)_S + \boldsymbol{\Omega} \times \boldsymbol{A} \tag{6.7}$$

To motivate Eq. (6.7), let's consider when $\boldsymbol{A}$ corresponds to the position vector $\boldsymbol{r}$. According to Eq. (6.7), the rate of change of the position vector in frame S_0 can be expressed as:

$$\left(\frac{d\boldsymbol{r}}{dt}\right)_{S_0} = \left(\frac{d\boldsymbol{r}}{dt}\right)_S + \boldsymbol{\Omega} \times \boldsymbol{r} \tag{6.8}$$

The left-hand side of Eq. (6.8) represents the velocity measured in the stationary reference frame S_0. The first term on the right-hand side corresponds to the velocity measured *in* the rotating reference frame S. We can show that the second term represents the velocity due to the rotation of the reference frame itself.

To illustrate this, consider a horizontally rotating turntable as shown in Figure 6.6A. Here, $\boldsymbol{r}$ is the position vector from point O and $\boldsymbol{v}$ is the tangential velocity at point P. From introductory physics, we know that the tangential speed on a horizontal turntable is related to the angular speed by the equation $v = \Omega r$.

Next, in Figure 6.6A, notice that $\boldsymbol{v}$ points in the $-\boldsymbol{j}$ direction, $\boldsymbol{r}$ points in the $-\boldsymbol{i}$ direction, and $\boldsymbol{\Omega}$ points in the $\boldsymbol{k}$ direction, according to right-hand rule. This

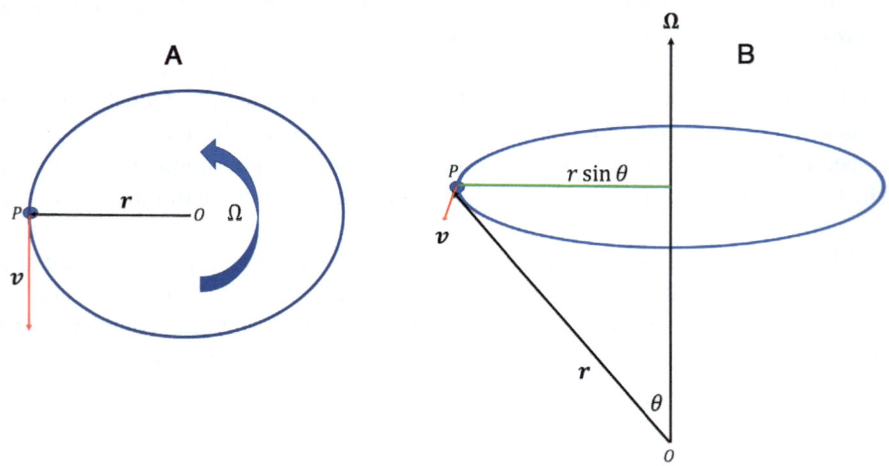

FIG. 6.6 Tangential vs angular velocity. The relationship between tangential velocity and angular velocity. (A) A rotating turntable in which the angular velocity points out of the page by the right-hand rule. (B) A point fixed on Earth rotating through a circle of latitude. Note that the velocity vector points out of the page at point P.

-No Permission Required

arrangement demonstrates that $\boldsymbol{v}$ is mutually perpendicular to both $\boldsymbol{r}$ and $\boldsymbol{\Omega}$. Additionally, it is important to note that $\boldsymbol{\Omega} \times \boldsymbol{r}$ must point in the same direction as $\boldsymbol{v}$, since $\boldsymbol{k} \times (-\boldsymbol{i}) = -\boldsymbol{j}$. These considerations show that $\boldsymbol{v} = \boldsymbol{\Omega} \times \boldsymbol{r}$. Formally stated: *for any object rotating with angular velocity* $\boldsymbol{\Omega}$ *about an axis through point* O, *the tangential velocity at any point* P *at position* $\mathbf{r}$ *fixed on the object is given by* $\mathbf{v} = \boldsymbol{\Omega} \times \mathbf{r}$.

We can demonstrate the same point by examining a point P fixed on the Earth with position $\boldsymbol{r}$ relative to the Earth's center, as shown in Figure 6.6B. As the Earth rotates on its axis, point P moves around a circle of latitude. The tangential speed at point P is related to the angular speed and the distance from the rotation axis such that $v = \Omega r \sin\theta$. Furthermore, note that $\boldsymbol{v}$ is perpendicular to both $\boldsymbol{\Omega}$ and $\boldsymbol{r}$, and by the right-hand rule, we conclude that $\boldsymbol{v} = \boldsymbol{\Omega} \times \boldsymbol{r}$, with $\boldsymbol{v}$ pointing out of the page.

Therefore the meaning of Eq. (6.8) can be expressed as follows:

The velocity measured in a stationary reference frame is equal to the velocity within the rotating frame plus the tangential velocity due to the rotation of the frame.

6.3 Newton's 2nd Law on the Rotating Earth

With these two results, we are now prepared to apply Newton's 2nd law of motion to an air parcel within Earth's rotating reference frame. We will begin by examining the effects of an accelerating reference frame in general, and then focus specifically on rotating reference frames.

6.3.1 Accelerating Reference Frames and Inertial Forces

To understand the effects of an accelerating reference frame, consider a passenger in a car traveling with a velocity $\boldsymbol{v}_c$ who throws a ball into the air. First, we will examine the motion of the ball relative to a reference frame anchored to the ground (e.g., a person standing by the roadside), which we will call reference frame S_O.

Relative to the stationary observer, the velocity of the ball can be expressed as:

$$\boldsymbol{v}_0 = \boldsymbol{v}_c + \boldsymbol{v}_p$$

where $\boldsymbol{v}_0$ is the velocity of the ball relative to the stationary reference frame S_O, and $\boldsymbol{v}_p$ is the velocity of the ball relative to the passenger in the car. Because the car moves horizontally (relative to the stationary observer) and the ball is thrown vertically (relative to the passenger), the stationary observer will see the ball follow a parabolic trajectory.

By differentiating with respect to time, we find that the acceleration relative to frame S_O is given by:

$$\boldsymbol{a}_0 = \boldsymbol{a}_p$$

where $\boldsymbol{a}_p$ is the acceleration of the ball relative to the passenger in the car. Because the only force acting on the ball while it is in the air is gravity, Newton's 2nd law implies that $\boldsymbol{F}_{net} = m\boldsymbol{a}_0$, where m is the mass of the ball and $\boldsymbol{a}_O$ is its acceleration relative to frame S_O. This result indicates that Newton's 2nd law holds in reference frame S_O. This reasoning can be extended to include any reference frame that moves with a constant velocity, confirming that Newton's 2nd law holds for all inertial reference frames.

Let's now examine the motion of the same ball relative to a reference frame that is moving with a velocity $\boldsymbol{v}'$ and an acceleration $\boldsymbol{a}'$ compared with frame S_O (e.g., when another vehicle accelerates past the original car). We will call this moving reference frame S.

The velocity of the ball in frame S_O is related to the velocity of the ball in frame S by the following equation:

$$\boldsymbol{v}_0 = \boldsymbol{v} + \boldsymbol{v}'$$

where $\boldsymbol{v}$ is the velocity of the ball in frame S. This reflects the familiar concept of relative velocity from introductory physics, which states that the velocity of an object in a stationary frame S_O is equal to the velocity of that object <u>in</u> frame S plus the velocity <u>of</u> frame S itself.

By differentiating this expression with respect to time, we obtain the corresponding relationship for acceleration:

$$\boldsymbol{a}_0 = \boldsymbol{a} + \boldsymbol{a}'$$

Applying Newton's 2nd law in frame S gives:

$$\boldsymbol{F}_{net} = m\boldsymbol{a} + m\boldsymbol{a}'$$

$$\Rightarrow \quad m\boldsymbol{a} = \boldsymbol{F}_{net} - m\boldsymbol{a}' \tag{6.9}$$

Notice that Eq. (6.9) has the same form as Newton's 2nd law, but includes an additional term of $-m\boldsymbol{a}'$. This means that we can use Newton's 2nd law in the non-inertial frame S <u>only if</u> we include this extra term, commonly referred to as <u>**inertial force**</u>.

Inertial forces experienced in accelerating reference frames are common in several everyday situations. Here are some examples:

- When you sit in an aircraft that is rapidly accelerating for takeoff, you feel as though a force is pushing you back into your seat. From your perspective within the accelerating reference frame, this apparent force is directed opposite to the aircraft's forward acceleration. This is an example of an inertial force.
- If you are standing in a bus and the driver suddenly applies the brakes, you feel yourself lurch forward. This apparent forward force - opposite the direction of the bus's deceleration - is an inertial force.
- When a car turns sharply around a corner, a net force must act toward the center of the curve. This is the centripetal force, which you apply by turning the steering wheel. However, within the car's accelerating reference frame, you feel

a force pushing you *outward,* away from the turn. This apparent outward force - opposite the direction of centripetal acceleration - is called the *centrifugal force*, which is a type of inertial force.

Because a rotating reference frame is also an accelerating reference frame, we must account for inertial forces when applying Newton's 2nd law to a rotating reference frame.

6.3.2 Derivation of the Coriolis and Centrifugal Forces

In a stationary reference frame S_0, the motion of an air parcel is governed by Newton's 2nd law expressed as $\boldsymbol{F}_{net} = m\boldsymbol{a}_0$, where $\boldsymbol{a}_0$ is the acceleration measured in the frame S_0. Recall from introductory physics that acceleration vector $\boldsymbol{a}$ is related to the velocity vector $\boldsymbol{v}$ and the position vector $\boldsymbol{r}$ through the following relationship:

$$\boldsymbol{a} = \frac{d\boldsymbol{v}}{dt} = \frac{d^2\boldsymbol{r}}{dt^2}$$

Therefore, we can write Newton's 2nd law in the stationary reference frame S_0 as:

$$m\left(\frac{d^2\boldsymbol{r}}{dt^2}\right)_{S_0} = \boldsymbol{F}_{net} \tag{6.10}$$

where the subscript S_0 indicates that the rate of change is measured within the stationary reference frame. Our goal is to rewrite Eq. (6.10) within Earth's rotating reference frame.

We can use Eq. (6.10) to obtain the acceleration of an air parcel in Earth's rotating reference frame S. Differentiating Eq. 6.8 with respect to time gives:

$$\left(\frac{d^2\boldsymbol{r}}{dt^2}\right)_{S_0} = \left(\frac{d}{dt}\right)_{S_0}\left[\left(\frac{d\boldsymbol{r}}{dt}\right)_S + \boldsymbol{\Omega}\times\boldsymbol{r}\right]$$

where the derivative is evaluated within the stationary reference frame and the subscript S indicates that the rate of change is measured within Earth's rotating reference frame. To simplify this result, we can apply Eq. (6.7) where $\boldsymbol{A} = (d\boldsymbol{r}/dt)_S + \boldsymbol{\Omega} \times \boldsymbol{r}$, which gives

$$\left(\frac{d^2\boldsymbol{r}}{dt^2}\right)_{S_0} = \left(\frac{d}{dt}\right)_{S}\left[\left(\frac{d\boldsymbol{r}}{dt}\right)_S + \boldsymbol{\Omega}\times\boldsymbol{r}\right] + \boldsymbol{\Omega}\times\left[\left(\frac{d\boldsymbol{r}}{dt}\right)_S + \boldsymbol{\Omega}\times\boldsymbol{r}\right]$$

$$\Rightarrow\left(\frac{d^2\boldsymbol{r}}{dt^2}\right)_{S_0} = \left(\frac{d}{dt}\right)_{S}\left[\left(\frac{d\boldsymbol{r}}{dt}\right)_S + \boldsymbol{\Omega}\times\boldsymbol{r}\right] + \boldsymbol{\Omega}\times\left(\frac{d\boldsymbol{r}}{dt}\right)_S + \boldsymbol{\Omega}\times(\boldsymbol{\Omega}\times\boldsymbol{r}) \tag{6.11}$$

Using the product rule for vector products given in Eq. (6.6), the first term on the right-hand side of Eq. (6.11) can be expressed as:

$$\left(\frac{d}{dt}\right)_{S}\left[\left(\frac{d\boldsymbol{r}}{dt}\right)_S + \boldsymbol{\Omega}\times\boldsymbol{r}\right] = \left(\frac{d^2\boldsymbol{r}}{dt^2}\right)_{S} + \left(\frac{d\boldsymbol{\Omega}}{dt}\right)_{S}\times\boldsymbol{r} + \boldsymbol{\Omega}\times\left(\frac{d\boldsymbol{r}}{dt}\right)_S$$

Since $\boldsymbol{\Omega}$ is constant, this result simplifies to:

$$\left(\frac{d}{dt}\right)_S\left[\left(\frac{d\boldsymbol{r}}{dt}\right)_S+\boldsymbol{\Omega}\times\boldsymbol{r}\right]=\left(\frac{d^2\boldsymbol{r}}{dt^2}\right)_S+\boldsymbol{\Omega}\times\left(\frac{d\boldsymbol{r}}{dt}\right)_S$$

Substituting this result into Eq. (6.11) gives:

$$\left(\frac{d^2\boldsymbol{r}}{dt^2}\right)_{S_0}=\left(\frac{d^2\boldsymbol{r}}{dt^2}\right)_S+\boldsymbol{\Omega}\times\left(\frac{d\boldsymbol{r}}{dt}\right)_S+\boldsymbol{\Omega}\times\left(\frac{d\boldsymbol{r}}{dt}\right)_S+\boldsymbol{\Omega}\times(\boldsymbol{\Omega}\times\boldsymbol{r})$$

$$\Rightarrow\left(\frac{d^2\boldsymbol{r}}{dt^2}\right)_{S_0}=\left(\frac{d^2\boldsymbol{r}}{dt^2}\right)_S+2\boldsymbol{\Omega}\times\left(\frac{d\boldsymbol{r}}{dt}\right)_S+\boldsymbol{\Omega}\times(\boldsymbol{\Omega}\times\boldsymbol{r}) \tag{6.12}$$

Substituting Eq. (6.12) into Eq. (6.10) gives

$$\boldsymbol{F}_{net}=m\left[\left(\frac{d^2\boldsymbol{r}}{dt^2}\right)_S+2\boldsymbol{\Omega}\times\left(\frac{d\boldsymbol{r}}{dt}\right)_S+\boldsymbol{\Omega}\times(\boldsymbol{\Omega}\times\boldsymbol{r})\right]$$

$$\Rightarrow m\left(\frac{d^2\boldsymbol{r}}{dt^2}\right)_S=\boldsymbol{F}_{net}-2m\left[\boldsymbol{\Omega}\times\left(\frac{d\boldsymbol{r}}{dt}\right)_S\right]-m[\boldsymbol{\Omega}\times(\boldsymbol{\Omega}\times\boldsymbol{r})] \tag{6.13}$$

We have now reached our result: Eq. (6.13) represents Newton's 2nd law within Earth's rotating reference frame. Notice that Eq. (6.13) resembles Eq. (6.10), but with two additional terms on the right-hand side. These extra terms account for the inertial forces arising from Earth's acceleration. The second term is known as the **Coriolis force**, and the third term is referred to as the **centrifugal force** caused by Earth's rotation. We will now examine the characteristics of both the centrifugal force and the Coriolis force in more detail.

6.4 Properties of the Centrifugal Force

As mentioned in Ch. 6.3, the centrifugal force is an inertial force that arises from the rotating reference frame. Based on Eq. (6.13), the centrifugal force $\boldsymbol{F}_{cf}$ is given by:

$$\boldsymbol{F}_{cf}=-m[\boldsymbol{\Omega}\times(\boldsymbol{\Omega}\times\boldsymbol{r})]=m(\boldsymbol{\Omega}\times\boldsymbol{r})\times\boldsymbol{\Omega}$$

We start our discussion of the centrifugal force by applying a conceptual analysis.

6.4.1 Conceptual Analysis of the Centrifugal Force

We can determine the direction of the centrifugal force by considering a horizontally rotating turntable as shown in Figure 6.7A, where $\boldsymbol{r}$ is the position vector from point O. Since $\boldsymbol{\Omega}$ points in the $\boldsymbol{k}$ direction (i.e., out of the page), then $\boldsymbol{\Omega}\times\boldsymbol{r}$ can be expressed as:

$$\boldsymbol{\Omega}\times\boldsymbol{r}=\Omega r[\boldsymbol{k}\times(-\boldsymbol{i})]=\Omega r(-\boldsymbol{j})$$

Thus, $\boldsymbol{\Omega} \times \boldsymbol{r}$, which corresponds to the tangential velocity associated with the turntable's rotation, points in the $(-\boldsymbol{j})$ direction. Therefore, the centrifugal force for the rotating turntable is expressed as:

$$\boldsymbol{F}_{cf} = m(\boldsymbol{\Omega} \times \boldsymbol{r}) \times \boldsymbol{\Omega} = m(\boldsymbol{\Omega} r)(\boldsymbol{\Omega})[(-\boldsymbol{j}) \times \boldsymbol{k}] = m\boldsymbol{\Omega}^2 r(-\boldsymbol{i})$$

Therefore, $\boldsymbol{F}_{cf}$ points away from the axis of rotation of the turntable, which is why we refer to this inertial force as the centrifugal (i.e., "center-fleeing") force.

6.4.2 The Centrifugal Force on a Rotating Earth

We can now apply these concepts to Earth's rotation. Consider an air parcel located near Earth's surface, as shown in Figure 6.7B. If the air parcel is at rest relative to Earth's surface, it will follow a circular path along a line of latitude ϕ. Since $\boldsymbol{\Omega} \times \boldsymbol{r}$ represents the tangential velocity vector, it must be tangent to this circular path. Using the right-hand rule, the vector $(\boldsymbol{\Omega} \times \boldsymbol{r}) \times \boldsymbol{\Omega}$ points outward from Earth's rotation axis. According to the definition of the vector product, the magnitude of this vector is $\boldsymbol{\Omega}^2 r \sin\theta$. Since $\phi = 90° - \theta$, the magnitude of the centrifugal force associated with Earth's rotation can be expressed as follows:

$$F_{cf} = m\boldsymbol{\Omega}^2 r \sin(90° - \phi) = m\boldsymbol{\Omega}^2 r \cos\phi$$

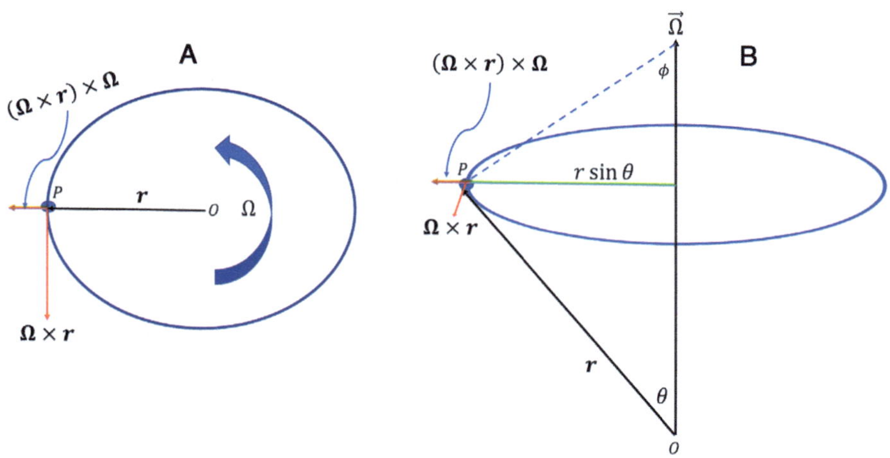

FIG. 6.7

Schematic of the centrifugal force. (A) A rotating turntable in which the angular velocity points out of the page by the right-hand rule. (B) A point fixed on Earth rotating through a circle of latitude ranging from 90° at the North Pole, 0° at the equator, and –90° at the South Pole. Note that the velocity vector points out of the page at point P.

-No Permission Required

Since ϕ ranges from 90° at the North Pole to −90° at the South Pole, note that $F_{cf} = 0$ at both poles, and F_{cf} obtains its maximum value at the equator where $\phi = 0°$. To summarize: for an air parcel in Earth's rotating reference frame, there exists a centrifugal force directed outward from Earth's rotation axis, with a magnitude $m\Omega^2 r \cos\phi$.

As mentioned in Ch. 5, air parcels are pulled toward Earth's surface due to gravitational attraction. However, when we analyze air parcels within Earth's rotating reference frame, we must also account for the centrifugal force arising from Earth's rotation. The combination of gravitational and centrifugal forces is referred to as **effective gravity**. As a result, all air parcels fall in the direction of effective gravity, which defines the local vertical direction.

Figure 6.8 shows a schematic of an air parcel subject to both gravitational and centrifugal forces. Due to the influence of the centrifugal force, the free-fall acceleration in the radial direction, denoted by $g_{eff,rad}$, is given by the equation:

$$g_{eff,rad} = \frac{F_g - F_{cf}\sin\theta}{m} = g - \left(\Omega^2 r \sin\theta\right)\sin\theta$$

$$\Rightarrow g_{eff,rad} = g - \Omega^2 r \cos^2\phi \tag{6.14}$$

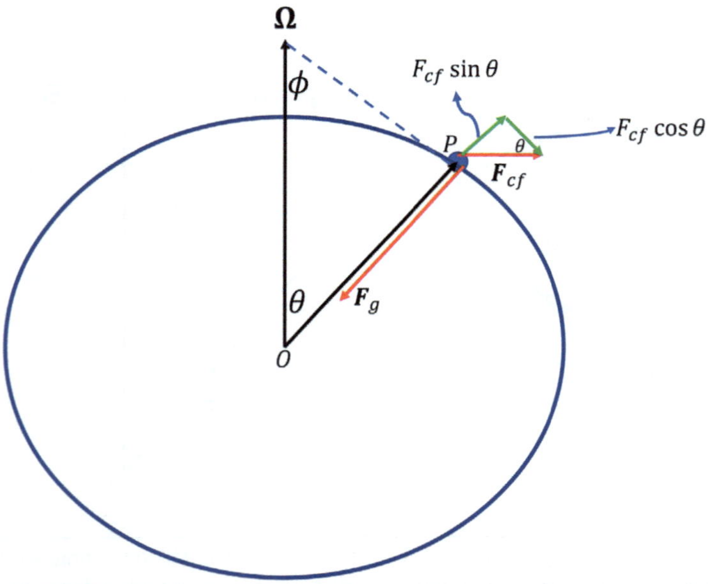

FIG. 6.8 An air parcel near Earth's surface under the influence of the gravitational force and the centrifugal force.

-No Permission Required

where $g = \frac{GM_E}{r^2}$ is the gravitational acceleration, ϕ is the latitude (where $\theta = 90° - \phi$), and r is the distance from the center of the Earth. Note that the second term on the right side of Eq. (6.14), which accounts for the impact of the centrifugal force, is zero at the poles (where $\phi = \pm 90°$), and reaches its maximum magnitude at the equator (where $\phi = 0°$).

At Earth's surface, the maximum magnitude of the centrifugal force associated with Earth's rotation can be calculated as:

$$\Omega^2 R = \left(7.29 \times 10^{-5} s^{-1}\right)^2 \times \left(6.3781 \times 10^6 m\right) = 0.0339 \frac{m}{s^2}$$

Since $g \approx 9.80 m/s^2$, the effect of the centrifugal force reduces the value of the free-fall acceleration by about 0.3% at the equator compared to the poles. Thus, the free-fall acceleration measured on Earth is be a weak function of latitude.

Figure 6.8 shows that the direction of free-fall acceleration in Earth's rotating frame differs from the direction of true gravitational acceleration. The tangential component of the effective gravitational acceleration arises from the centrifugal force and is given by:

$$g_{eff,tan} = F_{cf} \cos\theta = \left(\Omega^2 r \sin\theta\right) \cos\theta = \Omega^2 r \cos\phi \sin\phi$$

The maximum magnitude of $g_{eff,tan}$ at $\phi = 45°$, and its magnitude at Earth's surface is given by:

$$g_{eff,tan} = \Omega^2 R \cos(45.0°)\sin(45.0°) = \Omega^2 R \left(\frac{\sqrt{2}}{2}\right)^2 = \frac{\Omega^2 R}{2} = \frac{0.0339 \frac{m}{s^2}}{2} = 0.0170 m / s^2$$

Because the centrifugal force due to Earth's rotation has minimal impact, it is customary in synoptic meteorology to neglect the tangential component of the effective gravitational force and approximate $g_{eff,rad} \approx g$. However, the combined effect of the tangential and radial components of the centrifugal force results in a slight bulging at the equator. As a result, the equatorial radius (the distance from the Earth's center to the equator) is approximately 21 km greater than the polar radius (the distance from the Earth's center to the poles). Thus, Earth's shape is an ellipsoid rather than a sphere, which it would be if free-fall acceleration were determined solely by gravity.

6.5 Properties of the Coriolis Force

According to Eq. (6.13), the Coriolis force $\boldsymbol{F}_{cor}$ is given by:

$$\boldsymbol{F}_{cor} = -2m\left[\boldsymbol{\Omega} \times \left(\frac{d\boldsymbol{r}}{dt}\right)_S\right] = m(\boldsymbol{v} \times 2\boldsymbol{\Omega})$$

where $\boldsymbol{v}$ is the velocity of the air parcel as measured in Earth's rotating reference frame. Unlike the centrifugal force, which affects all objects in a rotating frame regardless of their motion, the Coriolis force depends on both the magnitude of $\boldsymbol{v}$ (i. e., the wind speed) and the direction of $\boldsymbol{v}$ relative to $\boldsymbol{\Omega}$. We will begin our discussion of the Coriolis force with a conceptual analysis.

6.5.1 Conceptual Analysis of the Coriolis Force

We can determine the direction of the Coriolis force by considering a horizontally rotating turntable, as shown in Figure 6.9. Because Earth rotates eastward, its rotation appears counterclockwise when viewed from the Northern Hemisphere in Figure 6.9A. Consequently, the angular velocity vector $\boldsymbol{\Omega}$ points in the $\boldsymbol{k}$ direction (out of the page). For an object moving with velocity $\boldsymbol{v} = u\boldsymbol{i}$, the Coriolis force is calculated as:

$$\boldsymbol{F}_{cor} = m(\boldsymbol{v} \times 2\boldsymbol{\Omega}) = 2m\boldsymbol{\Omega}u(\boldsymbol{i} \times \boldsymbol{k}) = 2m\boldsymbol{\Omega}u(-\boldsymbol{j})$$

Therefore, the object will be deflected to the south. For an object moving with velocity $\boldsymbol{v} = -v\boldsymbol{j}$, the Coriolis force is calculated as:

$$\boldsymbol{F}_{cor} = m(\boldsymbol{v} \times 2\boldsymbol{\Omega}) = 2m\boldsymbol{\Omega}v(-\boldsymbol{j} \times \boldsymbol{k}) = 2m\boldsymbol{\Omega}v(-\boldsymbol{i})$$

Therefore, the object will be deflected to the west. If we continue this process, we find that the Coriolis force remains perpendicular to the velocity vector.For a counterclockwise rotating turntable, the Coriolis force deflects objects to the right of their velocity vector (i.e., $\boldsymbol{F}_{cor}$ is directed 90° clockwise of $\boldsymbol{v}$).

When we view Earth's rotation from the Southern Hemisphere, it appears as a clockwise-turning turntable, as shown in Figure 6.9B. In this case, the angular

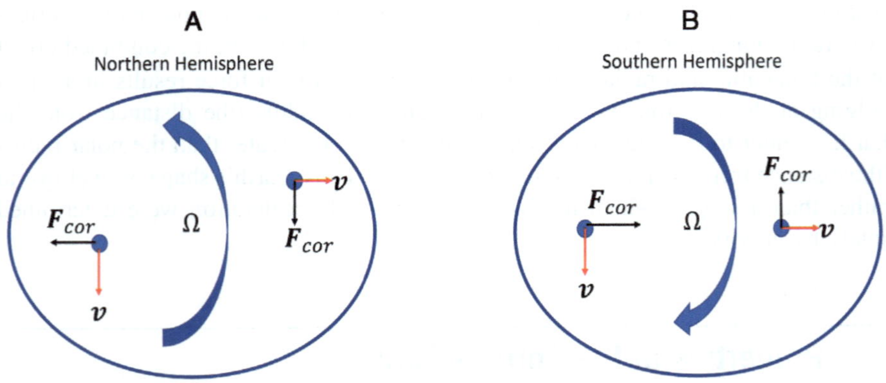

FIG. 6.9 Schematic of the Coriolis force on a rotating turntable. A rotating turntable with (A) counterclockwise rotation and (B) clockwise rotation. From the perspective of the Northern Hemisphere, Earth's rotation is counterclockwise, which produces an angular velocity that points out of the page. In contrast, from the perspective of the Southern Hemisphere, Earth's rotation is clockwise, which produces an angular velocity that points into the page.

-No Permission Required

velocity $\boldsymbol{\Omega}$ points in the $-\boldsymbol{k}$ direction (i.e., into the page). Based on this orientation, we can expect the Coriolis force to deflect moving objects to the left of their velocity vector (i.e., $\boldsymbol{F}_{cor}$ is directed 90° counterclockwise of $\boldsymbol{v}$) in this scenario.

For example, if an object is moving with a velocity $\boldsymbol{v} = u\boldsymbol{i}$ on a clockwise-rotating table, the Coriolis force is expressed as:

$$\boldsymbol{F}_{cor} = m(\boldsymbol{v} \times 2\boldsymbol{\Omega}) = 2m\boldsymbol{\Omega}u(\boldsymbol{i} \times -\boldsymbol{k}) = 2m\boldsymbol{\Omega}u(\boldsymbol{j})$$

Thus, the object will be deflected to the north, as expected. If we apply these conclusions to the Earth, we can predict specific behavior regarding air parcels.

- The Coriolis force is always perpendicular to the velocity of air parcels.
- In the Northern Hemisphere, $\boldsymbol{F}_{cor}$ is directed 90° clockwise of $\boldsymbol{v}$, whereas in the Southern Hemisphere, $\boldsymbol{F}_{cor}$ is directed 90° counterclockwise of $\boldsymbol{v}$. As a result, the Coriolis force deflects air parcels to the right of their velocity vector in the Northern Hemisphere and to the left in the Southern Hemisphere.

Because the Coriolis force is always perpendicular to the direction of motion, it cannot change the speed of an air parcel. To demonstrate this, recall from introductory physics that the work done by a force is given by:

$$W = \int \boldsymbol{F} \cdot d\boldsymbol{r}$$

where **F** is the applied force and **dr** is the displacement. By substituting $d\boldsymbol{r} = \boldsymbol{v}dt$, we can derive the work done by the Coriolis force as:

$$W_{cor} = m \int (\boldsymbol{v} \times 2\boldsymbol{\Omega}) \cdot (\boldsymbol{v} \quad dt)$$

Since the vector product $(\boldsymbol{v} \times 2\boldsymbol{\Omega})$ is perpendicular to $\boldsymbol{v}$, their scalar product is zero: $(\boldsymbol{v} \times 2\boldsymbol{\Omega}) \cdot \boldsymbol{v} = 0$. Therefore, the Coriolis force does no work on an air parcel and cannot change its speed. It can only deflect the parcel from its original path of motion consistent with the fact that this force acts on moving air parcels.

6.5.2 The Coriolis Force on the Rotating Earth

We now examine the Coriolis force within Earth's rotating reference frame. Consider an air parcel on Earth's surface, and establish a coordinate system such that the y-direction points north and the z-direction points upward, as shown in Figure 6.10A. With this choice of coordinates, the components of the air parcel's velocity vector and Earth's angular velocity vector are expressed as follows:

$$\boldsymbol{v} = u\boldsymbol{i} + v\boldsymbol{j} + w\boldsymbol{k}, \boldsymbol{\Omega} = (\boldsymbol{\Omega} \cos\phi)\boldsymbol{j} + (\boldsymbol{\Omega} \sin\phi)\boldsymbol{k}$$

where ϕ is the latitude, which ranges from 90° at the North Pole, 0° at the equator, and −90° at the South Pole, as shown in Figure 6.10B. Using the definition of the vector product, the acceleration generated by the Coriolis force is expressed as:

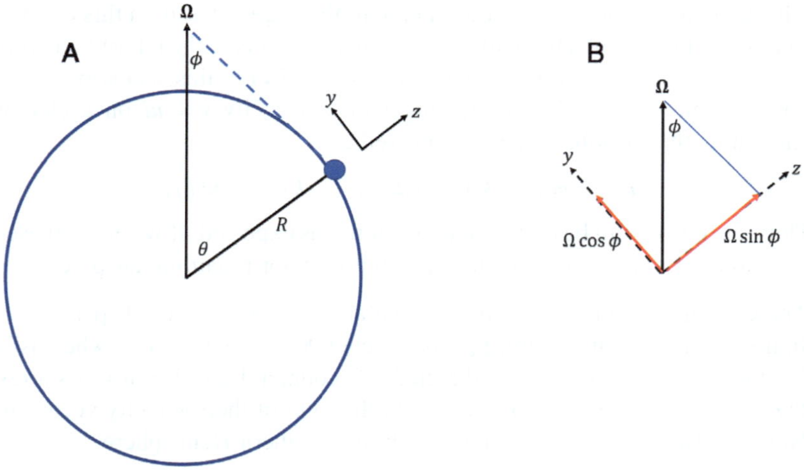

FIG. 6.10 Effect of the Coriolis force on an air parcel. An air parcel located near Earth's surface. (A) The coordinate system is defined such that the direction points upward from the air parcel and the direction points toward the north. (B) The components of Earth's angular velocity are given related to the coordinate system defined in (A).

-No Permission Required

$$\boldsymbol{a}_{cor} = \frac{\boldsymbol{F}_{cor}}{m} = \boldsymbol{v} \times 2\boldsymbol{\Omega} = \begin{vmatrix} \boldsymbol{i} & \boldsymbol{j} & \boldsymbol{k} \\ u & v & w \\ 0 & 2\boldsymbol{\Omega}\cos\phi & 2\boldsymbol{\Omega}\sin\phi \end{vmatrix}$$

$$\Rightarrow \boldsymbol{a}_{cor} = \boldsymbol{i}(2\boldsymbol{\Omega}v\sin\phi - 2\boldsymbol{\Omega}w\cos\phi) - \boldsymbol{j}(2\boldsymbol{\Omega}u\sin\phi) + \boldsymbol{k}(2\boldsymbol{\Omega}u\cos\phi)$$

For clarity, we can define $f = 2\boldsymbol{\Omega}\ sin\ \phi$ where f is known as the **Coriolis parameter**. This parameter f depends on both Earth's angular speed and the latitude of the air parcel. Specifically, $f > 0$ in the Northern Hemisphere (since $\sin\phi > 0$ for $0° < \phi < 90°$), and $f < 0$ in the Southern Hemisphere (since $\sin\phi < 0$ for $-90° < \phi < 0°$). Additionally, $f = 0$ at the equator, indicating that **there is no Coriolis deflection** along **the equator**.

Using this notation, the acceleration produced by the Coriolis force can be summarized as:

$$\boldsymbol{a}_{cor} = \boldsymbol{i}(fv - 2\boldsymbol{\Omega}\mathrm{w}\cos\phi) - \boldsymbol{j}(fu) + \boldsymbol{k}(2\boldsymbol{\Omega}\mathrm{u}\cos\phi) \quad (6.15)$$

Let's focus on the horizontal acceleration associated with the Coriolis force given in Eq. (6.15). In component form, the horizontal acceleration associated with Eq. (6.15) can be expressed as:

$$\left(\frac{\partial u}{\partial t}\right)_{cor} = fv - 2\boldsymbol{\Omega}w\cos\phi, \left(\frac{\partial v}{\partial t}\right)_{cor} = -fu$$

As we observed in Ch. 5, the vertical velocity component is typically much smaller than the horizontal velocity components for synoptic-scale processes. This implies that $2\Omega w \cos\phi \ll fu$. In other words, the Coriolis acceleration resulting from vertical motion is much weaker than that from horizontal motion. For this reason, we usually neglect this term in the context of synoptic-scale motion, allowing us to simplify the equations to:

$$\left(\frac{du}{dt}\right)_{cor} \approx fv, \left(\frac{dv}{dt}\right)_{cor} = -fu \tag{6.16}$$

We can use Eq. (6.16) to predict the direction of the deflection caused by the Coriolis force. For example, consider an air parcel moving eastward in the Northern Hemisphere. In this case, $u > 0$ and $v = 0$, which implies that $(dv/dt)_{cor} < 0$. Therefore, an eastward-moving air parcel will experience a southward acceleration, resulting in a deflection toward the south.

Consider an air parcel moving southward in the Northern Hemisphere. In this case, we have $v < 0$ and $u = 0$, which indicates that $(du/dt)_{cor} < 0$. Thus, a southward-moving air parcel will experience a westward acceleration, resulting in a deflection toward the west.

Continuing this reasoning, we conclude that the Coriolis force causes air parcels to deflect 90° to the right (i.e., 90° clockwise) of their velocity vector in the Northern Hemisphere. Conversely, in the Southern Hemisphere, air parcels are deflected 90° to the left (i.e., 90° counterclockwise) of their velocity vector. This observation is consistent with our earlier analysis of Figure 6.9.

To summarize, we have the following conclusions:

- Since Earth's angular velocity is constant, the magnitude of the Coriolis force depends on both latitude and the velocity of the air parcel.
- In the Northern Hemisphere, the Coriolis force vector is 90° to the right (i.e., clockwise) of the velocity vector.
- In the Southern Hemisphere, the Coriolis force vector is 90° to the left (i.e., counterclockwise) of the velocity vector. Additionally, Box 6.4 provides an example of the amount of deflection caused by the Coriolis force.

Having defined the major atmospheric forces acting on an air parcel in Chs. 5 and 6, we can now explore the dynamics of atmospheric flow patterns. We will begin this discussion in Ch. 7, where we examine atmospheric flow under balanced conditions within Earth's atmosphere.

Questions

1. In words, explain why the centrifugal force and the Coriolis force are called inertial forces.
2. What are the directions of the centrifugal and Coriolis forces on a person moving (a) south near the North pole, (b) east on the equator, and (c) south across the equator?

Box 6.4 Estimate of the Coriolis deflection

Suppose an air parcel is moving eastward with a constant speed of 25 m/s at 30° latitude. By how much is it deflected laterally due to Earth's rotation over a 1.00-minute span?

Answer: Since the air parcel is moving eastward, then by Eq. (6.15), it will have a southward deflection given by

$$\frac{dv}{dt} = -fu$$

If we integrate both sides with respect to time, we obtain

$$\int_0^t \frac{dv}{dt}dt = -\int_0^t fudt \quad (i)$$

We can solve the integral on the left-hand side of (*i*) by using the fundamental theorem of calculus

$$\int_0^t \frac{dv}{dt}dt = \int_{v_0}^{v} dv = v - v_0$$

Since u is constant, then integral on the right-hand side of (*i*) is given by

$$\int_0^t fudt = fut$$

Therefore, the integral in (*i*) gives $v - v_0 = -fut$. Since the air parcel is moving eastward, there is no meridional velocity component, and thus $v_0 = 0$. Thus, we have

$$v = -fut$$

Integrating both sides with respect to time gives

$$\int \frac{dy}{dt}dt = -\int futdt$$

$$\Rightarrow \int dy = -fu\int tdt$$

$$\Rightarrow \Delta y = -\frac{fut^2}{2} = -\frac{[2\Omega \sin\phi]ut^2}{2} = -\frac{\left[2\left(7.29\times10^{-5}\frac{rad}{s}\right)\sin 30.0^\circ\right]\left(25.0\frac{m}{s}\right)(60.0s)^2}{2} =$$

$$-3.28m$$

3. In Ch. 2, it was shown that mercury barometers require temperature corrections due to thermal expansion of the fluid.
   a. Explain why mercury barometers require corrections due to gravitation and due to Earth's rotation.
   b. If mercury barometers are calibrated based on sea-level pressure at 45° latitude, would the measured barometric pressure be an overestimate or an underestimate of the true pressure if the measurement were done at 30° latitude if no corrections were made? Defend your answer.
4. Regarding the Coriolis force,
   a. Explain why the Coriolis force vanishes for an air parcel that is stationary with respect to the Earth's surface, even though that air parcel has a large tangential velocity associated with Earth's rotation.
   b. Explain why the Coriolis force vanishes for an air parcel that is moving along the equator.

5. A geostationary satellite is a satellite whose orbital period is the same as Earth's rotational period, which means that the satellite appears to be fixed in the sky relative to observers at the ground.
   a. Relative to an inertial reference frame, what forces are acting on the geostationary satellite?
   b. Relative to a reference frame fixed on the geostationary satellite, what forces are acting on geostationary satellite?

6. Figure 6.11 shows an air parcel on Earth's surface from the perspective of an observer looking down at the North Pole (NP). Suppose that the air parcel is at rest on Earth's surface. Relative to the air parcel, what forces are acting upon the air parcel? Draw a free body diagram to defend your answer and use your free body diagram to explain why the air parcel must remain on the same parallel (i.e., line of constant latitude ϕ).

7. Let's return to Figure 6.11 and now suppose that the air parcel moves with a velocity northward (i.e., towards the North Pole). Relative to the air parcel, what forces are acting upon the air parcel? Draw a free body diagram to defend your answer and use your free body diagram to explain why the net force on the air parcel is to the right of the velocity vector.

8. Consider an eastward-moving air parcel with slow westerly winds to the west and fast westerly winds to the east. Use a conceptual argument to show that $\frac{\partial u}{\partial t} < 0$ at a fixed location.

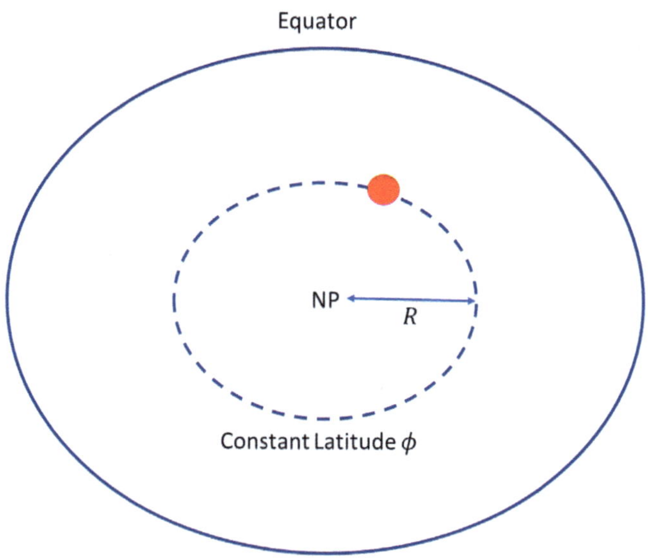

FIG. 6.11

An air parcel on Earth's surface relative to the north pole.

-No Permission Required

9. Consider the following three vectors:
$\boldsymbol{A} = -3.00\boldsymbol{i} + 3.00\boldsymbol{j} + 2.00\boldsymbol{k}$, $\boldsymbol{B} = -2.00\boldsymbol{i} - 4.00\boldsymbol{j} + 2.00\boldsymbol{k}$,
$\boldsymbol{C} = 2.00\boldsymbol{i} + 3.00\boldsymbol{j} + 1.00\boldsymbol{k}$
   a. Calculate the magnitude of $\boldsymbol{A} \times \boldsymbol{C}$
   b. Calculate the magnitude of $(\boldsymbol{A} \times \boldsymbol{B})\cdot(\boldsymbol{C} + \boldsymbol{A})$
   c. Determine the direction of $\boldsymbol{k} \times (\boldsymbol{A} + \boldsymbol{B} + \boldsymbol{C})$
10. Consider an air mass that moves northward at a rate of 10 m/s and suppose that the surface pressure increases toward the northwest at the rate of 5.00 Pa/km. Determine how the pressure at a nearby surface observation station will change over 3 hours if the pressure within the air parcel decreases at a rate of 1 hPa over 3 hours.
11. Consider an air mass with a temperature of 12°C along its southern side but smoothly increases in temperature to 15°C on its northern side. If this 100-km air mass is advecting toward the north at 25 km/h, determine the heating rate that can be attributed to thermal advection. For simplicity, you may assume that the air parcel is shaped as a square.
12. Indianapolis, IN, is about 175 miles west of Columbus, OH. The wind at Indianapolis is 14.5 m/s from 207°, and the wind at Columbus is 12.8 m/s from 173°. Determine the acceleration of the wind due to advection. You may assume that the wind is measured using the meteorological direction.
13. Answer the following questions regarding the Coriolis force:
   a. Assuming that the air parcel is located at a latitude $\phi = 33.0°$N, calculate the Coriolis acceleration in unit-vector notation if the wind is northerly at 15.0 m/s.
   b. Assuming that the air parcel is located at $\phi = 45.0°$N, calculate the Coriolis acceleration in terms of magnitude and meteorological direction if the wind speed is 10.0 m/s and the meteorological direction is from 285°.
14. Assuming that the air parcel is located at a latitude $\phi = 45.0°N$, calculate the wind speed and the wind direction if the Coriolis acceleration is
$\boldsymbol{a}_{COR} = (7.50 \times 10^{-4} \ s^{-1})\boldsymbol{i} + (3.70 \times 10^{-4} \ s^{-1})\boldsymbol{j}$

CHAPTER

Balance in synoptic-scale flow

7

Chapter outline

In Chapters 5 and 6 we showed that the forces governing the motion of an air parcel in Earth's rotating reference frame include the gravitational force, the pressure gradient force, the turbulent drag force, the Coriolis force, and the centrifugal force associated with Earth's rotation. Therefore, Newton's Second Law can be expressed in component form as:

$$\begin{aligned}\frac{Du}{Dt} &= fv - \frac{1}{\rho}\frac{\partial P}{\partial x} - \frac{1}{\rho}\frac{\partial \tau_x}{\partial z},\\ \frac{Dv}{Dt} &= -fu - \frac{1}{\rho}\frac{\partial P}{\partial y} - \frac{1}{\rho}\frac{\partial \tau_y}{\partial z},\\ \frac{dP}{dz} &= -\rho g\end{aligned} \tag{7.1}$$

where *D/Dt* is the Lagrangian rate of change, observed in the reference frame following the air parcel. Since Eq. (7.1) describes the evolution of the wind, it is

An Introduction to Synoptic-Dynamic Meteorology. https://doi.org/10.1016/B978-0-443-24858-0.00007-1

commonly referred to as the **equations of motion**. Because Eq. (7.1) forms a coupled set of equations, there are no simple analytic solutions that apply to all types of atmospheric flow. However, observations of synoptic-scale motion allow us to make useful approximations to Eq. (7.1).

In this chapter, we will analyze atmospheric motion by applying such simplifications, focusing on the midlatitude synoptic-scale atmosphere. This analysis will lead us to examine balanced flow, beginning with one of the most fundamental balance conditions: geostrophic balance.

7.1 Geostrophic balance

Figure 7.1 shows a 500-hPa isobaric chart of the continental United States on January 8, 2012. Along the western coast, we observe the formation of a rex block, characterized by a ridge over the Pacific Northwest and a trough over the Southwest. Three key features are worth emphasizing in this section.

First, there is significant meridional flow across California, where the wind moves parallel to the isohypses. Second, a strong zonal flow is evident across the central United States, with the wind again aligned with the isohypses. Third, wind speeds increase over the Northeast, as indicated by the decreasing spacing of the isohypses in that region.

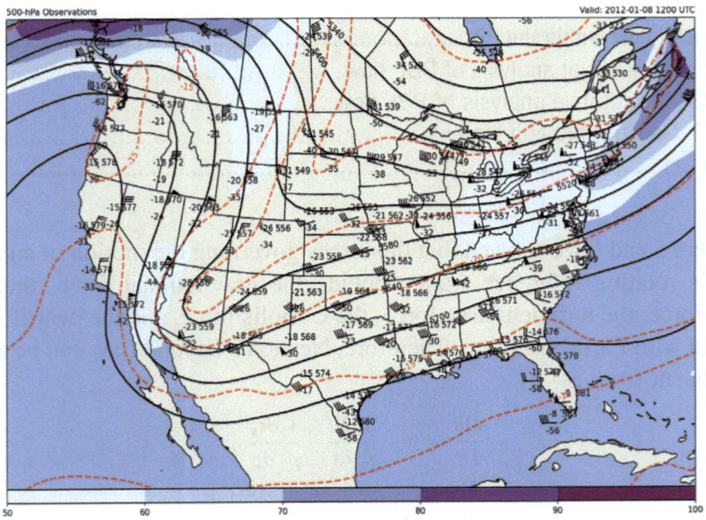

FIGURE 7.1 Observation of geostrophic balance

500-hPa isobaric chart of the continental United States from January 8, 2012, at 12 UTC. Solid black contours correspond to isohypses, dashed red contours correspond to isotherms, and contour color fill corresponds to isotachs.

No Permission Required

Looking at the chart as a whole, a recurring pattern emerges: in regions with straight isohypses, the wind flows parallel to them, and wind speed varies with isohypse spacing. This pattern reflects an important balance condition in the atmosphere known as **geostrophic balance**, which occurs when the Coriolis force is balanced by the horizontal pressure gradient force. The resulting flow is called **geostrophic flow**.

7.1.1 The geostrophic wind vector

We can now examine geostrophic flow mathematically. Above the planetary boundary layer (PBL), the turbulent drag force can be neglected. Thus, the zonal and meridional acceleration components in Eq. (7.1) can be expressed as:

$$\frac{D\mathbf{v}_H}{Dt} = -\frac{1}{\rho}\nabla_z P - f\mathbf{k} \times \mathbf{V}$$

where $\boldsymbol{V}_H$ is the horizontal velocity vector, and the subscript "z" indicates that the pressure gradient force is evaluated on a surface of constant geometric height. From Ch. 5, recall that the pressure gradient force can be rewritten in terms of the geopotential height gradient using the hydrostatic equation:

$$\frac{1}{\rho}\nabla_z P = g_0 \nabla_P Z$$

where the subscript "P" signifies that the geopotential height gradient is evaluated on an isobaric surface. Substituting this into the previous equation gives:

$$\frac{D\mathbf{v}_H}{Dt} = -g_0 \nabla_P Z - f\mathbf{k} \times \mathbf{V}$$

Geostrophic balance corresponds to equilibrium between the pressure gradient force and the Coriolis force. Thus, under Newton's Second Law:

$$\frac{D\boldsymbol{v}_H}{Dt} = 0 \Rightarrow -g_0 \nabla_P Z - f\mathbf{k} \times \mathbf{V} = 0$$

$$\Rightarrow f\mathbf{k} \times \mathbf{V} = -g_0 \nabla_P Z \tag{7.2}$$

Taking the cross product with **k**:

$$f[\mathbf{k} \times (\mathbf{k} \times \mathbf{V})] = -\mathbf{k} \times g_0 \nabla_P Z \tag{7.3}$$

The left side of Eq. (7.3) can be simplified as follows:

$$\mathbf{k} \times \mathbf{V} = \mathbf{k} \times [u\mathbf{i} + v\mathbf{j} + w\mathbf{k}] = u\mathbf{j} - v\mathbf{i}$$

$$\Rightarrow \mathbf{k} \times (\mathbf{k} \times \mathbf{V}) = \mathbf{k} \times [u\mathbf{j} - v\mathbf{i}] = -u\mathbf{i} - v\mathbf{j} = -\boldsymbol{V}_H$$

Substituting this result into Eq. (7.4) gives

$$-f\boldsymbol{V}_H = -\mathbf{k} \times g_0 \nabla_P Z$$

$$\Rightarrow \boldsymbol{V}_H = \left(\frac{g_0}{f}\right)(\mathbf{k} \times \nabla_P Z) \tag{7.4}$$

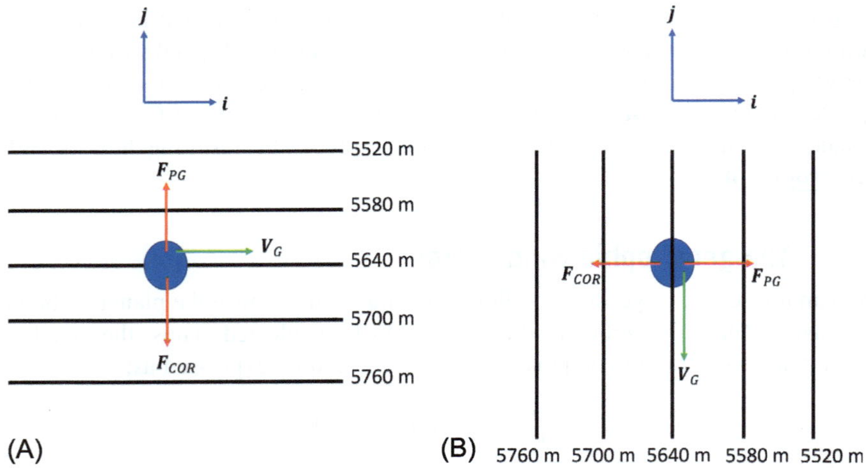

FIGURE 7.2 Schematic of geostrophic balance

Free-body diagrams associated with (A) westerly geostrophic flow and (B) northerly geostrophic flow within the Northern Hemisphere. The solid black contours correspond to isohypses associated with a 500-hPa isobaric chart. The red vector arrows correspond to the pressure gradient force F_{PG} and the Coriolis force F_{COR}. The green vector arrow corresponds to the geostrophic wind vector.

No Permission Required

The horizontal wind in isobaric coordinates, expressed in Eq. (7.4), is known as the **geostrophic wind vector**. Figure 7.2 illustrates the free-body diagram analysis for both westerly and northerly geostrophic flow. As discussed in Ch. 5, the pressure gradient force points toward lower isohypses. Because the flow is balanced, the Coriolis force must point toward higher isohypses. From Ch. 6, recall that the Coriolis force is directed 90° to the right of the velocity vector. Therefore, a southward-directed Coriolis force corresponds to westerly geostrophic flow (as shown in Figure 7.2A), while a westward-directed Coriolis force corresponds to southerly geostrophic flow (as shown in Figure 7.2B). In both cases, the geostrophic wind vector is parallel to the isohypses.

In component form within Cartesian coordinates, Eq. (7.4) can be written as:

$$u_g = -\frac{g_0}{f}\frac{\partial Z}{\partial y}, v_g = \frac{g_0}{f}\frac{\partial Z}{\partial x} \tag{7.5}$$

which emphasizes that the magnitude of the geostrophic wind is proportional to the isohypse spacing.

7.1.2 Observations of geostrophic flow

We can assess whether the atmospheric flow is in geostrophic balance by comparing the actual wind with the predicted geostrophic wind, as given in Eq. (7.5). The difference between the two is the **ageostrophic wind**, calculated as:

$$\boldsymbol{V}_{ag} = \boldsymbol{V} - \left(\frac{g_0}{f}\right)(\boldsymbol{k} \times \nabla_P Z) \tag{7.6}$$

From Eq. (7.6), the flow approaches geostrophic balance when $\boldsymbol{V}_{ag} \approx 0$. In this case, the pressure gradient force and the Coriolis force are nearly in balance. In contrast, significant ageostrophic flow indicates an imbalance between these forces, resulting in horizontal acceleration of the atmosphere.

We first examine the ageostrophic flow on upper-air charts, as shown in Figure 7.3. The subtropical jet extends across the central United States, while the polar jet lies across southern Canada. Over most of the upper-air chart, the flow is nearly in geostrophic balance, with ageostrophic wind speeds less than 10% of the total wind speed. However, jet streaks within the subtropical jet over Colorado and southern California produce strong ageostrophic flow. Jet streaks therefore represent localized accelerations of the airflow, and as will be shown in Ch. 9 are typically accompanied by vertical motion.

Figure 7.4 illustrates ageostrophic flow at 500 hPa. Much of this chart shows weak geostrophic flow (less than 10% of the actual wind), indicating that the atmosphere is close to geostrophic balance. Under these conditions, variations in isohypse spacing correspond to changes in wind speed at specific locations. For example, the 500-hPa winds over central Florida are weaker than those over the Carolinas due to differences in isopleth spacing. The atmosphere departs from geostrophic balance in regions where isopleths are curved. Modest ageostrophic flow appears over the Great Lakes while

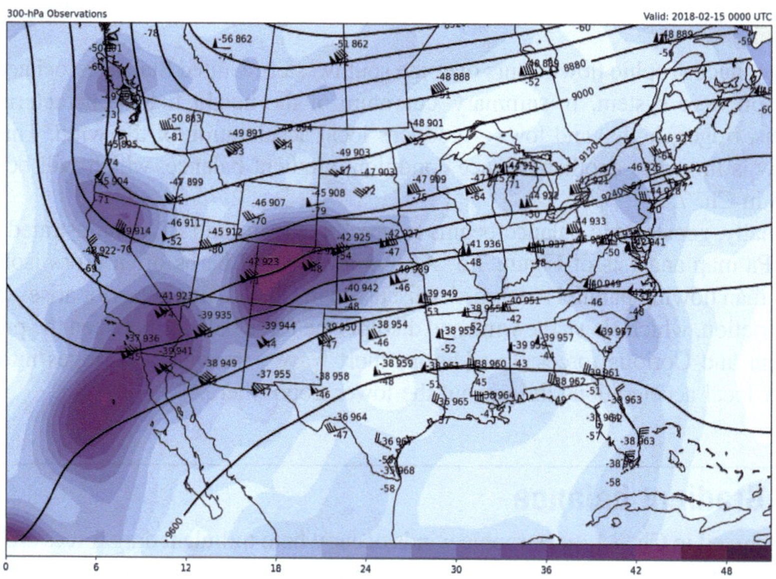

FIGURE 7.3 300-hPa jet stream analysis

300-hPa map analysis on February 15, 2018, at 00Z. Solid black contours correspond to isohypses at 120-m intervals and contour fill corresponds to ageostrophic wind speed in units of knots.

No Permission Required

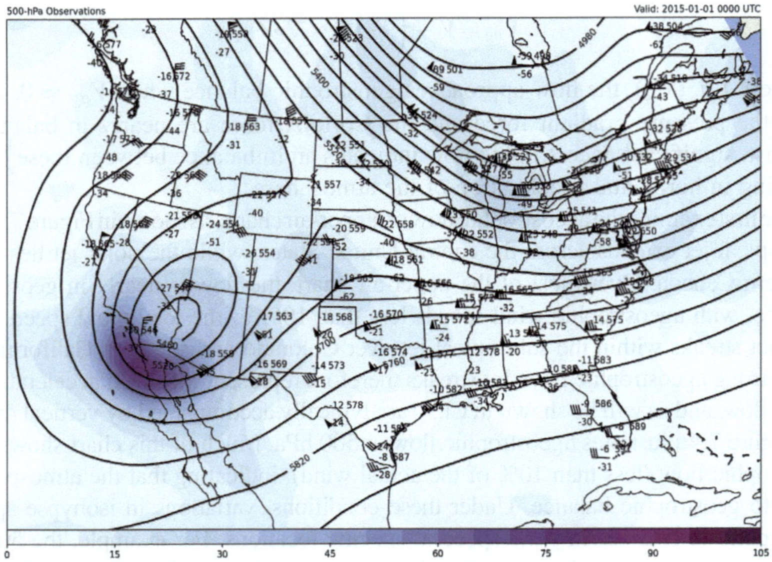

FIGURE 7.4 500-hPa map analysis showing geostrophic balance

500-hPa map analysis on January 1, 2015, at 00Z. Solid black contours correspond to isohypses at 60-m intervals and contour fill corresponds to ageostrophic wind speed in units of knots.

No Permission Required

stronger ageostrophic flow occurs over the southwestern United States, associated with a low-pressure system. In summary, curvature of the height field—characteristic of troughs, ridges, highs, and lows—produces local acceleration of the wind. This type of flow is linked to another balance condition, gradient balance, which will be introduced in Ch. 7.2.

Finally, geostrophic balance begins to break down in the PBL, as illustrated in the 700-hPa map analysis of Figure 7.5. At this level, wind barbs often cross isohypses rather than flowing parallel to them. As discussed in Ch. 7.4, this behavior arises mainly from friction, which slows the wind and disrupts the force balance between the pressure gradient and Coriolis forces. In addition, short waves, troughs, and ridges introduce further local accelerations throughout the lower troposphere.

7.2 Gradient balance

As discussed in Ch. 7.1, geostrophic winds occur where turbulent drag forces are negligible (ie, above the PBL) and where the isohypses are straight. However, in many cases, isohypses are curved. Figure 7.6 shows a 500-hPa isobaric chart of the continental United States from January 8, 2024 with a deep trough over the southwestern United States and a weaker ridge over the Northern Plains and the Midwest.

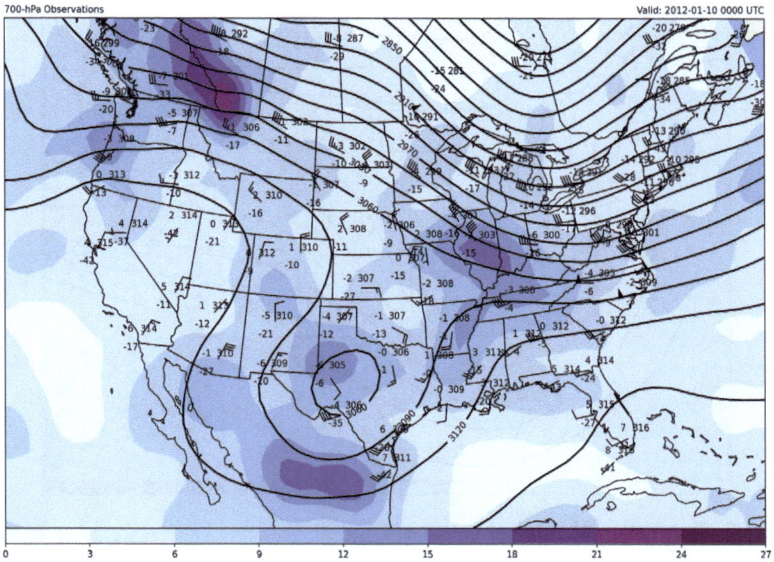

FIGURE 7.5 700-hPa map analysis showing geostrophic balance

700-hPa map analysis on January 1, 2012, at 00Z. Solid black contours correspond to isohypses at 30-meter intervals and contour fill corresponds to ageostrophic wind speed in units of knots.

No Permission Required

For this section, three key points should be noted. First, near the base of the trough, the wind flows parallel to the curved isohypses. Second, the wind also flows parallel to the curved isohypses near the peak of the trough. Third, air parcels are not in geostrophic balance as they travel through the base of the trough or the ridge. These features are commonly observed on upper-air charts: the wind flows parallel to curved isohypses with speeds that are ageostrophic.

This pattern is a consequence of another balance condition in the atmosphere known as **gradient balance**, which occurs when the horizontal components of the Coriolis force, the pressure gradient force, and the centrifugal force associated with curved motion are in equilibrium.[1] The resulting flow is known as **gradient flow**.

7.2.1 Gradient flow in the atmosphere

To understand the airflow along curved isohypses, consider an air parcel located south of a closed low-pressure system, as shown in Figure 7.7A. The pressure gradient force acts toward areas of low geopotential height, pointing toward the

[1] It is important to note that this is not the same centrifugal force discussed in Ch. 6. The centrifugal force discussed in Ch. 6 was caused by Earth's rotation. However, as we will show, the centrifugal force here is caused by the movement along curved isohypses.

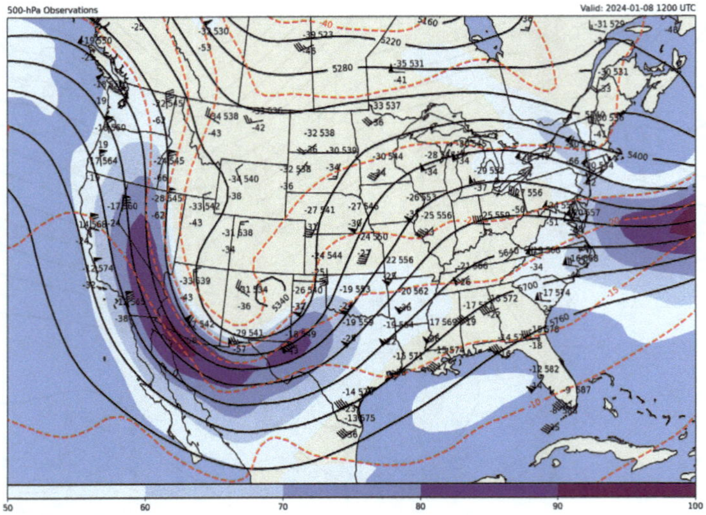

FIGURE 7.6 Observation of gradient balance

500-hPa isobaric chart of the continental United States from January 8, 2024, at 12 UTC. Solid black contours correspond to isohypses, dashed red contours correspond to isotherms, and contour color fill corresponds to isotachs.

No Permission Required

center of the low. In contrast, the Coriolis force deflects the parcel to the right of its motion, directing it away from the low. Since the Coriolis force vector is oriented clockwise 90° of the velocity vector, the parcel's velocity must point eastward, as shown in Figure 7.7A.

As the parcel moves along a curved path, its velocity vector changes direction, which requires an inward centripetal acceleration. However, in the rotating reference frame of the air parcel, the parcel does not experience acceleration; instead, the forces acting on it must balance. The centrifugal force, directed outward from the center of the parcel's circular path, balances the inward-directed pressure gradient and Coriolis forces. This balance allows the air parcel to flow parallel to the curved isohypses, producing gradient flow.

As the air parcel moves eastward, it shifts slightly away from the low-pressure center. The centrifugal force decrease in magnitude since it varies inversely with the distance from the air parcel to the low-pressure center. If the pressure gradient force remains constant, the reduced centrifugal force creates a net inward force, pushing the parcel back to its original radius and maintaining the flow along the isohypses.

A similar analysis applies to high-pressure systems, as shown in Figure 7.7B. Here, the pressure gradient force points radially outward, while the Coriolis force deflects the parcel to the right (i.e. toward the west in the figure), pointing radially inward. In this case, the centrifugal force acts outward, in the same direction as the

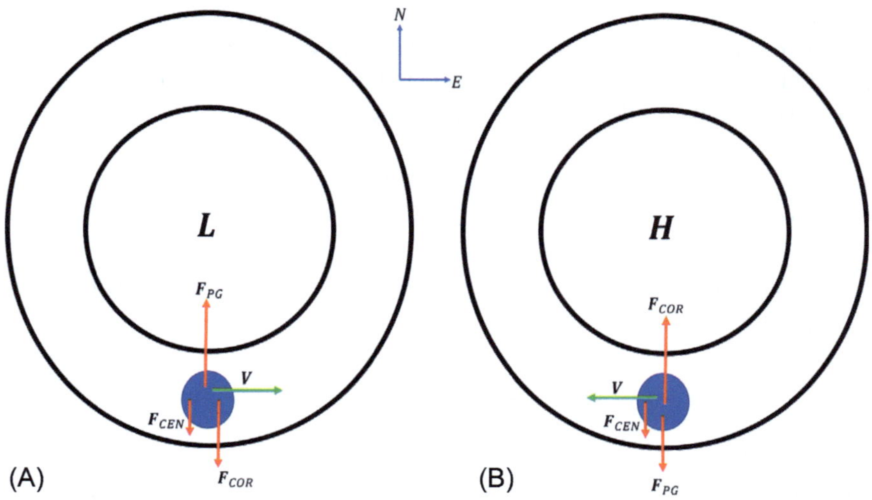

FIGURE 7.7 Schematic of gradient balance

(A) Schematic of a low-pressure system in the Northern Hemisphere. (B) Schematic of a high-pressure system in the Northern Hemisphere.

No Permission Required

pressure gradient force. To maintain gradient flow around a high, 90° the Coriolis force must be stronger than the pressure gradient force.

Consequently, the gradient wind is typically stronger around a high-pressure center than around a low-pressure center with the same pressure gradient. If the air parcel moves further from the high-pressure center, the centrifugal force decreases, producing a net force directed toward the center and restoring the parcel to its original curved path.

We can now quantify the previous discussion by deriving a relationship for the gradient wind around a low-pressure center. We examine the motion of an air parcel in the parcel's rotating reference frame. Because gradient balance involves a three-way equilibrium between the Coriolis force, the horizontal pressure gradient, and the centrifugal force, we find that $dV/dt = 0$, where V is the air parcel's speed.

Referring to Figure 7.7A and applying Newton's Second Law of motion, this balance can be expressed as:

$$F_{PG} + F_{COR} + F_{CEN} = 0$$

$$\Rightarrow g_0\left|\frac{\partial Z}{\partial r}\right| - fV_{gr} - \frac{V_{gr}^2}{R} = 0$$

$$\Rightarrow V_{gr}^2 + fRV_{gr} - g_0R\left|\frac{\partial Z}{\partial r}\right| = 0 \tag{7.7}$$

where V_{gr} is the gradient wind, R is the distance from the center of the low-pressure center (also known as the **radius of curvature**), and $|\partial Z/\partial r|$ is the magnitude of the radial geopotential height gradient. Equation (7.7) is a quadratic equation in V_{gr}, which can be solved using the quadratic formula:

$$V_{gr} = \frac{-fR \pm \sqrt{(fR)^2 - 4(1)\left(-g_0 R\left|\frac{\partial Z}{\partial r}\right|\right)}}{2} = -\frac{fR}{2} \pm \sqrt{\left(\frac{fR}{2}\right)^2 + g_0 R\left|\frac{\partial Z}{\partial r}\right|}$$

Choosing the positive root ensures that $V_{gr} > 0$. Thus, the gradient wind around a low-pressure center is given by:

$$V_{gr} = -\frac{fR}{2} + \sqrt{\left(\frac{fR}{2}\right)^2 + g_0 R\left|\frac{\partial Z}{\partial r}\right|} \tag{7.8}$$

Because low-pressure systems are cyclones, the counterclockwise gradient flow around these cyclones described by Eq. (7.8) is often termed **cyclonic flow**.

A similar analysis applies to high-pressure centers. Referring to Figure 7.7B and applying Newton's Second Law of motion gives:

$$F_{PG} + F_{COR} + F_{CEN} = 0$$

$$\Rightarrow -g_0\left|\frac{\partial Z}{\partial r}\right| + fV_{gr} - \frac{V_{gr}^2}{R} = 0$$

$$\Rightarrow V_{gr}^2 - fRV_{gr} + g_0 R\left|\frac{\partial Z}{\partial r}\right| = 0$$

Solving this equation and choosing the positive root[2] gives:

$$V_{gr} = \frac{-(-fR) \pm \sqrt{(-fR)^2 - 4(1)\left(g_0 R\left|\frac{\partial Z}{\partial r}\right|\right)}}{2}$$

$$\Rightarrow V_{gr} = \frac{fR}{2} + \sqrt{\left(\frac{fR}{2}\right)^2 - g_0 R\left|\frac{\partial Z}{\partial r}\right|} \tag{7.9}$$

Eq. (7.9) yields a real solution only when the term inside the radical is positive, which requires:

$$\left(\frac{fR}{2}\right)^2 - g_0 R\left|\frac{\partial Z}{\partial r}\right| > 0$$

[2] For many cases, choosing the negative sign will also lead to a positive answer for the gradient wind for a closed high. This solution is known as an anomalous high pressure. As we will show in Ch. 11, the solution for anomalous high pressure is connected to negative absolute vorticity.

$$\Rightarrow \left|\frac{\partial Z}{\partial r}\right| < \frac{\left(\frac{fR}{2}\right)^2}{g_0 R}$$

$$\Rightarrow \left|\frac{\partial Z}{\partial r}\right| < \frac{f^2 R}{g_0}$$

This highlights a key distinction between high- and low-pressure centers. Near low-pressure centers, strong radial pressure gradients with small curvature radii can exist, resulting in strong winds. In contrast, near high-pressure centers, only weak radial pressure gradients with large curvature radii are possible, producing relatively light winds. High-pressure systems, also known as anticyclones, exhibit a clockwise flow, referred to as **anticyclonic flow**, as described in Eq. (7.9).

7.2.2 Observations of gradient flow

The dynamics presented in Ch. 7.2 lead to two testable conclusions regarding gradient flow. First, low-pressure centers are characterized by strong pressure gradients and small radii of curvature, whereas high-pressure centers exhibit weak pressure gradients and large radii of curvature. These patterns are clearly visible in atmospheric blocking patterns, as shown in Figure 7.8. For example, the cutoff low in Figure 7.8C has a relatively small radius of curvature and strong winds near its center, indicating a strong radial pressure gradient force. By contrast, the cutoff high in Figure 7.8D exhibits a much larger radius of curvature with weaker winds, reflecting a weaker and more diffuse pressure gradient. This behavior is also evident in the omega block shown in Figure 7.8A: the lows contain numerous closed isohypses, indicating a large pressure gradient, whereas the high lacks closed contours.

The second testable conclusion is that the geostrophic wind tends to *overestimate* the actual flow around a low-pressure center, while it *underestimates* the actual flow around a high-pressure center. To demonstrate this mathematically, recall that geostrophic balance between the pressure gradient force and the Coriolis force gives:

$$fv_g = g_0 \left|\frac{\partial Z}{\partial r}\right|$$

where v_g represents the geostrophic flow associated with this system. Substituting this expression into Eqs. (7.8) and (7.9) yields:

$$V_{low} = \frac{fR}{2}\left[-1 + \sqrt{1 + \frac{4V_g}{fR}}\right], \qquad V_{high} = \frac{fR}{2}\left[1 - \sqrt{1 - \frac{4V_g}{fR}}\right] \tag{7.10}$$

Both expressions in Eq. (7.10) share a common ratio V_g/fR, known as the **curvature Rossby number** Ro_c. Rewriting Eq. (7.10) in terms of the curvature Rossby number gives:

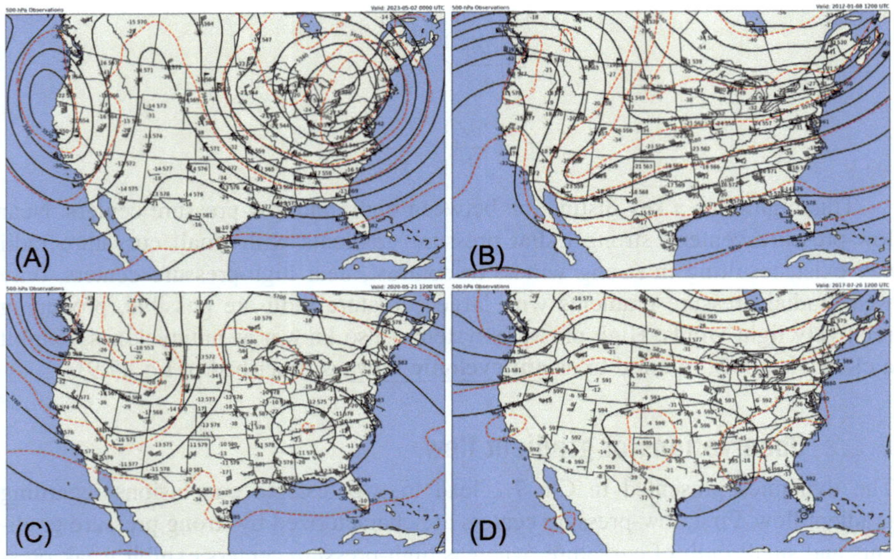

FIGURE 7.8 Comparing lows and highs

Four examples of atmospheric blocking patterns at the 500-hPa level: (A) Omega block on May 2, 2023; (B) Rex block on January 8, 2012; (C) Cutoff low on May 21, 2020; and (D) Cutoff high on July 20, 2017. For each map, the solid black contours correspond to isohypses at intervals of 60 meters, and the dashed red contours correspond to isotherms at intervals of 5°C.

No Permission Required

$$V_{low} = \frac{V_g}{2Ro_c}\left[-1 + \sqrt{1 + 4Ro_c}\right], \qquad V_{high} = \frac{V_g}{2Ro_c}\left[1 - \sqrt{1 - 4Ro_c}\right] \tag{7.11}$$

Figure 7.9 plots V_{low} and V_{high} (relative to V_g) as a function of the curvature Rossby number. Three key points emerge:

First, both $V_{low} \approx V_g$ and $V_{high} \approx V_g$ when $Ro_c \to 0$. Since $V_g$remains constant in Figure 7.9, the condition $Ro_c \to 0$ occurs when $R \to \infty$. In other words, a small curvature Rossby number corresponds to nearly straight isohypses.

Second, $V_{low} \leq V_g$for all values of Ro_c. This means the gradient wind around a low is **subgeostrophic** (ie, it is weaker than the geostrophic winds). Thus, the geostrophic winds overestimate the actual winds. Moreover, as shown in Figure 7.9, V_{low} decreases as Ro_c increases. This can be understood in terms of force balance: holding V_g constant keeps the pressure gradient fixed. As Ro_c increases (smaller radius of curvature), the centrifugal force must increase. To maintain balance with the pressure gradient force, the Coriolis force - and therefore the wind speed - must decrease.

Third, $V_{high} \geq V_g$ for all values of Ro_c. Thus, the gradient wind around a high-pressure system is **supergeostrophic (ie, it is stronger than the geostrophic winds). Thus, the** geostrophic flow underestimates the actual winds around a

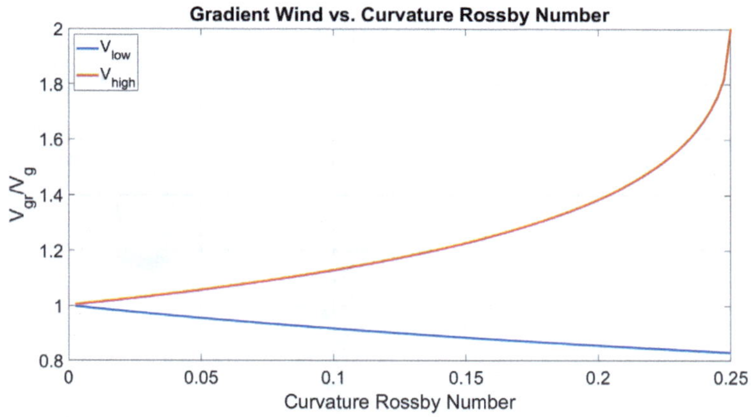

FIGURE 7.9 Gradient wind vs curvature Rossby number

Plot of the gradient wind V_{gr} (relative to the geostrophic wind V_g) as a function of the curvature Rossby number Ro_c for $V_g = 50.0$ m/s. In this figure, V_{low} corresponds to the gradient wind associated with a low-pressure system and V_{high} corresponds to the gradient wind associated with a high-pressure system. Since V_g is constant, an increase in Ro_c corresponds to a decrease in the radius of curvature of the flow.

No Permission Required

high. As shown in Figure 7.9, V_{high} increases rapidly with increasing Ro_c. Again, holding V_g constant fixes the pressure gradeint force. As Ro_c decreases, the centrifugal force increases. Since the Coriolis force balances both the pressure gradient force and the centrifugal force, the Coriolis force (and hence the wind speed) must increase to maintain this balance.

The relationship between gradient and geostrophic flow can be seen clearly in the omega block shown in Figure 7.10. The low-pressure systems in the eastern portion of the United States are associated with geostrophic winds exceeding 70 knots, with some regions surpassing 100 knots in the Southeast. However, the actual winds are weaker than the geostrophic estimate, consistent with subgeostrophic flow. Similarly, the lows in the western United States feature strong winds, but again the geostrophic wind overestimates the actual flow. By contrast, in the central United States, the high-pressure system has weak winds, while the geostrophic wind is nearly zero - illustrating that the geostrophic wind underestimates the true flow in high-pressure systems.

7.3 Thermal wind balance

In Ch. 5, it was shown that for synoptic-scale motions, vertical motion is governed by hydrostatic balance. This implies that the vertical velocity component is much

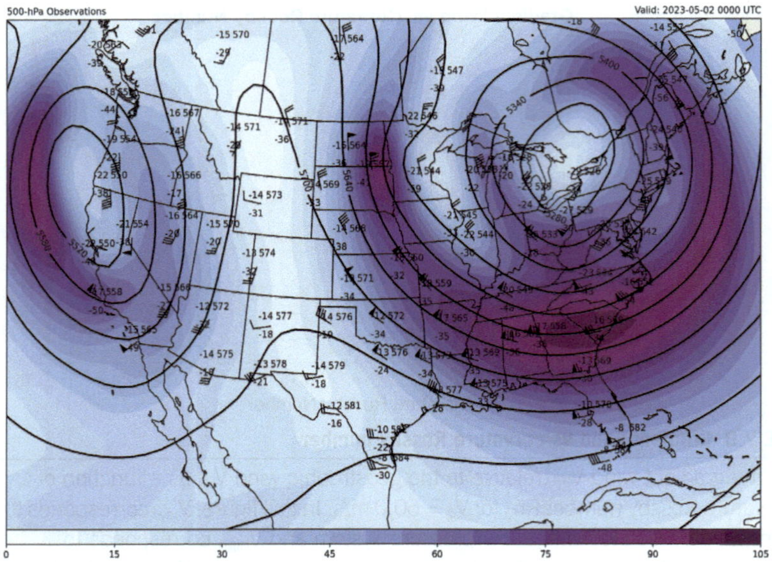

FIGURE 7.10 500-hPa map analysis of omega block

500-hPa map analysis on January 1, 2015, at 00Z. Solid black contours correspond to isohypses at 60-meter intervals and contour fill corresponds to geostrophic wind speed in units of knots.

No Permission Required

smaller than the horizontal wind components. Additionally, Ch. 7.1 explained that synoptic-scale horizontal motion is approximately geostrophic above the PBL, meaning that horizontal winds are largely balanced by the geopotential height field. Therefore, these observations indicate that the midlatitude synoptic-scale atmosphere is in an approximate state of hydrostatic and geostrophic balance.

In this section, we will show that hydrostatic balance and geostrophic balance can be combined to yield a single atmospheric balance condition, known as **thermal wind balance**. Understanding thermal wind balance serves two purposes. First, it reinforces a central theme of this textbook: thermodynamic and dynamical processes are fundamentally coupled in atmospheric motion. Second, it provides a diagnostic tool for understanding the evolution of midlatitude weather systems.

Thermal wind balance establishes a connection between the horizontal thickness gradient and the vertical shear of the geostrophic wind. Consider a warm column of air adjacent to a cold column of air, as shown in Figure 7.11. Recall from Ch. 5 that the thickness between any two isobaric surfaces is proportional to the layer-averaged virtual temperature. Therefore, the thickness in the warmer air column is greater than in the colder air column. In Figure 7.11, the thickness between the 1013-hPa and 850-hPa surfaces is larger in the warm column than in the cold column, implying that the 850-hPa surface slopes downward toward the cold air. Similarly, the thickness between the 700-hPa and 500-hPa surfaces is also larger in the warm column,

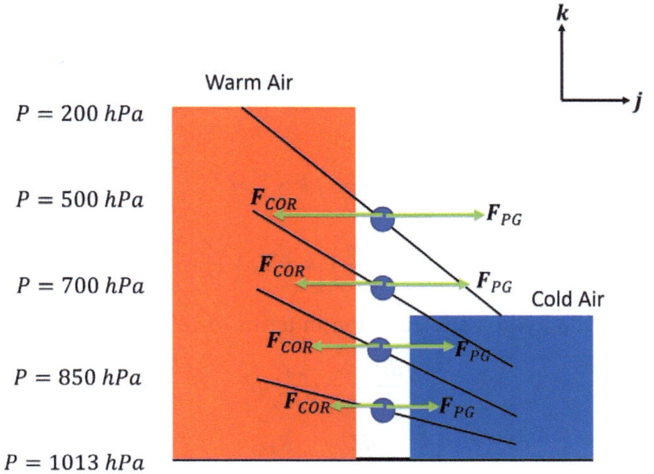

FIGURE 7.11 Schematic of thermal wind balance

Schematic of a warm air column next to a cold air column. In this figure, the horizontal direction points to the north and the vertical direction points vertically upward from the surface. The thick black contours correspond to five isobaric surfaces (1013 hPa, 850 hPa, 700 hPa, 500 hPa, and 200 hPa). The green vectors correspond to the Coriolis force (F_{COR}) and the pressure gradient force (F_{PG}).

No Permission Required

producing an even steeper downward slope of the 500-hPa surface. In other words, the slope of the isobaric surfaces *increases* with height in the presence of a horizontal gradient in layer-averaged virtual temperature.

As discussed in Ch. 7.1, the pressure gradient force (per unit mass) can be expressed as $-g_0\nabla_P Z$, where the subscript "P" indicates that the gradient is evaluated on an isobaric surface. This shows that the pressure gradient force on an isobaric surface is directly related to the geopotential height gradient along that surface.

In Figure 7.11, the slope of the isobaric surface toward the colder air corresponds to the geopotential height gradient along the isobaric surface. Therefore, if the slope of the isobaric surface *increases* with height, the horizontal pressure gradient force must also increase. Since the geostrophic wind is proportional to the magnitude of this horizontal pressure gradient force, it follows that the geostrophic wind must also increase with height.

This leads to the key conclusion:

Horizontal gradients in thickness throughout the atmosphere must be accompanied by vertical changes in the geostrophic wind.

The relationship between thickness gradients and geostrophic flow links the wind, height, and thermal fields of the atmosphere. Conceptually, this is the essence of thermal wind balance. In the next section, we will formalize this relationship by deriving an expression for the vertical shear of the geostrophic wind in terms of the virtual temperature gradient.

7.3.1 Derivation of the thermal wind equation

We begin by determining the vertical shear of the geostrophic wind vector. This can be done by taking a vertical derivative of Eq. (7.4) with respect to pressure, which gives:

$$\boldsymbol{V}_g = \left(\frac{g_0}{f}\right)[\boldsymbol{k} \times \nabla_P Z]$$

$$\Rightarrow \frac{\partial \boldsymbol{V}_g}{\partial P} = \left(\frac{g_0}{f}\right)\left[\boldsymbol{k} \times \nabla_P \left(\frac{\partial Z}{\partial P}\right)\right] \tag{7.12}$$

Rearranging the hydrostatic equation in terms of $\partial Z / \partial P$ gives:

$$\frac{dP}{P} = -\frac{g_0}{R_d T_v} dZ$$

$$\Rightarrow \frac{dZ}{dP} = -\frac{R_d T_v}{g_0 P}$$

Substituting this result into Eq. (7.12) gives us:

$$\frac{\partial \boldsymbol{V}_g}{\partial P} = \left(\frac{g_0}{f}\right)\left[\boldsymbol{k} \times \nabla_P \left(-\frac{R_d T_v}{g_0 P}\right)\right]$$

Because the gradient is evaluated on an isobaric surface, P can be treated as a constant. Thus, we obtain:

$$\frac{\partial \boldsymbol{V}_g}{\partial P} = -\left(\frac{R_d}{fP}\right)(\boldsymbol{k} \times \nabla_P T_v) \tag{7.13}$$

To determine the vertical shear of the geostrophic wind, we integrate both sides of Eq. (7.13) from pressure P_1 to pressure P_2 (where $P_2 < P_1$) to obtain:

$$\int_{P_1}^{P_2} \frac{\partial \boldsymbol{V}_g}{\partial P} dP = -\frac{R_d}{f} \int_{P_1}^{P_2} (\boldsymbol{k} \times \nabla_P T_v) \frac{dP}{P}$$

For simplicity, we assume that $T_v \approx \overline{T}_v$ where $\overline{T}_v$ is the layer-averaged virtual temperature between P_1 and P_2. Under this approximation, the integrals can be simplified, yielding the following expression for the vertical shear of the geostrophic wind:

$$\int_{P_1}^{P_2} \frac{\partial \boldsymbol{V}_g}{\partial P} dP = -\frac{R_d}{f}(\boldsymbol{k} \times \nabla_P \overline{T}_v) \int_{P_1}^{P_2} \frac{dP}{P}$$

$$\Rightarrow \Delta \boldsymbol{V}_g = \frac{R_d}{f} \ln\left(\frac{P_1}{P_2}\right)(\boldsymbol{k} \times \nabla_P \overline{T}_v) \tag{7.14}$$

where the order of the integration has been reversed and $\Delta \boldsymbol{V}_g = \boldsymbol{V}_g(\mathrm{P}_2) - \boldsymbol{V}_g(\mathrm{P}_1)$.

Eq. (7.14) establishes an explicit relationship between the vertical shear of the geostrophic wind and the horizontal virtual temperature gradient. Because this vertical shear depends on the temperature gradient, it is commonly referred to as the **thermal wind vector**. Consequently, Eq. (7.14), which allows us to calculate the thermal wind vector, is known as the **thermal wind equation**.

Defining $\boldsymbol{V}_T = \Delta \boldsymbol{V}_g$ as the thermal wind vector, the component form of Eq. (7.14) can be written as:

$$u_T = -\frac{R_d}{f}\frac{\partial \overline{T}_v}{\partial y}\ln\left(\frac{P_1}{P_2}\right), \qquad v_T = \frac{R_d}{f}\frac{\partial \overline{T}_v}{\partial x}\ln\left(\frac{P_1}{P_2}\right) \tag{7.15}$$

where u_T and v_T are the components of the thermal wind vector.

7.3.2 Properties of the thermal wind vector

Because of the importance of thermal wind balance in atmospheric dynamics, we consider three key applications. First, Eq. (7.15) shows that we can estimate the geostrophic wind at any level using the layer-averaged virtual temperature, provided the geostrophic wind vector is known at a single reference level. For example, if the geostrophic wind vector is known at 850 hPa and the average horizontal virtual temperature gradient in the 850- to 500-hPa layer is available, we can apply Eq. (7.15) to obtain the geostrophic wind at 500 hPa. Box 7.1 illustrates this procedure.

Second, Eq. (7.15) indicates that we can determine the average virtual temperature gradient within a layer if the geostrophic wind is known at the top and bottom of that layer. This is often easier in practice, as radiosonde data can provide the necessary wind observations. Box 7.2 demonstrates this approach.

Third, because the thickness of a layer is proportional to the layer-averaged virtual temperature, we can rewrite the thermal wind equation in terms of thickness. As discussed in Ch. 5, the thickness equation can be expressed as:

$$Z_T = \frac{R_d \overline{T}_v}{g_0}\ln\left(\frac{P_1}{P_2}\right)$$

where $Z_T = Z_2 - Z_1$ represents the thickness of the layer. By substituting the thickness equation into Eq. (7.15) we obtain:

$$\Delta \boldsymbol{V}_g = \frac{R_d}{f}\ln\left(\frac{P_1}{P_2}\right)\left(\boldsymbol{k} \times \nabla_P \overline{T}_v\right) = \frac{g_0}{f}\left(\boldsymbol{k} \times \nabla_P Z_T\right) \tag{7.16}$$

Eq. (7.16) carries two important implications regarding the thermal wind vector. First, the magnitude of the thermal wind vector is proportional to the thickness gradient between the two isobaric surfaces. In other words, a strong thickness gradient corresponds to a strong thermal wind, indicating a large vertical shear of the geostrophic wind. Second, the direction of the thermal wind vector is parallel to the thickness gradient. That is, the thermal wind "flows" parallel to thickness contours, with colder layers (lower thickness) to the left and warmer layers (greater thickness) to the right in the Northern Hemisphere. Because layer thickness is

Box 7.1 Calculating the geostrophic wind using the thermal wind vector

Suppose that the column-averaged virtual temperature in the layer at 43°N between 750 and 500-hPa decreases eastward by 3°C per 100 km. If the 750-hPa geostrophic wind is from the southeast at 20 m/s, let's calculate the geostrophic wind (in unit-vector notation) at 500 hPa.

First, we can convert the 750-mb geostrophic wind into component form by noting that

$$u_{750} = -M \sin\alpha = -\left(20.0\frac{\text{m}}{\text{s}}\right)\sin 135° = -14.1\frac{\text{m}}{\text{s}}$$

$$v_{750} = -M \cos\alpha = -\left(20.0\frac{\text{m}}{\text{s}}\right)\cos 135° = 14.1\frac{\text{m}}{\text{s}}$$

Let's define $\overline{T}_v$to be the average virtual temperature in the layer between pressure $P_1 = 750$ hPa and $P_2 = 500$ hPa. Since the mean temperature in the layer only changes eastward, then $\partial\overline{T}_v/\partial y = 0$, which implies that $u_T = 0$according to Eq. (7.15). The meridional component of the thermal wind is given by:

$$v_T = \frac{R_d}{f}\frac{\partial\overline{T}_v}{\partial x}\ln\left(\frac{P_1}{P_2}\right) = \frac{\left(287\frac{J}{\text{kg}\cdot K}\right)}{2(7.29\times10^{-5}\text{s}^{-1})\sin(43.0°)}\left(\frac{-3°\text{C}}{100\times10^3\ \text{km}}\right)\ln\left(\frac{750\ \text{hPa}}{500\ \text{hPa}}\right) = -34.5\frac{\text{m}}{\text{s}}$$

Therefore, for this problem,

$$u_T = u_{500} - u_{750}$$

$$\Rightarrow u_{500} = u_T + u_{750} = 0 + \left(-14.1\frac{\text{m}}{\text{s}}\right) = -14.1\frac{\text{m}}{\text{s}}$$

$$v_T = v_{500} - v_{750}$$

$$\Rightarrow v_{500} = v_{750} + v_T = 14.1\frac{\text{m}}{\text{s}} + \left(-34.5\frac{\text{m}}{\text{s}}\right) = -20.4\frac{\text{m}}{\text{s}}$$

Therefore, the geostrophic wind at 500 hPa is:

$$\mathbf{V}_g = \left(-14.1\frac{\text{m}}{\text{s}}\right)\mathbf{i} + \left(-20.4\frac{\text{m}}{\text{s}}\right)\mathbf{j}$$

proportional to the layer-averaged virtual temperature, the thermal wind can be used to diagnose the average virtual temperature within the layer, as illustrated in Box 7.2.

Finally, our conclusions regarding the thermal wind vector are summarized in Table 7.2. To fully understand these results, it is useful to compare the thermal wind vector in Eq. (7.15) with the geostrophic wind vector in Eq. (7.4), as both share similar mathematical forms. These concepts will now be applied to observe thermal wind balance in the atmosphere.

7.3.3 Observations of thermal wind balance

In this section, we will explore how thermal wind balance can be observed on upper-air weather charts. Figure 7.12 shows an upper-air analysis at 300 hPa and at 850 hPa for a frontal system that occurred on November 3, 2004, at 12 UTC. This system produced a zonal temperature gradient over Louisiana and eastern Texas, as highlighted in Figure 7.12B, where $\partial\overline{T}_v/\partial x > 0$. According to Eq. (7.15), this temperature gradient implies vertical shear in the geostrophic wind, with the meridional component increasing with height ($v_T > 0$). As a result, the observed wind speed near the Louisiana–Texas border increased from about 15 knots at 850 hPa to more than 85 knots at 300 hPa, as shown in Figure 7.12A.

Box 7.2 Calculate the horizontal temperature gradient using the thermal wind vector

Suppose that a vertical column of the atmosphere at 43°N is initially isothermal from 900 to 500 hPa. Suppose that radiosonde data shows that the geostrophic wind is 10 m/s from the south at 900 hPa, 10 m/s from the west at 700 hPa, and 20 m/s from the west at 500 hPa. Let's calculate the average virtual temperature gradient in the two layers 900 to 700 hPa and 700 to 500 hPa.

First, we can arrange the wind data in component form, as shown in Table 7.1.

Since the thermal wind vector is the vertical shear of the geostrophic wind within a given layer, the thermal wind in the 900- to 700-hPa layer is:

$$\boldsymbol{V}_T = \boldsymbol{V}_{g,700} - \boldsymbol{V}_{g,900} = \left(10.0\frac{\text{m}}{\text{s}}\right)\boldsymbol{i} - \left(10.0\frac{\text{m}}{\text{s}}\right)\boldsymbol{j}$$

Similarly, the thermal wind in the 700- to 500-hPa layer is:

$$\boldsymbol{V}_T = \boldsymbol{V}_{g,500} - \boldsymbol{V}_{g,700} = \left(20.0\frac{\text{m}}{\text{s}}\right)\boldsymbol{i} - \left(10.0\frac{\text{m}}{\text{s}}\right)\boldsymbol{i} = \left(10.0\frac{\text{m}}{\text{s}}\right)\boldsymbol{i}$$

Solving Eq. (7.15) for $\partial\overline{T}_v/\partial y$ in the 700- to 500-hPa layer gives:

$$\frac{\partial\overline{T}_v}{\partial y} = -\frac{fu_T}{R_d \ln\left(\frac{P_1}{P_2}\right)} = -\frac{2(7.29\times10^{-5}\text{s}^{-1})\sin(43.0°)\left(10.0\frac{\text{m}}{\text{s}}\right)}{\left(287\frac{J}{\text{kg}\cdot K}\right)\ln\left(\frac{700\text{ hPa}}{500\text{ hPa}}\right)} = -1.03\times10^{-5}\frac{\text{K}}{\text{m}} = -\frac{1.03°\text{C}}{100\text{ km}}$$

Solving Eq. (7.15) for $\partial\overline{T}_v/\partial x$ gives in the 700- to 500-hPa layer gives:

$$\frac{\partial\overline{T}_v}{\partial x} = \frac{fv_T}{R_d \ln\left(\frac{P_1}{P_2}\right)} = \frac{(10^{-4}\text{s}^{-1})\left(0\frac{\text{m}}{\text{s}}\right)}{\left(287\frac{J}{\text{kg}\cdot K}\right)\ln\left(\frac{700\text{ hPa}}{500\text{ hPa}}\right)} = 0$$

Therefore, the average virtual temperature gradient in the 700- to 500-hPa layer is:

$$\nabla_P\overline{T}_v = \left(-\frac{1.03°\text{C}}{100\text{ km}}\right)\boldsymbol{i}$$

Thus, the virtual temperature decreases from west to east in this layer.

Solving Eq. (7.15) for $\frac{\partial\overline{T}_v}{\partial y}$ in the layer 900 to 700 hPa gives:

$$\frac{\partial\overline{T}_v}{\partial y} = -\frac{fu_T}{R_d \ln\left(\frac{P_1}{P_2}\right)} = -\frac{2(7.29\times10^{-5}\text{s}^{-1})\sin(43.0°)\left(10.0\frac{\text{m}}{\text{s}}\right)}{\left(287\frac{J}{\text{kg}\cdot K}\right)\ln\left(\frac{900\text{ hPa}}{700\text{ hPa}}\right)} = -1.38\times10^{-5}\frac{\text{K}}{\text{m}} = -\frac{1.38°\text{C}}{100\text{ km}}$$

Solving Eq. (7.15) for $\partial\overline{T}_v/\partial x$ gives in the layer 900 to 700 hPa gives:

$$\frac{\partial\overline{T}_v}{\partial x} = \frac{fv_T}{R_d \ln\left(\frac{P_1}{P_2}\right)} = \frac{2(7.29\times10^{-5}\text{s}^{-1})\sin(43.0°)\left(-10.0\frac{\text{m}}{\text{s}}\right)}{\left(287\frac{J}{\text{kg}\cdot K}\right)\ln\left(\frac{900\text{ hPa}}{700\text{ hPa}}\right)} = -1.38\times10^{-5}\frac{\text{K}}{\text{m}} = -\frac{1.38°\text{C}}{100\text{ km}}$$

Therefore, the average virtual temperature gradient in the 900- to 700-hPa layer is:

$$\nabla_P\overline{T}_v = \left(-\frac{1.38°\text{C}}{100\text{ km}}\right)(\boldsymbol{i}+\boldsymbol{j})$$

Thus, the virtual temperature decreases from the southwest to northeast in this layer.

In contrast, away from the frontal zone, the vertical shear is much weaker. This explains why the 300-hPa winds over the southeastern United States are comparatively weaker, as shown in Figure 7.12. This observation confirms that strong horizontal temperature gradients, such as those that found in a frontal zone, are directly tied to strong vertical wind shear.

To examine this balance more formally, we analyze how the geostrophic flow changes near the frontal zone between 850 hPa and 300 hPa. Because the thermal wind vector is parallel to thickness contours, **thickness charts** are especially useful. Figure 7.13A shows the 850-hPa geostrophic flow, 850 to 300-hPa thickness, and 850-hPa isotherms for this event. By comparing Figure 7.12 with Figure 7.13, we see that the frontal zone corresponds to tightly packed thickness contours, which according to the thermal wind equation, requires a strong thermal wind- that is, strong vertical shear of the geostrophic wind.

Additionally, because the thermal wind vector is oriented parallel to the thickness contours, it points northeast across the study region. Comparing the isotherms and thickness contours in Figure 7.13A reveals that colder air lies to the northwest (left) of the thermal wind vector, while warmer air lies to the southeast (right). In other words, the top of the 850- to 300-hPa thickness layer slopes downward to the northwest.

As discussed in Ch. 7.3.2, the geostrophic wind at a higher level can be predicted if the geostrophic wind at a lower level and the thermal wind vector are known. In this case, the geostrophic wind at 300 hPa is determined by vector addition: $\boldsymbol{V}_{g,300} = \boldsymbol{V}_{g,850} + \Delta\boldsymbol{V}_g$.

Two general scenarios arise from this framework. First, if thickness contours are parallel to isohypses, the thermal wind vector must also be parallel to the geostrophic wind vector. In this case, geostrophic winds at different levels remain parallel, but they differ in magnitude. In contrast, if thickness contours intersect isohypses, the thermal wind vector cannot be parallel to the geostrophic wind. Here, *the geostrophic wind changes both speed and direction with height.*

In our study region in eastern Texas, as shown in Figure 7.13, the 850-hPa geostrophic wind crosses the thickness contours, indicating that it is not parallel to the vertical shear of the geostrophic wind. Adding the weak 850-hPa geostrophic wind to the strong thermal wind yields a much stronger 300-hPa geostrophic flow directed toward the northeast. Indeed, Figure 7.13B shows that the geostrophic flow both strengthens and rotates northeastward across this region.

In contrast, in northern Louisiana, as shown in Figure 7.13A, the 850-hPa geostrophic wind is parallel to the 850- to 300-hPa thickness. Consequently, the 300-hPa geostrophic wind, as shown in Figure 7.13B, remains parallel to the 850-hPa flow, differing only in magnitude.

Table 7.1 Geostrophic wind at 900 hPa, 700 hPa, and 500 hPa

Pressure (hPa)	u_g (m/s)	v_g (m/s)
900	0	10
700	10	0
500	20	0

No Permission Required

Table 7.2 Comparing the geostrophic wind with the thermal wind

Geostrophic Wind Vector	Thermal Wind Vector
The geostrophic wind vector is proportional to the geopotential height gradient on a single isobaric surface	The thermal wind vector is proportional to the thickness gradient between two isobaric surfaces
The geostrophic wind flows parallel to isohypses on a single isobaric surface	The thermal wind flows parallel to thickness contours between two isobaric surfaces
Lower isohypses are to the left of the geostrophic wind vector in the Northern Hemisphere	Lower thickness contours are to the left of the thermal wind vector in the Northern Hemisphere

No Permission Required

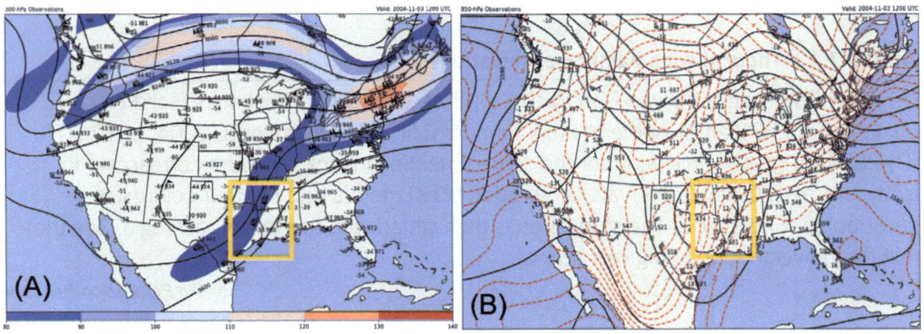

FIGURE 7.12 Thermal wind balance observations 1

(A) 300-hPa observations for November 3, 2004, at 12 UTC. Solid black contours correspond to isohypses at 120-meter intervals centered at 9000 meters, and contour color fill corresponds to isotachs. (B) 850-hPa observations for the same time. Solid black contours correspond to isohypses at 30-meter intervals centered at 1500 meters and dashed red contours correspond to isotherms at 2°C intervals centered at 0°C. The study area is highlighted with the yellow rectangle.

No Permission Required

7.4 Balanced flow within the PBL

So far, we have examined balanced flow in the free atmosphere above the PBL. As mentioned in Ch. 5.4, turbulence within the PBL mixes the slower moving near-surface air with the faster-moving air aloft, reducing wind speeds throughout the layer. Because the influence of turbulent drag decreases with height, wind speeds generally increase as we move upward through the PBL. In this section, we analyze how turbulent drag forces affect atmospheric motion within the PBL, beginning with a qualitative description.

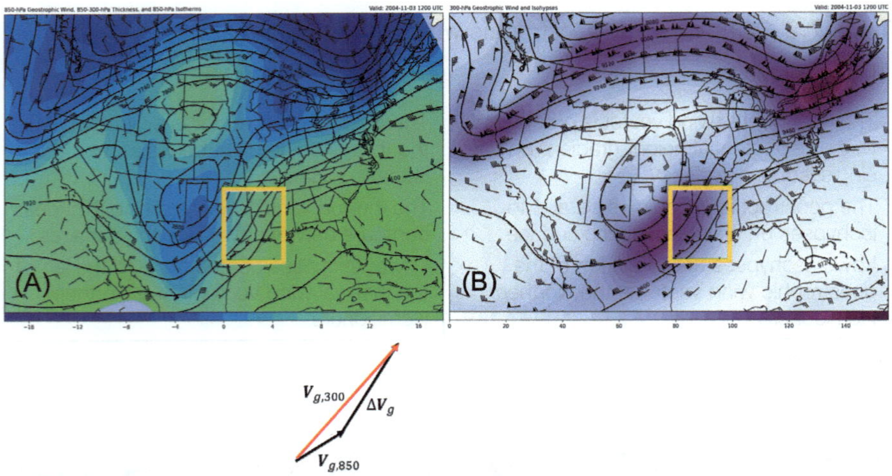

FIGURE 7.13 Thermal wind balance observations 2

(A) 850-hPa geostrophic wind (wind barbs), 850- to 300-hPa thickness (solid black contours), and 850-hPa isotherms (contour color fill) for November 3, 2004, at 12 UTC. (B) 300-hPa geostrophic wind for the same time. Solid black contours correspond to isohypses at 120-meter intervals centered at 9000 meters and contour color fill is geostrophic wind speed. The study area is highlighted with the yellow rectangle. See the text for the vector addition given below.

No Permission Required

7.4.1 Conceptual analysis of PBL flow

In Figure 7.14A, the wind above the PBL is nearly geostrophic- that is, it flows parallel to the isohypses, with the pressure gradient force balanced by the Coriolis force. However, near Earth's surface, the same pressure gradient does not produce the same wind response. As shown in Figure 7.14B,turbulent drag slows the wind, which reduces the Coriolis force. With a weaker Coriolis force, the pressure gradient force is no longer fully balanced, and the wind turns across the isohypses toward lower geopotential heights. Because the Coriolis force always acts 90° to the right of the wind in the Northern Hemisphere, its direction shifts as the wind turns, as shown in Figure 7.14B. Within the PBL, the pressure gradient force is therefore balanced jointly by turbulent drag force and the Coriolis force.

At the surface, this imbalance causes the wind to cross isobars at an angle, known as the **cross-isobar angle** (α). The size of α depends on surface characteristics such as terrain and roughness, but typically averages about 30°. With increasing height above the surface, the effect of friction weakens, α decreases, and the wind approaches the geostrophic direction (parallel to the isohypses, with lower geopotential heights on the left in the Northern Hemisphere). Thus, the cross-isobar angle reflects the combined influence of surface drag, wind speed, and elevation.

The existence of a nonzero cross-isobar angle means that PBL winds flow across isohypses, from higher toward lower pressure. Therefore, in the Northern Hemisphere, surface wind spiral counterclockwise and inward toward low-pressure systems, and clockwise and outward from high-pressure systems.

7.4.2 Quantitative analysis of PBL flow

Let's now quantify the conceptual discussion presented earlier. Rewriting Eq. (7.1) in terms of the horizontal wind vector within the PBL gives:

$$\frac{D\boldsymbol{V}_{\mathrm{PBL}}}{Dt} = -f(\boldsymbol{k} \times \boldsymbol{V}_{\mathrm{PBL}}) - \frac{1}{\rho}\nabla_z P - \frac{1}{\rho}\frac{\partial \boldsymbol{\tau}}{\partial z} \tag{7.17}$$

where $\boldsymbol{V}_{PBL}$ is the horizontal velocity vector within the PBL, and the pressure gradient force is evaluated on a surface of constant geometric height (such as a surface map). As shown in Ch. 5.4, the turbulent drag term can be approximated by expressing it in terms of the vertical wind shear within the PBL:

$$\frac{1}{\rho}\frac{\partial \boldsymbol{\tau}}{\partial z} = w_T \frac{\boldsymbol{V}_{\mathrm{PBL}}}{z_i} \tag{7.18}$$

where z_i is the PBL depth and $w_T$is the turbulent transport velocity. Substituting Eq. (7.18) into Eq. (7.17) gives:

$$\frac{D\boldsymbol{V}_{\mathrm{PBL}}}{Dt} = -f(\boldsymbol{k} \times \boldsymbol{V}_{\mathrm{PBL}}) - \frac{1}{\rho}\nabla_z P - w_T \frac{\boldsymbol{V}_{\mathrm{PBL}}}{z_i}$$

Assuming a steady-state condition (as illustrated in Figure 7.14B), the equation can be written in component form:

$$0 = f v_{PBL} - \frac{\partial P}{\partial x} - w_T \frac{u_{PBL}}{z_i}, \qquad 0 = -f u_{PBL} - \frac{\partial P}{\partial y} - w_T \frac{v_{PBL}}{z_i} \tag{7.19}$$

where u_{PBL} and v_{PBL} are the zonal and meridional components of the balanced PBL wind, respectively. Recall that the geostrophic wind can be expressed as:

$$u_g = -\frac{1}{\rho}\frac{\partial P}{\partial y}, \qquad v_g = \frac{1}{\rho}\frac{\partial P}{\partial x}$$

Rewriting Eq. (7.19) in terms of the geostrophic wind gives:

$$0 = f v_{PBL} - f v_g - w_T \frac{u_{PBL}}{z_i}, \qquad 0 = -f u_{PBL} + f u_g - w_T \frac{v_{PBL}}{z_i}$$

Therefore, the components of the balanced PBL wind can be written as:

$$u_{PBL} = u_g - \frac{w_T v_{PBL}}{f z_i}, \qquad v_{PBL} = v_g + \frac{w_T u_{PBL}}{f z_i} \tag{7.20}$$

The solution for Eq. (7.20) depends on the estimate of the turbulent transport velocity. As shown in Ch. 5.4, when thermal turbulence dominates (such as on sunny, light-wind days when surface heating drives strong convection), the turbulent trans-

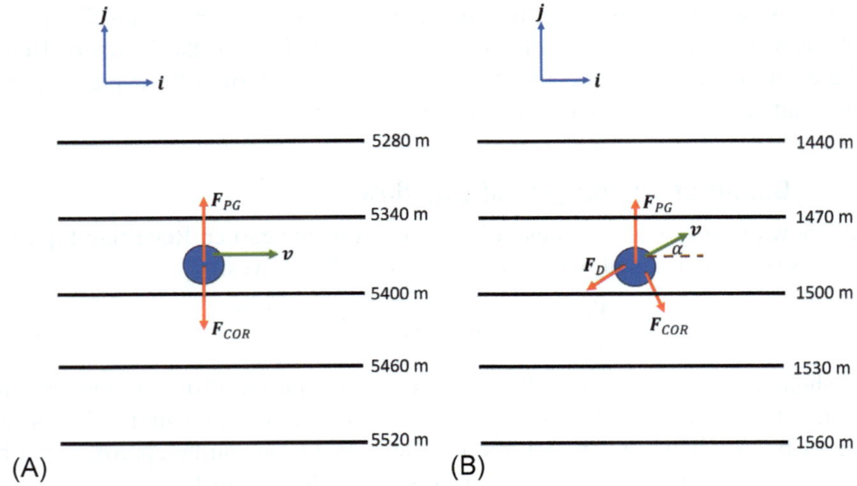

FIGURE 7.14 The impact of the turbulent drag force

(A) The atmospheric flow at the 500-hPa level where the viscous drag force is negligible. (B) The atmospheric flow at the 850-hPa level in the presence of the viscous drag force. For both sections, the green arrow corresponds to the velocity vector of the air parcel. The red vector arrows denote the forces acting on the air parcel, namely the pressure gradient force (F_{PG}), the Coriolis force (F_{COR}), and the turbulent drag force (F_D).

No Permission Required

port velocity can be estimated as $w_T = b_D w_B$, where $b_D = 1.83 \times 10^{-3}$ is a dimensionless quantity and w_B is the buoyancy velocity scale. Since w_B is independent of the steady-state PBL wind, a closed-form solution can be derived. It can be shown (see Problem 12) that the steady-state PBL wind is:

$$u_{PBL} = \frac{u_g - a v_g}{1 + a^2}, \qquad v_{PBL} = \frac{v_g + a u_g}{1 + a^2}, \qquad a = \frac{b_D w_B}{f z_i} \tag{7.21}$$

When mechanical turbulence dominates (such as on windy, overcast days), the turbulent transport velocity can be expressed as: $w_T = C_D V_H$ where C_D is the drag coefficient and V_H is the magnitude of the horizontal wind. Substituting this into Eq. (7.20) gives:

$$u_{PBL} = u_g - \frac{C_D v_{PBL}\sqrt{u_{PBL}^2 + v_{PBL}^2}}{f z_i}, \qquad v_{PBL} = v_g + \frac{C_D u_{PBL}\sqrt{u_{PBL}^2 + v_{PBL}^2}}{f z_i} \tag{7.22}$$

where $V_H = \sqrt{u_{PBL}^2 + v_{PBL}^2}$. Because Eq. (7.22) is a set of coupled nonlinear equations, it cannot be solved analytically, but it can be solved numerically (see Box 7.3 for an iterative method).

In summary, the turbulent drag within the PBL produces subgeostrophic flow. This drag force causes the winds to cross isobars at an angle α, giving the flow a component directed toward lower pressure.

7.4.3 Geostrophic adjustment

In the previous section, we found that air within the PBL eventually reaches a balanced state. An important question arises: *How* does the wind approach this balanced state? To answer this, we examine the time evolution of the wind within the PBL. For boundary layer flow with straight isobars, the equations of motion are:

$$\frac{Du_{PBL}}{Dt} = fv_{PBL} - \frac{1}{\rho}\frac{\partial P}{\partial x} - \frac{C_D u_{PBL}\sqrt{u_{PBL}^2 + v_{PBL}^2}}{z_i}, \qquad \frac{Dv_{PBL}}{Dt} = -fu_{PBL} - \frac{1}{\rho}\frac{\partial P}{\partial y} - \frac{C_D v_{PBL}\sqrt{u_{PBL}^2 + v_{PBL}^2}}{z_i}$$

Here, we assume mechanical turbulence dominates over thermal turbulence. Rewriting in terms fo the geostrophic wind, as shown in Eq. (7.23), gives:

$$\frac{Du_{PBL}}{Dt} = f\left(v_{PBL} - v_g\right) - \frac{C_D u_{PBL}\sqrt{u_{PBL}^2 + v_{PBL}^2}}{z_i}, \qquad \frac{Dv_{PBL}}{Dt} = -f\left(u_{PBL} - u_g\right) - \frac{C_D v_{PBL}\sqrt{u_{PBL}^2 + v_{PBL}^2}}{z_i}. \tag{7.23}$$

Rather than assuming a steady-state solution, we will use an iterative approach to solve Eq. (7.23). To adapt it for iteration, we will approximate the time derivatives in terms of difference quotients, resulting in the following expressions:

Box 7.3 Solving for the steady-state PBL wind using iteration

We can apply an iterative approach to Eq. (7.22) in the following manner.

First, we start with an initial value for $u_{PBL}$and v_{PBL}. One common approach would be $u_{PBL} = 0 = v_{PBL}$ since this would match the speed at Earth's surface. Another simple guess is $u_{PBL} = u_g$ and $v_{PBL} = v_g$ since this would match atmospheric flow above the boundary layer.

Solve for u_{PBL} using the initial values defined in Step 1.

Solve for v_{PBL}using the solution obtained from Step 2.

Repeat Steps 2 and 3 until the solution converges. Typically, this is done by comparing the current solution to the previous iteration. When the current solution differs by less than 10^{-6}, we can say that the solution has converged.

To see how this can be applied, let's suppose that $u_g = 20\ \mathrm{m/s}$, $v_g = 0\ \mathrm{m/s}$, $C_D = 0.003$, $z_i = 1.5$ km, and $f = 10^{-4}$. The MATLAB script to apply this iteration procedure is given below.

Continued

Box 7.3 Solving for the steady-state PBL wind using iteration—cont'd

```
f = 10^(-4);           % Coriolis parameter in s^(-1)
u_g = 20.0;            % Pressure gradient in x-direction
v_g = 0.0;             % Pressure gradient in y-direction in Pa/m
z_i = 1500;            % Depth of the layer in meters
C_D = 0.003;           % Drag coefficient

u_old = 0;             % The initial guess for the zonal wind
v_old = 0;             % The initial guess for the meridional wind
du = 1;                % Initial value of iteration parameter
dt = 60;               % Time step of 1 minute

N=0;
while du > 10^(-6)
    w_t = C_D*sqrt(u_old^2+v_old^2);
    u_new = u_old + dt*(f*(v_old-v_g) - w_t*u_old/z_i);
    v_new = v_old + dt*(-f*(u_old-u_g) - w_t*v_old/z_i);
    du = sqrt(abs(u_new-u_old)^2+(v_new-v_old)^2);
    u_old = u_new;
    v_old = v_new;
    N=N+1;
end

V_PBL = sqrt(u_new^2+v_new^2);
angle = atand(v_new/u_new);

disp(['The magnitude of the PBL wind is ', num2str(V_PBL), ' m/s.'])
disp(['The direction of the PBL wind is ', num2str(angle), ' degrees.'])
```

The above MATLAB script gives a PBL wind of 18.73 m/s with a cross-isobar angle of 20.5°.

$$\frac{\Delta u_{PBL}}{\Delta t} \approx f\left(v_{PBL} - v_g\right) - \frac{C_D u_{PBL}\sqrt{u_{PBL}^2 + v_{PBL}^2}}{z_i}, \qquad \frac{\Delta v_{PBL}}{\Delta t} \approx -f\left(u_{PBL} - u_g\right) - \frac{C_D v_{PBL}\sqrt{u_{PBL}^2 + v_{PBL}^2}}{z_i}$$

where $\Delta u_{PBL} = u_{PBL}(t+\Delta t) - u_{PBL}(t)$ and $\Delta v_{PBL} = v_{PBL}(t+\Delta t) - v_{PBL}(t)$.[3] Rearranging yields the time-stepping formulas:

$$u_{PBL}(t+\Delta t) \approx u_{PBL}(t) + \Delta t\left[f\left(v_{PBL} - v_g\right) - \frac{C_D u_{PBL}\sqrt{u_{PBL}^2 + v_{PBL}^2}}{z_i}\right], \qquad v_{PBL}(t+\Delta t) \approx v_{PBL}(t) + \Delta t\left[-f\left(u_{PBL} - u_g\right) - \frac{C_D v_{PBL}\sqrt{u_{PBL}^2 + v_{PBL}^2}}{z_i}\right]$$

[3] For simplicity, the effect of advection have been neglected.

These can be solved iteratively:

1. Begin with initial values for $u_{PBL}$and v_{PBL}. For simplicity, assume $u_{PBL} = 0 = v_{PBL}$, matching the surface winds.
2. Compute $u_{PBL}(t + \Delta t)$ and $v_{PBL}(t + \Delta t)$ using the initial values defined in Step 1.
3. Let $u_{PBL}(t) = u_{PBL}(t + \Delta t)$ and $v_{PBL}(t) = v_{PBL}(t + \Delta t)$. In other words, replace the old values with the newly calculated ones.
4. Repeat Steps 2 and 3 until the solution converges, typically defined when the difference between success iterations is less than 10^{-6},

For our toy model with $f = 10^{-4} s^{-1}$, $u_g = 15.0$ m/s, $v_g = 0$ m/s, $z_i = 1500$ m, and $C_D = 0.003$, the PBL converges to a magnitude of 9.81 m/s at a direction of 11.1°. The MATLAB script solving Eq. (7.23) is provided below:

```
f = 10^(-4);             % Coriolis parameter in s^(-1)
u_g = 10.0;              % Zonal geostrophic wind component in m/s
v_g = 0.0;               % Meridional geostrophic wind component in m/s
z_i = 1500;              % Depth of the PBL in meters
C_D = 0.003;             % Drag coefficient

u_old = 0;               % The initial guess for the zonal wind
v_old = 0;               % The initial guess for the meridional wind
du = 1;                  % Initial value of iteration parameter
dt = 60;                 % Time step of 1 minute

N=0;
while du > 10^(-6)
    w_t = C_D*sqrt(u_old^2+v_old^2);
    u_new = u_old + dt*(f*(v_old-v_g) - w_t*u_old/z_i);
    v_new = v_old + dt*(-f*(u_old-u_g) - w_t*v_old/z_i);
    du = sqrt(abs(u_new-u_old)^2+(v_new-v_old)^2);
    u_old = u_new;
    v_old = v_new;
    N=N+1;
end

V_PBL = sqrt(u_new^2+v_new^2);      % Magnitude of converged PBL wind
angle = atand(v_new/u_new);          % Direction of converged PBL wind

disp(['The magnitude of the PBL wind is ', num2str(V_PBL), ' m/s.'])
disp(['The direction of the PBL wind is ', num2str(angle), ' degrees.'])
```

Figure 7.15 illustrates how the PBL evolves toward balance for different drag coefficients. For small drag coefficients, the PBL wind oscillates around the geostrophic value before settling into balance. This behavior reflects Newton's laws: an air parcel accelerates toward low pressure under the pressure gradient force, while turbulent drag damps the acceleration and the Coriolis force deflects the flow to the right. Momentum causes the parcel to overshoot, then undershoot until drag

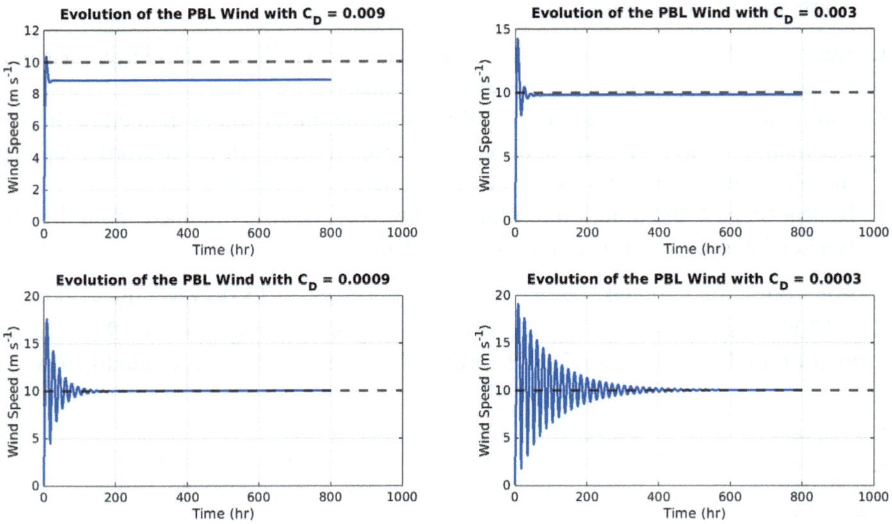

FIGURE 7.15 Evolution of PBL flow as a function of the drag coefficient

The evolution of the PBL flow for four different drag coefficients. The PBL wind at a given time is the blue curve, whereas the geostrophic wind is displayed with the black dashed lines.

No Permission Required

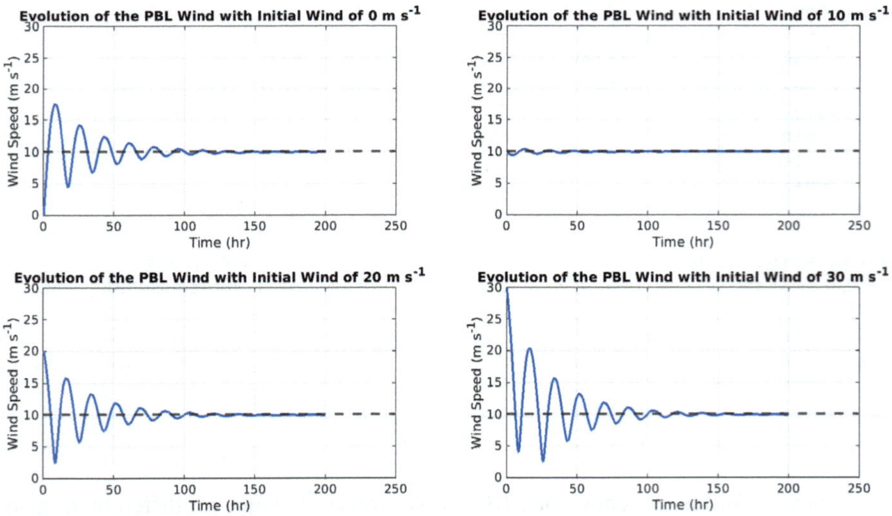

FIGURE 7.16 Evolution of PBL flow as a function of the initial velocity

The evolution of the PBL flow for four different initial velocities. The PBL wind at a given time is the blue curve, whereas the geostrophic wind is displayed with the black dashed lines.

No Permission Required

and the Coriolis force gradually damp the oscillations. The number of oscillations and the adjustment time depend on the drag coefficient.

For all drag coefficients, the wind eventually reaches balance with the imposed pressure gradient. This tendency holds regardless of initial wind conditions, as shown in Figure 7.16. When the initial flow is near geostrophic, little adjustment is required. When the flow is strongly unbalanced, the adjustment is larger and slower. If the initial flow is supergeostrophic, the process involves a significant energy loss, whereas if it is subgeostrophic, energy is added as the pressure gradient force performs work on the air.

The toy model describes only half the total response of the real atmosphere. In reality, both the pressure gradient and the flow adjust simultaneously. For example, suppose the wind is initially geostrophic and the horizontal pressure gradient increases. The pressure gradient force temporarily exceeds the Coriolis force, producing a net acceleration toward low pressure. The resulting ageostrophic component redistributes mass, weakening the geopotential height field and reducing the geostrophic wind. As winds respond to the new pressure gradient, both the flow and height field adjust. These processes are collectively known as **geostrophic adjustments and play a central role in how the atmosphere responds to changes in pressure and temperature**.

Questions

1. The Buys Ballot's law provides a rule of thumb regarding wind and pressure aloft in the Northern Hemisphere: "If we stand with our backs to the direction from which the clouds are moving, lower pressure aloft will always be to our left and higher pressure to our right."
   a. Use free-body diagram analysis to justify the Buys Ballot's law. How would this statement be rewritten for the Southern Hemisphere?
   b. Suppose we observe that the winds aloft are blowing from the northwest in the Northern Hemisphere. According to the Buys Ballot's law, what would be the pressure pattern aloft? Defend your answer.
   c. Can the Buys Ballot's law be applied within the PBL? Defend your answer.
2. Use the thermal wind equation to answer the following questions:
   a. Use the thermal wind equation to prove that the presence of a poleward-directed temperature gradient across the globe must be reflected by the presence of westerly vertical shear in both hemispheres.
   b. Use thermal wind balance to explain why there should be strong geostrophic winds directly above a cold front.
3 Consider the 500-hPa chart shown in Figure 7.17.
   a. Estimate the magnitude of the geostrophic wind (in knots) at Charleston, SC. Note that Charleston, SC, is located at a latitude of 32.78°N.

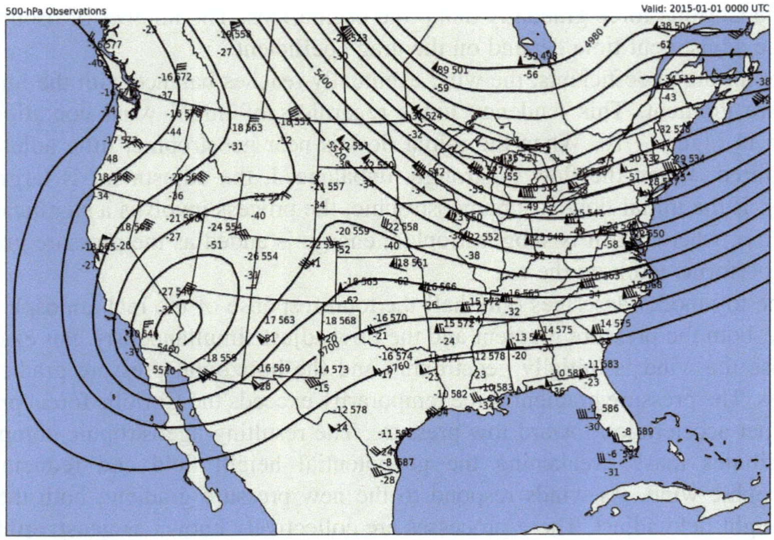

FIGURE 7.17 500-hPa observations

500-hPa observations of the continental United States on January 1, 2015, at 00Z.

No Permission Required

b. Estimate the magnitude of the geostrophic wind (in knots) at Dayton, OH. Note that Dayton, OH, is located at a latitude of 39.76°N.

4. Consider a closed low on a 500-hPa isobaric chart with a radius of curvature of 600 km. Suppose there is an air parcel located north of the closed high.

a. Draw a free-body diagram of the forces acting on the air parcel (i) from the reference frame of the air parcel and (ii) from the reference frame associated with the center of the low.

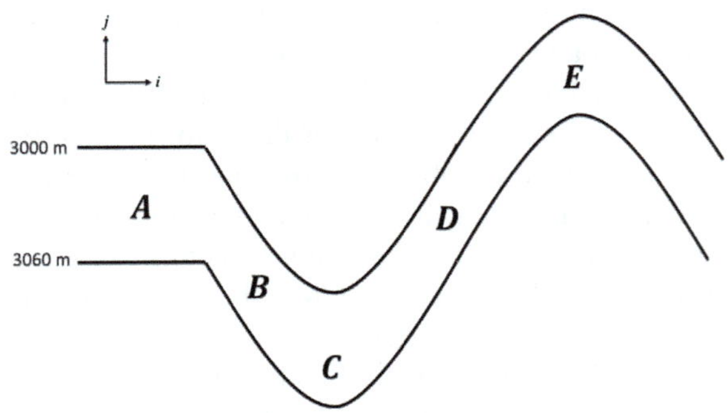

FIGURE 7.18

A trough–ridge couplet.

No Permission Required

b. Suppose that the geopotential height gradient is 100 meters per 1000 km (ie, the geopotential height increases by 100 meters for every 1000 km away from the center of the low). Calculate the geostrophic wind speed and the gradient wind speed at the location of the air parcel. Explain why the gradient wind is smaller than the geostrophic wind. You may assume that $f = 10^{-4}\ \text{s}^{-1}$.

5. Consider a closed high on a 500-hPa isobaric chart with a radius of curvature of 500 km. Suppose there is an air parcel located west of the closed high.

a. Draw a free-body diagram of the forces acting on the air parcel (i) from the reference frame of the air parcel and (ii) from the reference frame associated with the center of the high.

b. If the Coriolis parameter is $f = 10^{-4}\ \text{s}^{-1}$, calculate the largest possible pressure gradient for this system. Express your answer in terms of meters per 100 km.

c. Suppose that the geopotential height gradient is 100 meters per 1000 km (ie, the geopotential height decreases by 100 meters for every 1000 km away from the center of the high). Calculate the geostrophic wind speed and the gradient wind speed at the location of the air parcel. Explain why the gradient wind is larger than the geostrophic wind. Assume that $f = 10^{-4}\ \text{s}^{-1}$.

6. Consider the schematic of a trough-ridge couplet at the 500-hPa level with a radius of curvature of 3000 km as shown in Figure 7.18. Assume that the zonal component of the geostrophic wind is 20.0 m/s at point A and that the magnitude of the geostrophic wind is 25.0 m/s from points B through E. Assume that $f = 10^{-4}\ \text{s}^{-1}$.

a. If the wind is in geostrophic balance at point A, determine the isohypse spacing between the 3,000 and 3,060 meters isohypses at point A and at point B. Express your answer in kilometers.

b. If the wind is in gradient balance at point C, calculate the ageostrophic wind vector (in m/s) at point C and express your answer in unit-vector notation. As the air parcel moves from point B to point C, is the air parcel decelerating or accelerating?

c. If the wind is in gradient balance at point E, calculate the ageostrophic wind vector (in m/s) at point E and express your answer in unit-vector notation. As an air parcel moves from point D to point E, is the air parcel decelerating or accelerating?

7. A hurricane can be pictured as a closed low-pressure system in which the winds are in gradient balance. Suppose that an Atlantic hurricane is located over the Caribbean Sea at a latitude of 26°N. Balanced wind speed of 60 m/s is found at 950-hPa at a station 200 km from the center. Based on this information, calculate the ageostrophic wind speed at this station. Are the winds highly supergeostrophic or subgeostrophic?

Table 7.3 Radiosonde data

TTAA	52121	72208	99022	03656	33503	00190	03058	00521	92818
00258	02523	85494	01558	33524	70034	05979	32525	50559	21784
33045	40719	34769	31058	30915	43775	29076	25037	47982	28596
20183	50381	27596	15368	54580	28077	10623	59379	27560	88130
59579	26575	77188	28103	42017	31313				

No Permission Required

8. Consider the radiosonde data from Charleston, SC (at a latitude of 32.78°N) as shown in Table 7.3. Suppose that the wind is in geostrophic balance between 700-hPa and 500-hPa.

a. Calculate the thermal wind vector in this layer. Express your answer in terms of wind speed (in m/s) and in meteorological direction.

b. Calculate the mean horizontal virtual temperature gradient in this layer. Express your answer in unit-vector notation and in terms of °C per 100 km.

9. Suppose that a vertical column of the atmosphere at 43°N between 850-hPa and 700-hPa. Suppose that the geostrophic wind is 10 m/s from the east and that the average virtual temperature in this layer is: $\nabla \overline{T}_v = \left(\frac{4.50\ K}{1.00\ \text{km}}\right)(-\boldsymbol{i} + \boldsymbol{j})$

Calculate the geostrophic wind at 700-hPa in m/s. Express your answer in terms of wind speed (in m/s) and in meteorological direction.

10. A radiosonde measurement at an upper-air station indicates that the 925-hPa geostrophic wind is 10 m/s from 225° and the 700-hPa geostrophic wind is 30.0 m/s from 185°. Calculate the layer-averaged horizontal virtual temperature gradient. Express your answer in magnitude in terms of °C per 100 km and in meteorological direction.

11. Consider an air parcel within the PBL in the presence of straight isobars. Assume that the PBL is dominated by convective turbulence such that $w_T = b_D w_B$ where $b_D = 1.83 \times 10^{-3}$ is a dimensionless coefficient and wB is the buoyancy velocity scale.

a. Derive Eq. (7.21) $u_{PBL} = \frac{u_g - a v_g}{1 + a^2}$, $v_{PBL} = \frac{v_g + a u_g}{1 + a^2}$, $a = \frac{b_D w_B}{f z_i}$.

b. Suppose that the PBL depth is 1.5 km, the Coriolis parameter f = $10^{-4}\ \text{s}^{-1}$, the buoyancy velocity scale is 50 m/s, and $\boldsymbol{V}_g = \left(10.0\frac{\text{m}}{\text{s}}\right)\boldsymbol{i} + \left(3.00\frac{\text{m}}{\text{s}}\right)\boldsymbol{j}$.

Determine the steady-state PBL flow and the cross-isobar angle.

12. Consider a closed high on a surface chart with a radius of curvature R. Suppose there is an air parcel located south of the surface high pressure.

a. Show that the equations of motion for the zonal and meridional wind components can be written as:

$$\frac{Du}{Dt} = fv - \frac{1}{\rho}\frac{\partial P}{\partial x} - w_T\frac{u}{z_i} + \frac{v}{R}\sqrt{u^2 + v^2}, \quad \frac{Dv}{Dt} = -fu - \frac{1}{\rho}\frac{\partial P}{\partial y} - w_T\frac{v}{z_i} - \frac{u}{R}\sqrt{u^2 + v^2}$$

b. If we approximate $Du/Dt \approx \Delta u/\Delta t$ and $Dv/Dt \approx \Delta v/\Delta t$, show that the above set of equations can be written as:

$$u(t + \Delta t) = u(t) + \Delta t\left[fv(t) - \frac{1}{\rho}\frac{\partial P}{\partial x} - w_T\frac{u(t)}{z_i} + \frac{v(t)}{R}\sqrt{u(t)^2 + v(t)^2}\right] \quad v(t + \Delta t) = v(t) + \Delta t\left[-fu(t) - \frac{1}{\rho}\frac{\partial P}{\partial y} - w_T\frac{v(t)}{z_i} - \frac{u(t)}{R}\sqrt{u(t)^2 + v(t)^2}\right]$$

c. Consider an air parcel that is located south of the closed low. Suppose that the geostrophic wind has a magnitude of 10.0 m s^{-1}, the radius of curvature is 400 km, the PBL depth is 1 km, the Coriolis parameter is $f = 10^{-4}\ \text{s}^{-1}$, and the drag coefficient is 0.02. Use the iterative method to find the PBL wind speed and meteorological direction. Investigate how the PBL wind vector changes as (i) the drag coefficient decreases and (ii) as the PBL depth increases.

CHAPTER

Thermodynamic processes in the atmosphere

8

Chapter outline

In Chs. 5 to 7, we applied Newtonian mechanics to atmospheric motion. Chs. 5 and 6 showed that the main forces affecting atmospheric flow are the gravitational force, the pressure gradient force, the turbulent drag force, the Coriolis force, and the centrifugal force due to Earth's rotation. Ch. 7 demonstrated that synoptic-scale flow is approximately geostrophic in the horizontal and hydrostatic in the vertical. Together, these two balance conditions yield thermal wind balance, which links the wind, height, and temperature fields and thus connects thermodynamic and dynamic processes in the atmosphere. Having examined the dynamics of atmospheric motion, we now turn to the thermodynamic processes that accompany them. We begin with the 1st law of thermodynamics, which extends the principle of energy conservation to include heat transfer.

An Introduction to Synoptic-Dynamic Meteorology. https://doi.org/10.1016/B978-0-443-24858-0.00008-3

8.1 The 1st law of thermodynamics

In introductory physics, the mechanical energy of a system consists of kinetic energy (the energy due to the system's overall motion) and potential energy (the energy due to forces acting on the system). Because every physical system is made of a large collection of molecules and atoms, we can also describe the system's energy in terms of the kinetic and potential energy of its constituent molecules. This combined energy is called **internal energy**, defined as the sum of the total molecular kinetic energy and molecular potential energy within a system.

Molecular kinetic energy includes all forms of molecular motion, primarily translational kinetic energy (from the random linear motion of molecules), rotational kinetic energy (from the rotation of molecules), and vibrational kinetic energy (from the vibration of molecules about their centers of mass).[1] Molecular potential energy consists of **chemical energy** stored in the bonds *within* a molecule and **latent energy** stored in the bonds *between* molecules. In atmospheric science, latent energy is the most important form of molecular potential energy because it is directly linked to the phase changes of substances, such as the transition between water vapor, liquid water, and ice.

The 1st law of thermodynamics describes the mechanisms that can change a system's internal energy. It can be expressed as follows:

Changes in the internal energy of a system are due to energy transfer via heat and/or work.

Mathematically:

$$\Delta E = W + Q \tag{8.1}$$

where ΔE is the change in internal energy, W is the work done **on** the system, and Q is the heat transferred **to** the system. Two key points should be emphasized:

1. Internal energy refers to the energy that a system *possesses*, whereas heat and work are *transfers* of energy between the system and its environment.
2. The signs of Q and W follow physical conventions. Thus, $Q > 0$ indicates that the system gains energy via heat (i.e. heat is absorbed), whereas $Q < 0$ signifies that the system loses energy via heat (i.e. heat is released). Similarly, $W > 0$ indicates that the system gains energy from work done on it by the surroundings, whereas $W < 0$ implies that the system loses energy from work it does on the surroundings.

In the next section, we will examine the properties of internal energy, work, and heat in greater detail, focusing on their application to gases in Earth's atmosphere.

---

[1]The molecular kinetic energy also includes contributions associated with electron motion, electron spin, and nuclear spin. However, these contributions can be considered negligible for our considerations.

8.1.1 Internal energy

In synoptic meteorology, the physical system of interest is Earth's atmosphere. Therefore, it is useful to determine the internal energy of the gases within it. As discussed in Ch. 1, the kinetic theory of gases describes an ideal gas as a collection of point-like particles in random motion. The kinetic theory of gases shows that the energy associated with an ideal gas can be expressed as:

$$E_{mol} = \frac{1}{2}Mv_{rms}^2, \qquad v_{rms} = \sqrt{\frac{3RT}{M_{mol}}}$$

where M is the total mass of the gas, R is the universal gas constant, M_{mol} is the molar mass of the gas, and T is the temperature in Kelvin.

Since the kinetic theory of gases assumes no molecular interactions except during collisions, the only contribution to the molecular energy comes from the translational kinetic energy. In other words, for a gas composed of a collection of noninteracting particles, the total energy is entirely due to the kinetic energy of their straight-line motion. Moreover, according to the kinetic theory of gases, the internal energy of the gas depends directly on its temperature.

The kinetic theory model presented in Ch.1 applies to monatomic gases such as helium and argon. However, over 98% of Earth's atmosphere consists of nitrogen (N_2)and oxygen (O_2), which are diatomic gases. Fortunately, the results from kinetic theory can be generalized for polyatomic gases using the **equipartition theorem**. Although its derivation is beyond the scope of this text, the equipartition theorem states:

Each degree of freedom associated with molecular motion contributes $k_BT/2$ ***to the total energy of a thermodynamic system***

where:

$$k_B = \frac{R}{N_A} = \frac{8.314\,J\,mol^{-1}\,K^{-1}}{6.022 \times 10^{23}\,mol^{-1}} = 1.38 \times 10^{-23}\frac{J}{K}$$

is known as **Boltzmann's constant** and a **degree of freedom** refers to an independent way in which a molecule can possess energy. For example, for a monatomic ideal gas such as helium, the molecular energy can be expressed as:

$$E_{mol} = \frac{1}{2}Mv_{rms}^2, \qquad v_{rms} = \overline{v^2}^{1/2}$$

Since $v_{rms} = \overline{v_x^2} + \overline{v_y^2} + \overline{v_z^2}$, we can write:

$$E_{mol} = \frac{1}{2}M\left(\overline{v_x^2} + \overline{v_y^2} + \overline{v_z^2}\right)$$

This shows that a monatomic molecule has three degrees of freedom: motion in the x-direction, motion in the y-direction, and motion in the z-direction, as shown in Figure 8.1A.

When applying this to the gases in Earth's atmosphere, we find that diatomic molecules have additional modes of energy storage beyond translational motion.

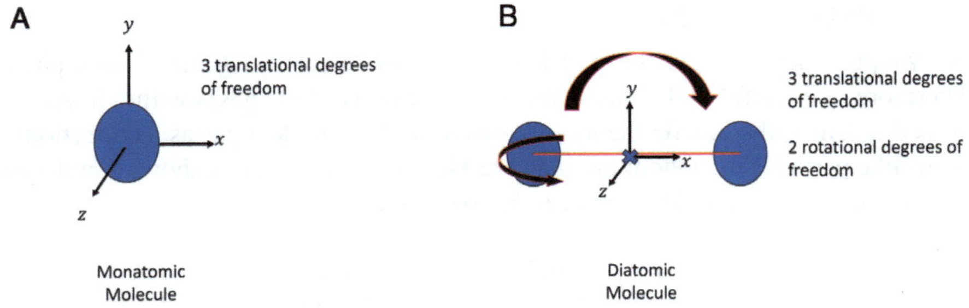

FIGURE 8.1 Degrees of freedom for monatomic and diatomic molecules

(A) A monatomic molecule (i.e. a single gas atom), such as helium and argon. Since there are three independent directions of motion, there are three degrees of freedom for this atom. (B) A diatomic molecule (such as nitrogen and oxygen) composed of two atoms joined by a molecular bond (shown in red) with the center of mass between the atoms (shown in blue). For typical tropospheric temperatures, there are three translational degrees of freedom associated with the translational motion of the center of mass, and there are two rotational degrees of freedom associated with rotation about the center of mass.

No Permission Required

These include rotational kinetic energy about their center of mass, and at very high temperatures, vibrational energy of the molecular bonds. At typical tropospheric temperatures, diatomic molecules exhibit five degrees of freedom: three translational and two rotational[2] as shown in Figure 8.1B.

According to the equipartition theorem, the internal energy of atmospheric gases in the troposphere is therefore:

$$E = \frac{5}{2} N k_B T$$

where N is the number of gas molecules. Therefore, for a diatomic ideal gas, internal energy is directly proportional to temperature.

8.1.2 Thermodynamic work

In introductory physics, the work W done on an object by a force is defined as

$$W = \int \boldsymbol{F} \cdot d\boldsymbol{r}$$

[2]A curious student may be interested on why there are only two rotational degrees of freedom associated with a diatomic gas. In a diatomic molecule, the bond holding the two atoms together is normally quite stiff, so it can be depicted as two nuclei held together by a rigid rod. If we neglect translational motion, then the configuration of the system in space depends only on two angles: the polar angle (which is defined from the positive z-axis) and the azimuthal angle (which is defined from the positive x-axis).

where $\boldsymbol{F}$ is the force vector and $\boldsymbol{dr}$ is the infinitesimal displacement vector. The scalar (dot) product in this definition shows that work is a scalar quantity that can be either positive or negative. From the properties of the scalar product, $W < 0$ when the force is directed opposite to the displacement. For example, the work done by turbulent drag force acting on an air mass is negative, which leads to its deceleration. Conversely, $W > 0$ when the force is directed parallel to the displacement. An example of this is the work done by the pressure gradient force acting on an air mass, which accelerates the air mass toward lower pressure.

Using the definition of the scalar product, the expression for work can also be written as:

$$W = \int F_r \, dr$$

where F_r is the component of the force that is parallel to the displacement.

A common form of work in the atmosphere involves the compression or expansion of air parcels as they move through their environment. This process changes the parcel's volume, implying that work is done during the process.

From Ch. 1, pressure on an air parcel is defined as the force acting perpendicular to the parcel's surface per unit area. Therefore, the work done can be written as:

$$W = \int F_r dr = -\int PAdr$$

The negative sign indicates that compression of the parcel increases its energy. Since compression or expansion changes its volume, then $dV = A\,dr$. Therefore, the work done can be rewritten as:

$$W = -\int_{V_i}^{V_f} P\,dV \tag{8.2}$$

Because this form of work depends on the thermodynamic properties of the system, it is called **thermodynamic work**.

Thermodynamic work commonly occurs when air parcels rise or sink in the atmosphere. For example, as an air parcel rises, it moves into regions of lower pressure, causing it to expand to match the surrounding pressure as shown in Figure 8.2. This expansion means the parcel does work on the environment so $W < 0$. Conversely, when an air parcel sinks, the higher surrounding pressure compresses it, meaning work is done on the parcel by the environment so $W > 0$ as shown in Figure 8.2. In the next section, we will examine energy transfer through heat.

8.1.3 Sensible heat and latent heat

As noted at the beginning of this section, heat is a form of energy transfer between a system and its environment. Whereas work involves the direct application of forces

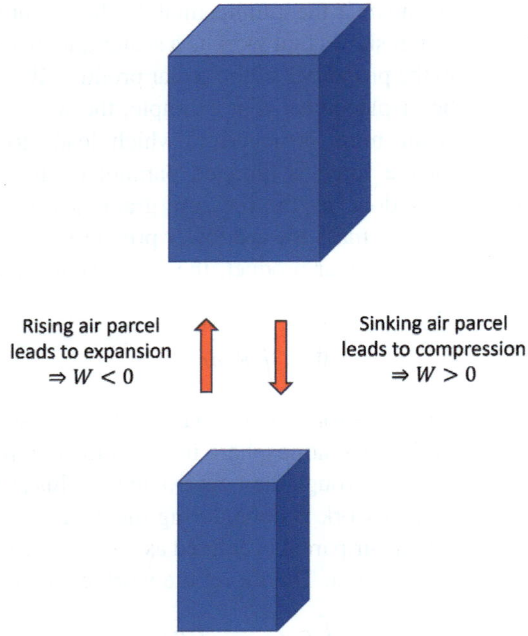

FIGURE 8.2 Schematic of rising air parcels and sinking air parcels

An air parcel rising in Earth's atmosphere and one sinking in Earth's atmosphere. Notice that the rising air parcel expands (indicating that $W < 0$), whereas the sinking air parcel is compressed (indicating that $W > 0$).

No Permission Required

on a system, heat transfer occurs for two reasons: (1) a temperature difference between the system and its environment (referred to as **sensible heating**) and (2) a phase change within the system (referred to as **latent heating**).

Regarding sensible heating, we know from common experience that some substances require more heat to produce a given temperature change. The total temperature change depends on the type of substance; for example, more heat is required to raise the temperature of water than an equal mass of air. The total temperature change also depends upon the mass; for example, a gallon of water requires more heat to warm than a cup of water.

An object's sensitivity to heat transfer is quantified by its **specific heat *c*,** defined as the amount of energy required to raise the temperature of 1 gram of a substance by 1°C. This gives the expression:

$$c = \frac{1}{m}\frac{dQ}{dT} \tag{8.3}$$

where Q is the heat exchange (in Joules), m is the mass of the object (in kilograms), and T is the temperature (in Kelvin). Thus, the units of specific heat are J kg^{-1} K^{-1}. Rearranging Eq. 8.3 and integrating gives:

$$dQ = mcdT$$

$$\Rightarrow Q = \int_{T_i}^{T_f} mc\, dT$$

For atmospheric gases, c can be treated as constant over a wide temperature range, simplifying to:

$$Q = mc(T_f - T_i) = mc\Delta T \quad (8.4)$$

Latent heat is associated with phase transitions. During a phase change, the temperature remains constant, yet energy must still be transferred into or out of the system for the change to occur. This can be best demonstrated by examining the properties of water.

In ice, the molecules are locked in a rigid lattice. Melting requires energy absorption from the environment to break the lattice bonds, while freezing requires the release of energy as molecules slow and bond into the lattice. A similar principle applies to the liquid-vapor transition. In liquid water, molecules move more freely than in a solid but are held together by cohesive forces. Evaporation requires energy to weaken these forces, allowing molecules to escape into the vapor phase. Condensation releases energy as vapor molecules lose kinetic energy and re-establish cohesive bonds.

In both melting and evaporation, energy is absorbed from the environment, cooling it. In freezing and condensation, energy is released into the environment, warming it. In all cases, the change in the system's latent energy is referred to as latent heating.

Latent heat is defined as:

$$L = \frac{Q}{m} \quad (8.5)$$

with units J/kg. The amount of latent heat varies depending on the specific type of phase transition and the properties of the substance involved. The term **latent heat of fusion** refers to the solid-liquid phase transition, while the **latent heat of vaporization** refers to the liquid-gas phase transition.

Figure 8.3 summarizes the phase transitions of water and the latent heat associated with each phase transition. In synoptic meteorology, the most common phase transitions are freezing, melting, evaporation, and condensation. However, under certain weather conditions – such as low relative humidity, dry winds, and high altitudes – snow and ice may bypass the liquid phase and turn directly into vapor (a process known as **sublimation**), or vapor may form ice or snow directly (a process known as **deposition**). The latent heat of sublimation for water is approximately 2838 kJ/kg.

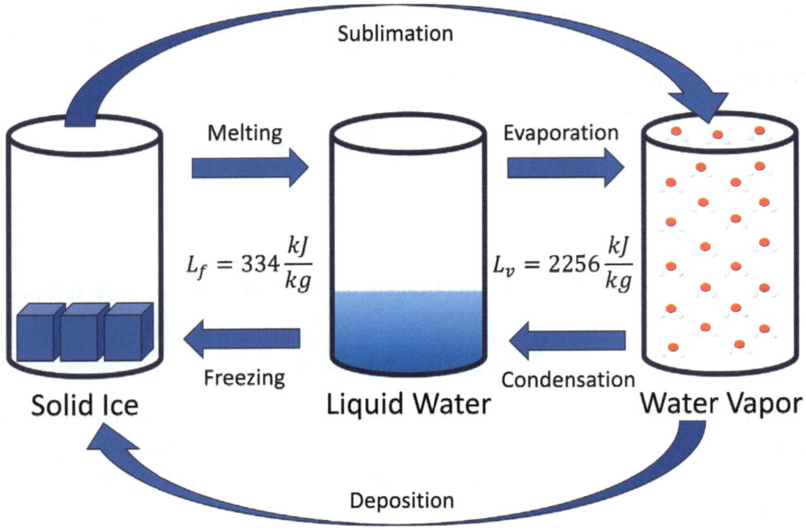

FIGURE 8.3 Phase transitions of water

Schematic of the phase transitions associated with water. Heat is absorbed by the substance for melting, evaporation, and deposition, whereas heat is released by the substance for freezing, condensation, and deposition. The latent heat of fusion and the latent heat of vaporization for water are given at standard temperature (0°C) and standard pressure (100 kPa).

No Permission Required

8.2 Common thermodynamic processes

In the previous section, we defined the 1st law of thermodynamics, along with internal energy, work, and heat. We now apply these principles to an air mass in the atmosphere. When external forces act on the air mass through its interaction with the environment, its internal energy changes as a result of work done on it. If the air mass is treated as an open system, energy can also be exchanged with the environment in the form of heat. Therefore, according to the 1st law of thermodynamics, the change in internal energy of the air mass equals the sum of the work done on the air mass and the heat exchanged with its surroundings. In this section, we apply the 1st law of thermodynamics to thermodynamic processes that are common in synoptic meteorology, focusing on isobaric and adiabatic processes.

8.2.1 Isobaric processes

An **isobaric process** is one in which the pressure remains constant. Such processes are common in Earth's atmosphere because air parcels adjust to the environmental pressure as they move. Rising parcels expand, while sinking parcels contract to match the surrounding pressure. As shown in Section 8.1, the change in internal energy for a diatomic ideal gas is given by:

$$\Delta E = \frac{5}{2} N k_B \Delta T$$

The work done on the gas is:

$$W = -\int_{V_i}^{V_f} P dV = -P(V_f - V_i) = -P\Delta V$$

For an ideal gas undergoing an isobaric process, volume is proportional to temperature, allowing us to write:

$$W = -P\Delta V = -N k_B \Delta T$$

By the 1st law of thermodynamics, the heat exchange between the gas and its environment is given by:

$$\Delta E = Q + W$$

$$\Rightarrow Q = \Delta E - W$$

$$\Rightarrow Q = \frac{5}{2} N k_B \Delta T - (-N k_B \Delta T) = \frac{7}{2} N k_B \Delta T$$

Comparing with Eq. (8.4), the specific heat for a diatomic ideal gas undergoing an isobaric process is:

$$m c_P = \frac{7}{2} N k_B$$

$$\Rightarrow c_P = \frac{7}{2} \frac{N k_B}{m} \tag{8.6}$$

Here, the subscript P indicates that this specific heat is evaluated at constant pressure, which is why we refer to c_P as the **isobaric-specific heat**.

Using $k_B = R/N_A$ and $N = nN_A$, the isobaric-specific heat can be written as:

$$c_P = \frac{7}{2} \frac{N k_B}{m} = \frac{7}{2} \frac{N}{m} \left(\frac{R}{N_A} \right) = \frac{7}{2} \frac{nR}{m}$$

Since the molar mass of a substance is defined as $M_{mol} = m/n$, we can rewrite this as:

$$c_P = \frac{7}{2} \frac{R}{M_{mol}} = \frac{7}{2} \frac{\left(8.314 \frac{J}{mol \cdot K} \right)}{\left(0.02897 \frac{kg}{mol} \right)} = 1004 \frac{J}{kg \cdot K}$$

8.2.2 Adiabatic processes

For our second application of the 1st law of thermodynamics, we examine how the temperature of an air parcel changes as it moves through the atmosphere. In this case, air parcels are assumed to be insulated from their surroundings, meaning there is no heat exchange between the parcel and the environment. Such a process is known as an **adiabatic process**. Applying the 1st law of thermodynamics in differential form to an adiabatic process gives:

$$dE = dW = -PdV$$

$$\Rightarrow \Delta E = -\int_{V_i}^{V_f} P\,dV$$

For a dry air parcel, $\Delta E = (5/2)Nk_B\Delta T$, so the above equation can be rewritten as:

$$\left(\frac{5}{2}\right)Nk_B\Delta T = -\int_{V_i}^{V_f} P\,dV$$

$$\Rightarrow \Delta T = -\frac{2}{5Nk_B}\int_{V_i}^{V_f} P\,dV \tag{8.7}$$

To interpret this result, consider an air parcel rising within the troposphere, as shown in Figure 8.2. As discussed in Section 8.1.2, rising air parcels must expand to adjust to decreasing environmental pressure. From Eq. (8.7), this adiabatic expansion must result in a temperature decrease, meaning air parcels cool as they rise. From an energy perspective, the parcel uses its internal energy to perform work on its surroundings during expansion Conversely, sinking air parcels undergo adiabatic compression, which increases their temperature.

Up to this point, we have seen that pressure remains constant during isobaric processes. One might ask whether there is a variable that remains constant during an adiabatic process. The answer is yes. To identify this variable, we apply the 1st law of thermodynamics to a dry air parcel undergoing an adiabatic process:

$$dE = -P\,dV$$

For dry air, we have:

$$dE = \frac{5}{2}Nk_B dT$$

Using the product rule, we can express $d(PV)$, as:

$$d(PV) = PdV + VdP$$

Therefore, we can rewrite PdV as:

$$P\,dV = d(PV) - V\,dP$$

$$\Rightarrow P\,dV = Nk_B dT - \frac{Nk_B T}{P}\,dP$$

where the ideal gas law has been used. By the 1st law of thermodynamics, we have:

$$\frac{5}{2}Nk_BdT = -\left(Nk_BdT - \frac{Nk_BT}{P}\,dP\right)$$

$$\Rightarrow \frac{7}{2}Nk_BdT = \frac{Nk_BT}{P}\,dP$$

$$\Rightarrow \frac{7}{2}\frac{dT}{T} = \frac{dP}{P}$$

Let's assume the air parcel has an initial temperature $T$and pressure P, and it undergoes an adiabatic process to a new pressure P_0 and a new temperature θ. Integrating the above expression yields:

$$\int_T^{\theta}\frac{7}{2}\frac{dT}{T} = \int_P^{P_0}\frac{dP}{P}$$

$$\Rightarrow \frac{7}{2}\ln\left(\frac{\theta}{T}\right) = \ln\left(\frac{P}{P_0}\right)$$

$$\Rightarrow \theta = T\left(\frac{P_0}{P}\right)^{2/7} \tag{8.8}$$

This temperature θ is referred to as the **potential temperature**, defined as the temperature an air parcel would have if it were adiabatically brought to a standard reference pressure P_0. In meteorology, this reference pressure is typically 1000 hPa. According to Eq. (8.8), an air parcel's temperature and pressure will change during an adiabatic process, but its potential temperature will remain constant. This property has two important applications in synoptic meteorology.

1. Because potential temperature is conserved as air parcels move throughout the atmosphere, it allows for meaningful comparisons between air parcels at different elevations. For example, consider an air parcel at 500 hPa with a temperature of −10°C and another air parcel at 850 hPa with a temperature of 5°C. Using Eq. (8.8), we find that the air parcel at 500 hPa has a potential temperature of 321 K, whereas the air parcel at 850 hPa has a potential temperature of 291 K. After accounting for the difference in elevation, this means the 500-hPa parcel is significantly warmer than the 850-hPa parcel. For this reason, potential temperature is often a more useful quantity than the actual temperature, especially when evaluating atmospheric stability.
2. Because potential temperature remains constant as air parcels move, we can label air parcels according to their potential temperatures, as in the example above. This leads to the concept of lines of constant potential temperature, known as **dry adiabats**, which represent natural pathways for the movement of air parcels. We will revisit this concept when discussing surface fronts in Ch. 11.

8.3 Thermodynamic energy equation

With our overview of the 1st law of thermodynamics, we can now derive an expression for the time evolution of temperature in an air parcel. First, we will rewrite the 1st law of thermodynamics in terms of temperature and pressure in differential form:

$$dE = dQ + dW$$
$$\Rightarrow dE = dQ - PdV \tag{8.9}$$

By the product rule of calculus, we note that

$$d(PV) = P\,dV + V\,dP$$
$$\Rightarrow P\,dV = d(PV) - V\,dP$$
$$\Rightarrow P\,dV = Nk_B\,dT - V\;dP$$

where the ideal gas law has been used. Substituting this expression into Eq. (8.9) gives us:

$$dE = dQ - (Nk_B\,dT + V\,dP)$$

Since $dE = (5/2)Nk_B\,dT$ for a dry air parcel, we can rewrite our equation as:

$$\frac{5}{2}Nk_B\,dT = dQ - (Nk_B\,dT + V\,dP)$$
$$\Rightarrow \frac{7}{2}Nk_B\,dT = dQ + V\,dP$$

Using the definition of isobaric-specific heat given in Eq. (8.6), we can rewrite the above expression as:

$$mc_P\,dT = dQ + V\,dP$$
$$\Rightarrow dT = \frac{dq}{c_P} + \frac{dP}{\rho c_P}$$

Here, we used the definition of density $\rho = m/V$, and note that dq represents the amount of heat transferred per unit mass. Applying the hydrostatic equation $dP = -\rho g\,dz$ to this expression gives us:

$$dT = \frac{dq}{c_P} - \frac{g}{c_P}dz$$

Since we are examining the evolution of the air parcel itself, we take a time derivative on both sides in the Lagrangian reference frame, giving:

$$\frac{DT}{Dt} = -\frac{g}{c_p}\frac{Dz}{Dt} + \frac{1}{c_p}\frac{Dq}{Dt} \tag{8.10}$$

Eq. (8.10) is known as the **thermodynamic energy equation** in the Lagrangian framework, describing the temperature budget of an air parcel as it moves. As noted in Ch. 6, it is preferable to examine temperature changes relative to a fixed location, which corresponds to the Eulerian reference frame. As discussed in Ch. 6, the rate of change in the Eulerian frame equals the rate of change in the Lagrangian reference frame, plus the effects of advection. Therefore, we can express Eq. (8.10) as follows:

$$\frac{\partial T}{\partial t} = -\boldsymbol{V}\cdot\nabla_z T - \frac{g}{c_p}\frac{Dz}{Dt} + \frac{1}{c_p}\frac{Dq}{Dt}$$

Here, $\boldsymbol{V}$ represents the full three-dimensional velocity vector, and the subscript "z" indicates that the gradient is applied on a surface of constant geometric height. Since $w = Dz/Dt$ is the vertical velocity component, we can rewrite the above equation as:

$$\frac{\partial T}{\partial t} = -\boldsymbol{V}_{\mathrm{H}}\cdot\nabla_z T - \left(w\frac{\partial T}{\partial z} + \frac{g}{c_p}\right) + \frac{1}{c_p}\frac{Dq}{Dt} \tag{8.11}$$

In this equation, $\boldsymbol{V}_{\mathrm{H}}$ corresponds to the horizontal wind vector. Eq. (8.11) presents the thermodynamic energy equation in the Eulerian reference frame. The following sections discuss the meaning of each term in Eq. (8.11).

8.3.1 Horizontal thermal advection

As shown in Ch. 6, the first term on the right-hand side of Eq. (8.11) represents thermal advection caused by the horizontal wind. Based on the properties of the scalar product, the magnitude of thermal advection depends on three factors: (1) the magnitude of the horizontal wind speed ($\boldsymbol{V}_H$); (2) the spacing of isotherms (∇T); and (3) the alignment of the horizontal wind vector with the temperature gradient. Thermal advection is maximized when strong winds blow perpendicular to closely spaced isotherms.

Since the synoptic-scale atmosphere remains close to geostrophic balance, thermal advection by the horizontal wind is approximately equal to that by the geostrophic wind- a concept known as **geostrophic thermal advection**. Thus, the horizontal thermal advection term can be approximated as:

$$-\boldsymbol{V}_{\mathrm{H}}\cdot\nabla_z T \approx -\boldsymbol{V}_{\mathrm{g}}\cdot\nabla_z T$$

This approximation is useful because it allows us to estimate horizontal thermal advection directly from upper-air charts. In Ch. 7, we showed that the magnitude of the geostrophic wind is proportional to the spacing of isohypses on an isobaric chart. Therefore, the intersection of isotherms and isohypses reflects geostrophic thermal advection. Strong geostrophic thermal advection occurs where tightly packed isohypses (indicating strong geostrophic flow) intersect tightly packed isotherms (indicating a strong temperature gradient), that is, where the distance between crossing isohypses and isotherms is smallest.

For example, in the 850-hPa observations shown in Figure 8.4, the strongest geostrophic warm advection occurs in eastern Texas and eastern Oklahoma, where closely spaced isohypses intersect multiple isotherms. Conversely, the strongest geostrophic cold advection is found near the New Mexico-Texas border.

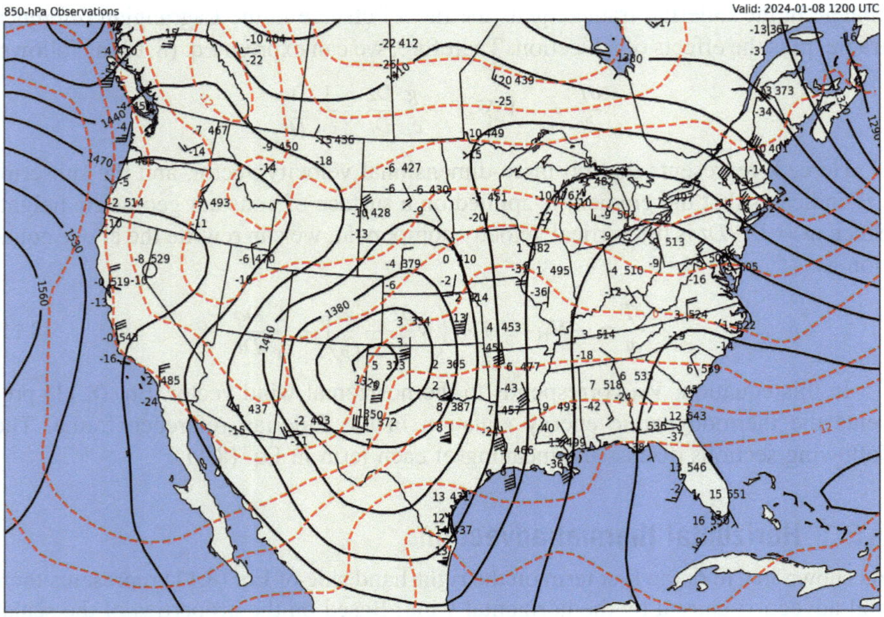

FIGURE 8.4 Observations of geostrophic thermal advection.

850-hPa observations for January 8, 2024, at 12 UTC. Solid black contours correspond to isohypses and dashed red contours correspond to isotherms. Notice that the area between cross isohypses and isotherms is smallest in eastern Texas and eastern Oklahoma.

No Permission Required

Thermal advection typically extends through an entire layer of the atmosphere rather than being confined to a single isobaric layer. Fortunately, we can assess average thermal advection within an entire layer by using the thermal wind vector. As noted in Ch. 7, the thermal wind flows parallel to isotherms corresponding to the layer's average temperature. This allows us to use the thermal wind equation to determine whether the average temperature in the layer is increasing or decreasing. In the Northern Hemisphere, cold air lies to the left of the thermal wind vector. Because the thermal wind is defined as the difference between the geostrophic wind at two levels, examining the geostrophic wind at those levels is sufficient to diagnose warm or cold advection.

Consider the schematic in Figure 8.5. With cold air to the north (and thus to the left of the thermal wind vector), the thermal wind must be westerly. For cold advection to occur, strong geostrophic flow must come from the north. If there is northerly geostrophic flow (as shown in Figure 8.5A), then from the relation $\boldsymbol{V}_{g,2} = \boldsymbol{V}_T + \boldsymbol{V}_{g,1}$, it follows that $\boldsymbol{V}_{g,2}$ must have be northwesterly – a

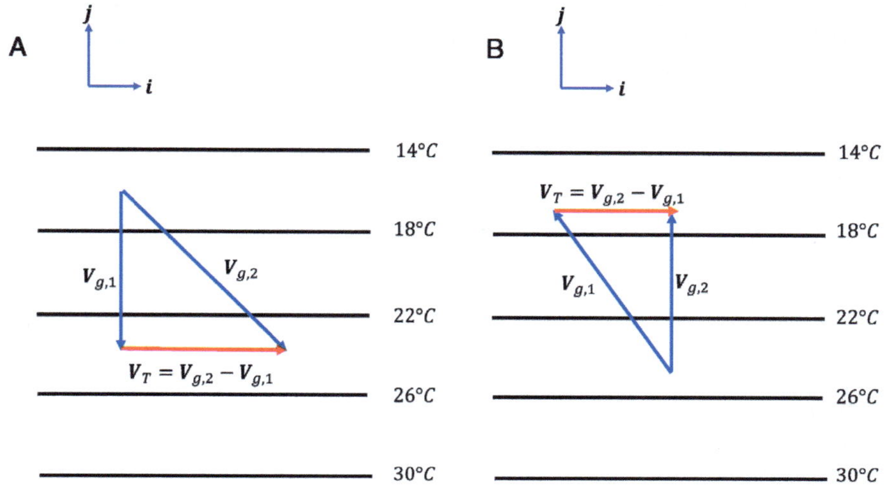

FIGURE 8.5 Schematic of thermal wind and thermal advection

The relationship between thermal advection and the thermal wind vector where $V_{g,1}$ corresponds to the geostrophic wind vector at a lower atmospheric level, $V_{g,2}$ corresponds to the geostrophic wind vector at a higher atmospheric level, and V_T corresponds to the thermal wind vector. (A) Geostrophic winds rotate counterclockwise with increasing height, leading to cold advection. (B) Geostrophic winds rotate clockwise with increasing height, leading to warm advection.

No Permission Required

counterclockwise rotation from $\boldsymbol{V}_{g,1}$. Thus, geostrophic winds that rotate counterclockwise with height, commonly referred to as **backing winds**, are associated with cold advection.

Similarly, for warm advection (as illustrated in Figure 8.5B), strong geostrophic flow must come from the south. If the lower-level geostrophic flow is southeasterly, then $\boldsymbol{V}_{g,2}$ must be southerly - a clockwise rotation from $\boldsymbol{V}_{g,1}$. Therefore, geostrophic winds that rotate clockwise with increasing height, commonly called **veering winds**, are associated with warm advection.

The relationship between thermal advection and the thermal wind vector makes it possible to estimate horizontal temperature advection at a location using only wind data from a radiosonde launch, as illustrated in Box 8.1.

8.3.2 Vertical thermal advection and adiabatic cooling

The second term on the right side of Eq. (8.11) represents temperature changes by vertical motion. An important distinction should be made here: when vertical motion transports heat due to buoyancy, the process is called **convection**. In contrast, when vertical motion transports heat due to non-buoyancy effects, the process is called

Box 8.1 Estimating horizontal temperature advection from radiosonde data

Suppose that a radiosonde measurement at an upper-air station indicates that the 700-hPa wind is 25 knots from 285° and the 500-hPa wind is 45 knots from 315°. Let us estimate the horizontal temperature advection at this site.

First, we will compute the components of the wind vector at 500 hPa and 700 hPa:

$$u_{500} = -V_{500}\sin(\alpha_{500}) = -(45.0\,kts)\sin(315°) = 31.8\,kts = 16.4\frac{m}{s}$$

$$v_{500} = -V_{500}\cos(\alpha_{500}) = -(45.0\,kts)\cos(315°) = -31.8\,kts = -16.4\frac{m}{s}$$

$$u_{700} = -V_{700}\sin(\alpha_{700}) = -(25.0\,kts)\sin(285°) = 24.1\,kts = 12.4\,\frac{m}{s}$$

$$v_{700} = -V_{700}\cos(\alpha_{700}) = -(25.0\,kts)\cos(285°) = -6.47\,kts = -3.33\,\frac{m}{s}$$

The PBL is usually below the 700-hPa level; thus the wind is approximately geostrophic at 700 hPa and 500 hPa. Therefore, we can use the thermal wind equation from Ch. 7 to estimate the horizontal temperature gradient within the 700- to 500-hPa layer. The meridional component of the virtual temperature gradient is given by:

$$u_T = -\frac{R_d}{f}\ln\left(\frac{P_1}{P_2}\right)\frac{\partial T_v}{\partial y}$$

$$\Rightarrow \frac{\partial T_v}{\partial y} = \frac{-fu_T}{R_d\ln\left(\frac{P_1}{P_2}\right)} \approx \frac{-f(u_{500}-u_{700})}{R_d\ln\left(\frac{P_1}{P_2}\right)} = \frac{-(10^{-4}\,s^{-1})\left[\left(16.4\frac{m}{s}\right)-\left(12.4\,\frac{m}{s}\right)\right]}{\left(287\,\frac{J}{kg\cdot K}\right)\ln\left(\frac{700\,hPa}{500\,hPa}\right)} = -4.14\times10^{-6}\,\frac{K}{m}$$

Similarly, the zonal component of the virtual temperature gradient is given by:

$$v_T = \frac{R_d}{f}\ln\left(\frac{P_1}{P_2}\right)\frac{\partial T_v}{\partial x}$$

$$\Rightarrow \frac{\partial T_v}{\partial x} = \frac{fv_T}{R_d\ln\left(\frac{P_1}{P_2}\right)} \approx \frac{f(v_{500}-v_{700})}{R_d\ln\left(\frac{P_1}{P_2}\right)} = \frac{(10^{-4}\,s^{-1})\left[\left(-16.4\frac{m}{s}\right)-\left(-3.33\,\frac{m}{s}\right)\right]}{\left(287\,\frac{J}{kg\cdot K}\right)\ln\left(\frac{700\,hPa}{500\,hPa}\right)} = -1.35\times10^{-5}\,\frac{K}{m}$$

The mean virtual temperature gradient within the layer can be written as:

$$-\mathbf{V}\cdot\nabla T_v = -\left(\frac{\mathbf{V}_{500}+\mathbf{V}_{700}}{2}\right)\cdot\nabla T_v$$

Therefore, we have

$$-\mathbf{V}\cdot\nabla T_v = -\frac{1}{2}\left[\left(16.4\frac{m}{s}\right)i+\left(-16.4\frac{m}{s}\right)j+\left(12.4\frac{m}{s}\right)i+\left(-3.33\frac{m}{s}\right)j\right]\cdot\left[\left(-1.35\times10^{-5}\frac{K}{m}\right)i+\left(-4.14\times10^{-6}\frac{K}{m}\right)j\right]$$

$$\Rightarrow -\mathbf{V}\cdot\nabla T_v = -\left[\left(14.4\frac{m}{s}\right)i+\left(-9.87\frac{m}{s}\right)j\right]\cdot\left[\left(-1.35\times10^{-5}\frac{K}{m}\right)i+\left(-4.14\times10^{-6}\frac{K}{m}\right)j\right]$$

$$\Rightarrow -\mathbf{V}\cdot\nabla T_v = -\left[\left(14.4\frac{m}{s}\right)\left(-1.35\times10^{-5}\frac{K}{m}\right)+\left(-9.87\frac{m}{s}\right)\left(-4.14\times10^{-6}\frac{K}{m}\right)\right] = 1.54\times10^{-4}\,\frac{K}{s} = 0.554\,\frac{K}{hr}$$

We obtain a positive value for thermal advection, consistent with the fact that the winds are veering. If this advection occurs at the same rate over the next 12 hours, the resulting temperature change within this layer will be:

$$\Delta T \approx (-\mathbf{V}\cdot\nabla T_v)\Delta t = \left(0.554\frac{K}{hr}\right)(12\,hr) = 6.63\,K$$

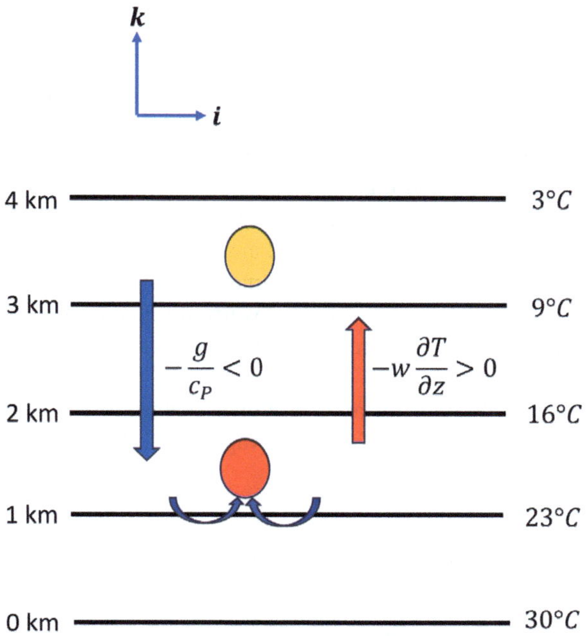

FIGURE 8.6 Schematic of vertical advection plus adiabatic cooling

Impacts of vertical temperature advection and adiabatic expansion on an air parcel. The air parcel is initially located at 850 hPa (as shown in red), and it is lifted upward. Since the temperature is warmer near the surface, the vertical temperature advection transports warmer air upward. However, as air rises, it undergoes adiabatic expansion and cooling. Thus, vertical temperature advection partially offsets adiabatic cooling.

No Permission Required

vertical temperature advection. For synoptic-scale processes, we focus on the latter.

To understand this term, consider an air parcel rising from a height of 1.5 km, as shown in Figure 8.6. Typically, the environmental temperature in the troposphere decreases with increasing height, so that $-w\partial T/\partial z > 0$. This implies that vertical temperature advection is bringing warmer air into this location. However, as the parcel rises, it undergoes adiabatic expansion, which causes cooling. According to Eq. (8.11), the cooling rate due to adiabatic expansion is:

$$\frac{g}{c_p} \approx \frac{9.80\ \frac{m}{s^2}}{1004\ \frac{J}{kg\cdot K}} = 9.76 \times 10^{-3}\frac{K}{m} = 9.76\frac{K}{km}$$

In this case, warm vertical temperature advection partly offsets the cooling from adiabatic expansion. Similarly, vertical advection of cold air can partially offset the warming caused by adiabatic compression in sinking air.

8.3.3 Diabatic heating rate

The third term on the right-hand side of Eq. (8.11) is referred to as the **diabatic heating rate term**, which encompasses all other processes that can cause heat transfer. It is important to note that this term addresses the heating rate, not just the total amount of heating. For synoptic-scale processes, two processes are most relevant: (1) latent heating and (2) radiational cooling.

For synoptic-scale processes, latent heating is usually associated with precipitation and condensation. Consider the vertical air column in Figure 8.7A. If the air in the center becomes saturated, water vapor condenses into clouds, warming the column's core as depicted in Figure 8.7B. This warming increases the column's thickness relative to its surroundings, creating a pressure gradient force directed toward the cooler air.

Now suppose precipitation forms from this condensation. If some precipitation evaporates before reaching the surface, the evaporation absorbs latent heat, cooling

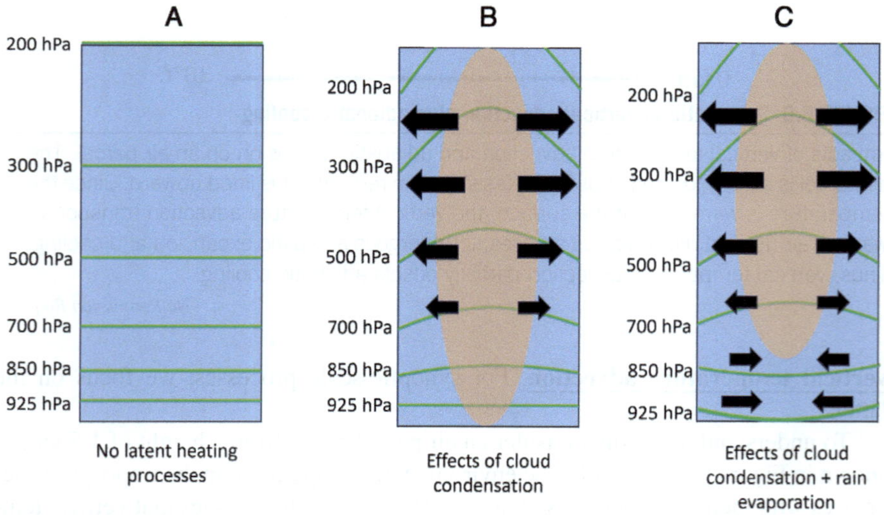

FIGURE 8.7 Effects of latent heating on atmospheric dynamics.

(A) A vertical column of air in thermal wind balance where green lines correspond to isobars. (B) Cloud condensation causes warming within the air column, which increases the thickness in the center of the air column. The thickness gradient produces pressure gradient forces (shown in black arrows) that point toward the cooler air. (C) Some of the precipitation evaporates below the cloud base, which decreases the thickness in the center of the air column. The thickness gradient produces gradient forces (shown in black arrows) that point toward the center of the column.

No Permission Required

the air column below the cloud base as shown in Figure 8.7C. The precipitation that reaches the surface is related to the latent heating rate from condensation, which can be expressed as:

$$\left(\frac{1}{c_P}\frac{Dq}{Dt}\right)_{LH} = \frac{L_v}{c_P}\frac{\rho_{liq}}{\overline{\rho}_{air}}\frac{RR}{\Delta z}$$

where L_v is the latent heat of vaporization for water, c_P is the isobaric-specific heat, $\rho_{liq} = 1000\ kg/m^3$ is the density of liquid water, ρ_{air} is the layer-averaged air density for depth Δz, and RR is the rainfall rate in mm/s (Stull, 2017). The rainfall rate can be estimated using rain gauges, such as those at ASOS sites, and is reported in METAR data.

The second diabatic heating process, radiational cooling, arises from the thermal motion of particles emitting electromagnetic waves. Since the microscopic components of all matter are in constant motion, all objects radiate energy. Two fundamental laws govern this process:

1. **Stefan-Boltzmann law**: Total radiated power is proportional to T^4, where the temperature is measured in Kelvin.
2. **Wien's displacement law**: The wavelength at which the peak power is emitted is inversely proportional to the temperature.

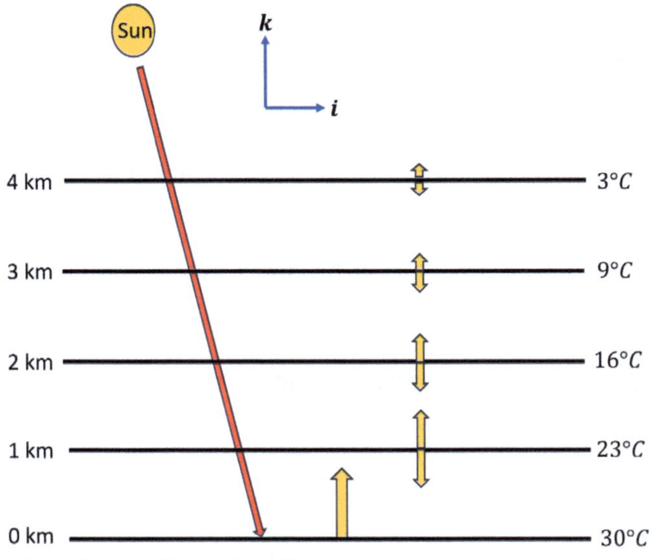

FIGURE 8.8 Schematic of radiational cooling

Radiation within Earth's atmosphere. The red arrow corresponds to solar radiation, which is absorbed by the Earth's surface. The yellow arrows correspond to longwave radiation from Earth's surface and throughout the atmosphere. See the text for additional details.

No Permission Required

Applying these laws, we can provide a conceptual understanding of the diabatic heating rate from radiation, as depicted in Figure 8.8. Solar energy, emitted at high temperatures, peaks at short wavelengths, so it is referred to as **shortwave radiation**. Observations show that most shortwave radiation passes through the atmosphere with minimal absorption and is absorbed mainly at Earth's surface – hence, direct solar heating of the air is often neglected.

Earth's surface, being cooler than the Sun, emits less total radiation, peaking at longer wavelengths, so it is referred to as **longwave radiation**. While the atmosphere is mostly transparent to shortwave radiation, it absorbs much of the surface's longwave emission and re-emits it in all directions,[3] with emission that is proportional to T^4 according to the Stefan-Boltzmann law.

However, because temperature decreases with height, more longwave radiation is lost upward from the warmer lower troposphere than is returned downward from the cooler air aloft, as shown in Figure 8.8. This imbalance causes a net cooling of roughly $-0.1°$C per hour.[4] In summary, the combined effects of latent heating and radiational cooling give the total diabatic heating term as:

$$\frac{1}{c_P}\frac{Dq}{Dt} \approx \frac{L_v}{c_P}\frac{\rho_{liq}}{\overline{\rho}_{air}}\frac{RR}{\Delta z} - \left(0.1\frac{K}{hr}\right)$$

8.4 Introduction to atmospheric stability

In the last section, we discussed vertical heat transport caused by non-buoyancy effects, known as vertical temperature advection. Here, we introduce vertical heat transport due to buoyancy effects, known as convection. For synoptic-scale processes, the vertical movement of air parcels, and thus the vertical heat transport by convection, is linked to the **static stability** of the environment. Static stability is the ability of a fluid at rest to accelerate away from its original position solely due to buoyancy. We will begin by examining lapse rates.

8.4.1 Introduction to atmospheric lapse rates

In Ch. 1, we showed that the buoyancy of an air parcel is based on Archimedes' principle, which states that the buoyant force on a fully submerged object equals the

[3]This is a consequence of Kirchhoff's law of thermal radiation, which states that the fraction of radiation absorbed at a given wavelength (known as the absorptivity) is equal to the fraction of radiation that is emitted at the same wavelength (known as emissivity).

[4]Observations show that horizontal temperature gradients are generally weak compared to vertical temperature gradients. For example, near a frontal boundary, the temperature gradient across the front is of the order of 10°C per 100 km, whereas the vertical temperature gradient is of the order 10°C per kilometer.

weight of the displaced fluid. The same idea applies to air parcels in the atmosphere. Since an air parcel's properties differ from those of its surroundings, we can treat it like an object fully submerged in a fluid. Ch. 1 Archimedes' principle implies that an air parcel will rise if its density (ρ_{par}) is less than the density of the surrounding air (ρ_{env}). Since $\rho \propto T_v^{-1}$, this means an air parcel will rise in the atmosphere if $T_{v,par} > T_{v,env}$. Conversely, air parcels with a lower virtual temperature than their surroundings will sink.

Consider an air parcel with a virtual temperature greater than that of its environment. As explained in Section 8.2, the air parcel will rise, undergoing adiabatic expansion and cooling, as shown in Figure 8.9. From Section 8.3.2, the rate of adiabatic cooling for dry air is 9.76 K/km, known as the **dry adiabatic lapse rate** (Γ_d).

Our first-level approximation of stability is straightforward: if an air parcel starts warmer than its surroundings, buoyancy exceeds its weight, and it accelerates upward. As it rises, cooling at Γ_d, it may remain warmer than the environment – continuing to accelerate – or cool below the environment's temperature, slowing its ascent.

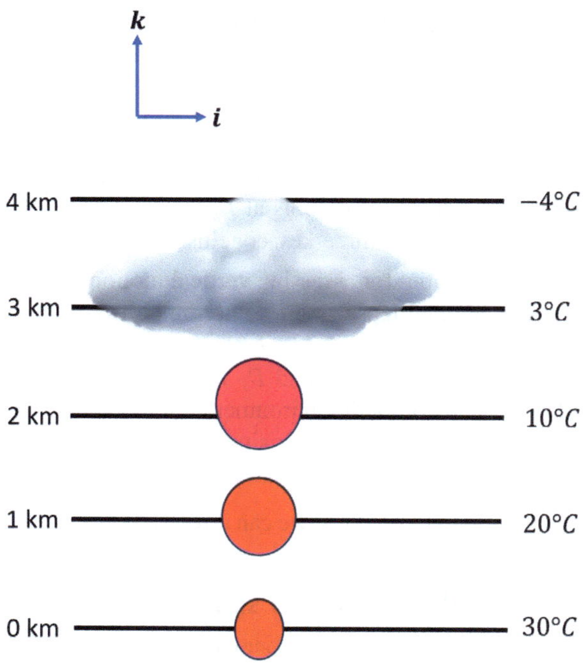

FIGURE 8.9 Schematic of a rising air parcel

A moist rising air parcel. When the rising parcel is unsaturated, its temperature decreases according to the dry adiabatic lapse rate. When the rising parcel becomes saturated, its temperature decreases according to the saturated adiabatic lapse rate.

No Permission Required

If the rising parcel contains water vapor but remains unsaturated, it cools at a slightly lower rate than Γ_d. In cold or dry conditions, the difference is negligible, and the cooling approaches Γ_d. Isolated from its environment, such a parcel undergoes a **moist adiabatic process**, cooling at the **moist adiabatic lapse rate** (Γ_m).

The difference between Γ_d and Γ_m arises because the isobaric-specific heat of water is about 85% greater than that of dry air. Although the amount of water vapor is small, empirical measurements show that $c_{P,m} \approx c_{P,d}(1 + 0.856r)$, where $c_{P,d}$ is the isobaric-specific heat for dry air and r is the mixing ratio (Markowski & Richardson, 2010). Therefore, we can approximate the moist adiabatic lapse rate as:

$$\Gamma_m = \frac{g}{c_{P,m}} \approx \frac{g}{c_{P,d}(1 + 0.856r)} = \frac{\Gamma_d}{1 + 0.856r}$$

Since $r \ll 1$, then $c_{P,m} \approx c_{P,d}$, so moisture effects are minor for unsaturated parcels.

Major moisture effects occur once the air parcel becomes saturated. As the air parcel rises, cooling continues until saturation is reached, condensation occurs, and latent heat release partially offsets adiabatic cooling, as shown in Figure 8.9. This is known as a **saturated adiabatic process**, and it produces a lapse rate lower than Γ_m, called the **saturated adiabatic lapse rate Γ_s**. While a full derivation is beyond our scope, Γ_s depends on the parcel's moisture. Observations indicate Γ_s ranges from 4 K/km near the ground in warm, humid air to 7 K/km in the mid-troposphere (Markowski and Richardson, 2010).

The vertical motion of a moist air parcel can thus be summarized: initially, it cools at Γ_d until it becomes saturated (which occurs when the air parcel's temperature equals its dewpoint), and then it cools at Γ_s as latent heat release reduces the cooling rate. From this, we draw three key conclusions:

1. Two lapse rates are essential for assessing vertical motion: the parcel's lapse rate Γ_{par} and the environmental lapse rate Γ, determined from radiosonde temperature profiles.
2. Although Γ_d remains constant, Γ_m and Γ_s vary with moisture, meaning atmospheric stability depends on both environmental temperature and moisture .
3. Spontaneous convection requires the rising parcel to remain warmer than its surrounding environment, despite adiabatic cooling.

With knowledge of these lapse rates, we can directly assess atmospheric stability.

8.4.2 Lapse rates and atmospheric stability

Consider the case where $\Gamma < \Gamma_s$ (the environmental lapse rate is less than the saturated adiabatic lapse rate), as shown in Figure 8.10A. If an air parcel is forced to rise to the levels labeled A and B, its temperature (T_A) will be lower than the environmental temperature T_B. The parcel's weight will then exceed the buoyant force, causing it to sink back to its original level. In this situation, the parcel encounters a restoring force after being displaced, which suppresses upward motion. Therefore,

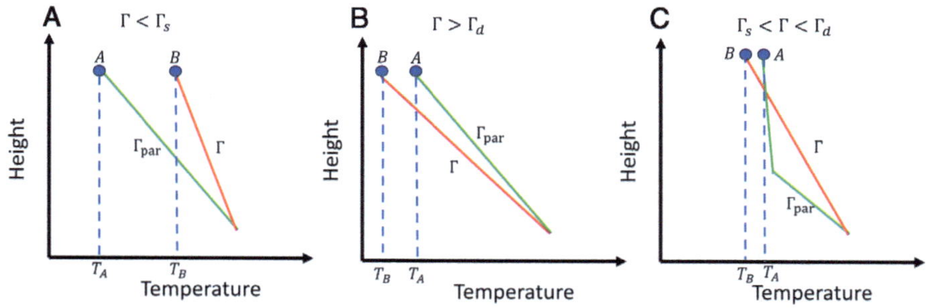

FIGURE 8.10 Schematic of stable vs. unstable environments

(A) An absolutely stable stratification. (B) An absolutely unstable stratification, (C) A conditionally unstable stratification. The green curve indicates the trajectory of an air parcel, whereas the red curve indicates the environmental temperature profile. Γ corresponds to the environmental lapse rate, Γ_d corresponds to the dry adiabatic lapse rate, and Γ_s corresponds to the saturated adiabatic lapse rate.

No Permission Required

$\Gamma < \Gamma_s$ indicates **absolute stability**. In such a stratification, the atmosphere strongly resists upward vertical motion and favors horizontal spreading. The strength of this stability depends on the difference $\Gamma_s - \Gamma$: the larger the difference, the stronger the restoring force.

A stable atmosphere occurs when the environmental lapse rate is small – that is, when the temperature difference between the surface air and air aloft is minimal. Warm advection aloft (i.e. the replacement of upper-level air with warmer air) reduces the lapse rate and increases stability, especially if the surface temperatures remain unchanged. Cooling of the lower troposphere has the same effect and can occur due to:

1. Nighttime radiational cooling of the surface (discussed in Section 8.5),
2. Cold air advection at the surface, such as behind a cold front,
3. Warm air moving over a cold surface, such as southerly winds over spring snow cover.

On most days, stability is strongest in the early morning, near sunrise, when surface temperatures are lowest. In stable conditions, surface air saturation can lead to persistent haze or fog.

Another stabilizing process occurs when an entire layer of air sinks. In a large-scale subsidence event, an unsaturated layer more than 100 m thick warms through adiabatic compression. The upper portion of the layer sinks farther and warms more than the lower part, as shown in Figure 8.11. This can produce an **inversion**, where temperature increases with height. Inversions formed by slow, large-scale sinking are called **subsidence inversions**. These are most often observed aloft in association with large high-pressure systems, where sinking air dominates. Inversions inhibit

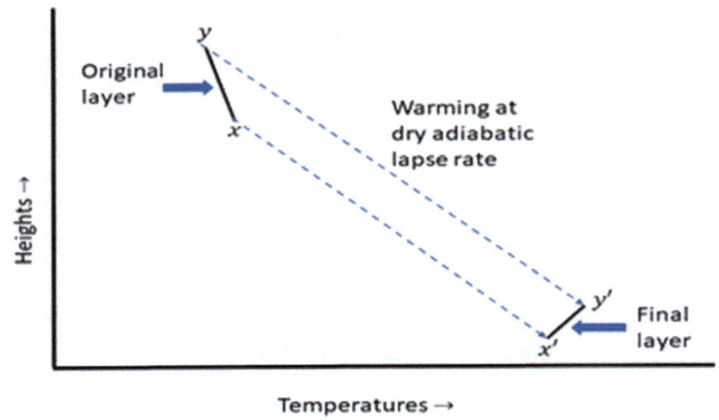

FIGURE 8.11 Schematic of subsidence inversion

A sinking layer of air forming a subsidence inversion. See the text for more details.

No Permission Required

vertical motion, and when near the surface, keep stratus clouds, fog, and haze trapped close to the ground.

Now consider the case where $\Gamma > \Gamma_d$, as shown in Figure 8.10B. When lifted to points A and B, the parcel's temperature (T_A) exceeds the environmental temperature (T_B), so the buoyant force exceeds the parcel's weight. This causes it to accelerate upward, and the magnitude of this acceleration depends on the difference $\Gamma - \Gamma_d$. Thus, $\Gamma > \Gamma_d$ indicates **absolute instability**. However, deep atmospheric layers are rarely absolutely unstable. This condition usually occurs only in a shallow near-surface layer on hot, sunny days, when the lapse rate exceeds the dry adiabatic lapse rate – producing a **superadiabatic** layer. Under these conditions, spontaneous convection and overturning occur.

The atmosphere becomes more unstable as the environmental lapse rate steepens, that is, when temperature decreases more rapidly with height. This can result from:

(1) cold advection aloft (such as when a midlatitude cyclone passes overhead);
(2) radiational cooling at cloud tops;
(3) warm advection near the surface (such as ahead of a cold front); or
(4) cold air flowing over a warm surface.

A common example is the combination of cold air aloft and warm surface air, which steepens the environmental lapse rate and enhances instability. This process is linked to **differential horizontal temperature advection** - a decrease in warm advection with height or an increase in cold advection with height. We will revisit this in Ch. 12 when discussing midlatitude cyclones.

Finally, consider the case where $\Gamma_s < \Gamma < \Gamma_d$, as shown in Figure 8.10C. Initially, an unsaturated parcel rises and cools at the dry adiabatic lapse rate Γ_d. Because $\Gamma < \Gamma_d$, the parcel will be cooler than its environment, mimicking absolute stability. However, once the parcel becomes saturated, it cools more slowly at the saturated adiabatic lapse rate Γ_s. Since $\Gamma > \Gamma_s$, the saturated parcel can eventually become warmer than its surroundings, enabling upward acceleration. In this regime, an unsaturated air parcel will always be cooler than its environment, but a saturated air parcel *may* become warmer. Therefore, stability depends on moisture content, and the environment is said to be **conditionally unstable**. On average, the Earth's environmental lapse rate is approximately 6.5 K/km, meaning the atmosphere is often in a state of conditional instability. Water vapor thus plays a central role in determining atmospheric stability and the potential for vertical motion.

8.5 Turbulent heat transport

So far, we have discussed several processes that can change the thermal structure of the atmosphere at a specific location. Thermal advection, driven by horizontal and vertical winds, transports warm or cold air into a region. In areas with active precipitation, latent heating plays a major role: condensation within clouds releases heat into the environment, while rain evaporation cools the surrounding air. Outside precipitation regions, radiative heating and cooling also modify the thermal structure.

In addition to these processes, turbulence mixes the thermal, moisture, and kinematic properties of the air within the planetary boundary layer (PBL). The rate of mixing depends on turbulence intensity, which varies in space and time. To illustrate, consider a daytime scenario in which solar heating generates thermal turbulence. As sunlight warms Earth's surface, the surface temperature rises, transferring heat to the air immediately above it. The resulting vertical temperature gradient near the surface produces buoyant thermals – rising bubbles of warm air. This generates an upward heat flux expressed as:

$$F_H = b_H w_B \left(\Theta_{sfc} - \Theta_{ML}\right), \qquad b_H = 5 \times 10^{-4}$$

where b_H is known as the **convective transfer coefficient**, w_B is the buoyancy velocity scale (introduced in Ch. 5), Θ_{sfc} is the surface potential temperature, and Θ_{ML} is the average potential temperature of the mixed layer (Stull, 2017). The convective circulations mix the PBL until the potential temperature and mixing ratio become nearly uniform throughout, creating a mixed PBL. Observations show that turbulent heat fluxes within the mixed layer decrease linearly with height such that:

$$\frac{\partial F_H}{\partial z} = -\frac{1.2 F_H}{z_i}$$

where z_i is the PBL depth (Stull, 2017). Therefore, the vertical heat transport generated by turbulence can be written as:

$$\left(\frac{\partial T}{\partial t}\right)_{turb} = -\left(\frac{\partial F_H}{\partial z}\right) = \begin{cases} \dfrac{1.2 b_H w_B \left(\Theta_{sfc} - \Theta_{ML}\right)}{z_i}, & z \leq z_i \\ 0, & z > z_i \end{cases}$$

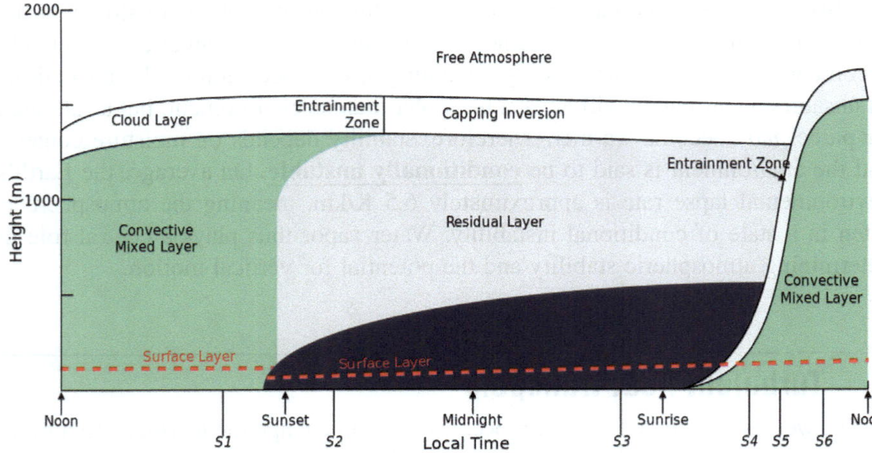

FIGURE 8.12 Schematic of the diurnal evolution of the PBL

The evolution of the atmospheric boundary layer during fair weather over land.

The vertical heat transport by turbulence, combined with radiative processes, produces a well-observed diurnal cycle in the PBL. Figure 8.12 illustrates the typical summer evolution of the PBL over land under fair-weather conditions. By the end of the night, the PBL often consists of a shallow and stable layer near the surface, known as the **nocturnal boundary layer**, and an overlying, nearly neutral **residual layer**. A strong capping inversion separates the PBL from the free atmosphere, trapping turbulence and moisture below it and preventing surface friction from influencing the air above. For this reason, the PBL is usually shallow and stable during the morning hours.

After sunrise, solar heating produces buoyant thermals, which rise and generate intense mixing, forming a mixed layer. When these thermals encounter the capping inversion, their momentum allows them to overshoot briefly into the free atmosphere before sinking back, bringing some free-atmosphere air into the mixed layer. This process, known as **entrainment**, occurs within the **entrainment zone** and increases the depth of the PBL. As the mixed layer grows, it erodes the previous evening's inversion and rises through the residual layer. By midday or early afternoon, the mixed layer reaches the capping inversion, which is replaced by the entrainment zone for the rest of the day. Under fair-weather conditions without strong mechanical turbulence, the PBL is usually deep and unstable during the afternoon hours.

At sunset, solar heating ceases, removing the main source of instability in the PBL. Longwave radiative cooling then lowers the ground temperature until a temperature inversion forms. Two effects follow: (1) the decay of turbulence from the mixed layer forms a new residual layer, and (2) the development of a stable inversion produces a deep nocturnal boundary layer with strong static stability. This diurnal cycle repeats under fair-weather conditions but varies seasonally. For example, during the winter, longer nights favor a relatively deep nocturnal boundary layer, while shorter days lead to a shallower daytime mixed layer.

Having covered the thermodynamic processes and dynamical processes common to synoptic-scale motion, we can now apply these concepts to better understand fronts, jets, and cyclones - the focus of the third unit of this textbook.

Questions

1. Answer the following conceptual questions regarding the 1st law of thermodynamics:
   a. Explain why it is improper to state that a system possesses heat, whereas it is proper to state that a system possesses internal energy.
   b. Explain the differences between heat and work.
   c. If an air parcel rises isothermally (namely heat is added or subtracted to maintain constant temperature), what would happen to its potential temperature as it rises?
2. Answer the following questions regarding the properties of the PBL:
   a. Explain why the veering of the wind with height within the PBL is not necessarily an indication of warm advection. What other mechanisms could cause the winds to veer?
   b. Explain how daytime solar heating can produce a superadiabatic layer near the surface and increase the PBL depth. Explain how radiational cooling reduces the PBL depth and stabilizes the PBL.
3. As a radiosonde rises within the atmosphere, suppose that it passes through a layer of clouds.
   a. Suppose that the humidity sensor freezes while passing through the clouds. How will this affect measurements of relative humidity? What are some of the potential consequences of this?
   b. As the radiosonde emerges from the clouds, water or ice will evaporate from the surface of the temperature sensor. How will this affect temperature measurements? What are some of the potential consequences of this?
4. Consider the contoured 700-hPa chart shown in Figure 8.13.
   a. Mark the locations of longwave troughs, longwave ridges, and shortwave troughs using standard conventions.
   b. Using a red pencil, shade regions of warm advection and use a blue pencil to shade regions of cold advection.
5. Consider the contoured 850-hPa chart shown in Figure 8.14.
   a. Using a yellow pencil, shade regions of warm advection and use a blue pencil to shade regions of cold advection.
   b. Using a green pencil, shade regions of dry advection and use red pencil to shade regions of moist advection.

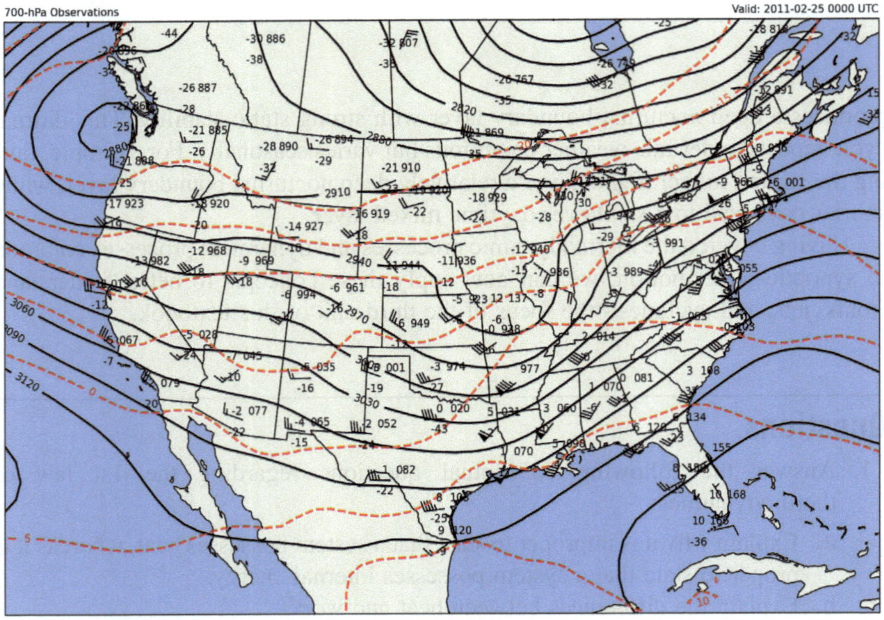

FIGURE 8.13 700-hPa observations

700-hPa observation for February 25, 2011, at 00 Z. Solid black contours correspond to isohypses, and dashed red contours correspond to isotherms.

No Permission Required

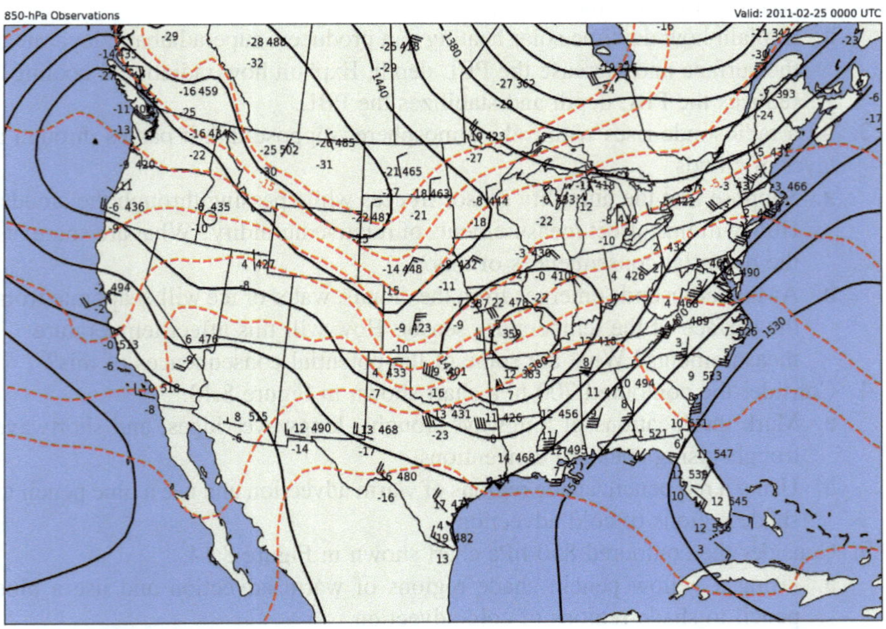

FIGURE 8.14 850-hPa observations

850-hPa observation for February 25, 2011, at 00 Z. Solid black contours correspond to isohypses, and dashed red contours correspond to isotherms.

No Permission Required

6. Consider a sunny day with no winds. If the PBL depth is 3 km, the mixed layer potential temperature is 290 K, and the surface potential temperature is 320 K, determine the heating rate due to turbulent transport. You may ignore the effects of moisture. Express your answer in K/hr.
7. Suppose that the 925-hPa temperature is 10°C, while the surface temperature is 5°C. If the vertical velocity is +1.50 cm/s, what will the temperature be at 925-hPa after 6 hours?
8. A 5-kg air parcel with an initial temperature of 5°C rises 1 km and loses 15 kJ of energy due to longwave radiation. What is the final temperature of the parcel?
9. Suppose a thunderstorm rains at rate 4 mm/hr. What is the average heating rate in the troposphere? You may assume a tropospheric depth of 11 km and a scale height of 7 km.
10. Suppose that an air parcel rises within the atmosphere.
    a. As the air parcel undergoes adiabatic expansion, show that $PV^{7/5} = constant$.
    b. Use the result from (a) to show that the work done by as the air parcel rises and expands is given by: $W = \frac{5P_iV_i^{7/5}}{2}\left[V_f^{-2/5} - V_i^{-2/5}\right]$
    c. Based on the result from (b), what must happen to the temperature of an air parcel as it rises?
11. In this question, you will derive a relationship between lapse rates and potential temperature.
    a. Use the definition of potential temperature to show that:

$$\frac{1}{\theta}\frac{d\theta}{dz} = \frac{1}{T}\frac{dT}{dz} - \frac{2}{7P}\frac{dP}{dz}$$

    [Hint: Take the natural logarithm of potential temperature, and differentiate your result as a function of height z]
    b. Use the hydrostatic equation, the ideal gas law, and the result from (a) to show that:

$$\frac{1}{\theta}\frac{d\theta}{dz} = \frac{1}{T}(\Gamma_d - \Gamma)$$

    [Hint: Note that $c_p/R_d = 7/2$].
    c. Based on the answer to (b), how should the potential temperature vary with height for an absolutely unstable environment? How should the potential temperature vary with height if the environmental lapse rate is equal to the dry adiabatic lapse rate?
12. Consider the mandatory-level radiosonde data from Charleston, SC, which is located at a latitude of 32.78°N, as shown in Table 8.1.
    a. Use the mandatory-level data to graph the environmental temperature (in Celsius) vs height (in km). Discuss the stability of the environmental sounding.

Table 8.1 Radiosonde data for Problem 13

TTAA	52121	72208	99022	03656	33503	00190	03058	00521	92818	00258	02523
85494	01558	33524	70034	05979	32525	21784	33045	40719	34769	31058	30915
43775	29076	25037	47982	28596	20183	50381	27596	15368	54580	28077	10623
59379	27560	88130	59579	26575	77188	28103	42017	31313	45408	81106	51515
10164	00091	10194	01021	32023							

No Permission Required

b. If the wind is in geostrophic balance between 700-hPa and 500-hPa, determine the layer-averaged thermal advection within this layer.

References

Markowski, P., Richardson, Y., 2010. Mesoscale Meteorology in Midlatitudes, first ed. John Wiley & Sons, Ltd. https://doi.org/10.1002/9780470682104.

Stull, R., 2017. Practical Meteorology: An Algebra-based Survey of Atmospheric Science, first ed. Univ. of British Columbia, British Columbia Univ. of British Columbia, British Columbia. Available from: https://www.eoas.ubc.ca/books/Practical_Meteorology/.

CHAPTER

Upper-tropospheric jets

9

Chapter outline

In the second unit of this textbook (Chs. 5–8), we introduced the physical laws and mechanisms necessary to describe the evolution of the atmosphere. As discussed in Ch. 5, the fundamental forces acting on air masses in the atmosphere are the pressure gradient force, the gravitational force, and the turbulent drag force. As discussed in Ch. 6, Newtonian mechanics assumes that atmospheric motion is examined from an inertial reference frame. However, the reference frame typically used when examining atmospheric motion on weather charts is fixed at a specific location, known as the Eulerian frame. To account for this shift, we must include advection into our analysis of atmospheric dynamics. In addition, because Earth rotates, observations made from its surface are taken from a non-inertial reference frame. To account for Earth's rotation, we must include two inertial forces: the centrifugal force and the Coriolis force. Although the equations of motion are complex, Ch. 7 demonstrated

An Introduction to Synoptic-Dynamic Meteorology. https://doi.org/10.1016/B978-0-443-24858-0.00009-5

that synoptic-scale motion tends to approach a balanced state known as thermal wind balance, which combines hydrostatic balance in the vertical direction with geostrophic balance in the horizontal direction. In Ch. 8, we examined the thermodynamic processes that also influence the evolution of synoptic-scale flow in the atmosphere.

In the third unit of this textbook (Chs. 9–12), we will apply the physical principles discussed in the previous unit to four major synoptic-scale systems: upper-tropospheric jets (Ch. 9), Rossby waves (Ch. 10), surface fronts (Ch. 11), and midlatitude cyclones (Ch. 12). We begin with the structure, dynamics, and evolution of upper-tropospheric jet streams, which are powerful, narrow, westerly air currents that encircle the globe. These jets play a central role in steering the movement of air masses, fronts, and surface pressure systems, all of which shape the weather we regularly experience. To understand their dynamics, we will introduce the concept of divergence, an important flow pattern associated with these jets.

9.1 Introduction to divergence

In Ch. 5, we introduced the mathematical definition of the gradient. For a function $f = f(x, y, z)$, the gradient is defined as follows:

$$\nabla f = \frac{\partial f}{\partial x}\boldsymbol{i} + \frac{\partial f}{\partial y}\boldsymbol{j} + \frac{\partial f}{\partial z}\boldsymbol{k}$$

where $\frac{\partial f}{\partial x}$, $\frac{\partial f}{\partial y}$, and $\frac{\partial f}{\partial z}$ are the partial derivatives in the x-, y-, and z-directions, respectively, and $\boldsymbol{i}, \boldsymbol{j}$, and $\boldsymbol{k}$ are the unit vectors in the x-, y-, and z-directions, respectively. As we discussed, the gradient is the generalization of the derivative to three dimensions. It represents a vector field that has both magnitude and direction at each point (x, y, z). At any point in space, the vector ∇f points in the direction of the maximum increase of the function f, and its magnitude $|\nabla f|$ gives the rate of increase in that direction.

Another way to understand the gradient is by viewing it as an operator. In mathematics, an operator acts on an input to generate an output. For the gradient, ∇ acts on the scalar function $f(x, y, z)$ and produces a vector field as the output. To see this, we can express the gradient of f as:

$$\nabla f = \left(\boldsymbol{i}\frac{\partial}{\partial x} + \boldsymbol{j}\frac{\partial}{\partial y} + \boldsymbol{k}\frac{\partial}{\partial z}\right)f \tag{9.1}$$

In this context, we refer to ∇ as the **gradient operator**. In summary, when the gradient operator ∇ acts on the scalar function $f\ (x, y, z)$, it produces a new vector field whose direction at each point in space indicates the direction of the greatest increase of $f(x, y, z)$, while its magnitude corresponds to rate of change of the function.

9.1.1 Definition of divergence

We see in Eq. (9.1) that the gradient operator can act on a scalar function. Additionally, it can also act on a vector function in two ways: through the scalar product and the vector product.

When the gradient operator acts on a vector function via the scalar product, we obtain:

$$\nabla \cdot \boldsymbol{v} = \left(\boldsymbol{i}\frac{\partial}{\partial x} + \boldsymbol{j}\frac{\partial}{\partial y} + \boldsymbol{k}\frac{\partial}{\partial z} \right) \cdot (v_x\boldsymbol{i} + v_y\boldsymbol{j} + v_z\boldsymbol{k})$$

where v_x, v_y, and v_z are the x-, y-, and z-components of the vector function $\boldsymbol{v}(x, y, z)$, respectively.

Performing the scalar product to each term gives us:

$$\begin{aligned}\nabla \cdot \boldsymbol{v} = {} & \boldsymbol{i}\cdot\frac{\partial}{\partial x}(v_x\boldsymbol{i}) + \boldsymbol{i}\cdot\frac{\partial}{\partial x}(v_y\boldsymbol{j}) + \boldsymbol{i}\cdot\frac{\partial}{\partial x}(v_z\boldsymbol{k}) + \boldsymbol{j}\cdot\frac{\partial}{\partial y}(v_x\boldsymbol{i}) + \boldsymbol{j}\cdot\frac{\partial}{\partial y}(v_y\boldsymbol{j}) \\ & + \boldsymbol{j}\cdot\frac{\partial}{\partial y}(v_z\boldsymbol{k}) + \boldsymbol{k}\cdot\frac{\partial}{\partial z}(v_x\boldsymbol{i}) + \boldsymbol{k}\cdot\frac{\partial}{\partial z}(v_y\boldsymbol{j}) + \boldsymbol{k}\cdot\frac{\partial}{\partial z}(v_z\boldsymbol{k})\end{aligned}$$

Since each unit vector is a constant, we can factor them out of the parentheses, resulting in the following:

$$\begin{aligned}\nabla \cdot \boldsymbol{v} = {} & (\boldsymbol{i}\cdot\boldsymbol{i})\frac{\partial v_x}{\partial x} + (\boldsymbol{i}\cdot\boldsymbol{j})\frac{\partial v_y}{\partial x} + (\boldsymbol{i}\cdot\boldsymbol{k})\frac{\partial v_z}{\partial x} + (\boldsymbol{j}\cdot\boldsymbol{i})\frac{\partial v_x}{\partial y} + (\boldsymbol{j}\cdot\boldsymbol{j})\frac{\partial v_y}{\partial y} + (\boldsymbol{j}\cdot\boldsymbol{k})\frac{\partial v_z}{\partial y} + (\boldsymbol{k}\cdot\boldsymbol{i})\frac{\partial v_x}{\partial z} \\ & + (\boldsymbol{k}\cdot\boldsymbol{j})\frac{\partial v_y}{\partial z} + (\boldsymbol{k}\cdot\boldsymbol{k})\frac{\partial v_z}{\partial z}\end{aligned}$$

From our previous discussion on the scalar product, we recall that:

$$\boldsymbol{i}\cdot\boldsymbol{i} = 1 = \boldsymbol{j}\cdot\boldsymbol{j} = \boldsymbol{k}\cdot\boldsymbol{k}, \boldsymbol{i}\cdot\boldsymbol{j} = 0 = \boldsymbol{j}\cdot\boldsymbol{i}, \boldsymbol{j}\cdot\boldsymbol{k} = 0 = \boldsymbol{k}\cdot\boldsymbol{j}, \boldsymbol{i}\cdot\boldsymbol{k} = 0 = \boldsymbol{k}\cdot\boldsymbol{i}$$

Applying these relations to the above expression yields:

$$\nabla \cdot \boldsymbol{v} = \frac{\partial v_x}{\partial x} + \frac{\partial v_y}{\partial y} + \frac{\partial v_z}{\partial z} \tag{9.2}$$

The vector operation in Eq. (9.2) is known as the **divergence**, denoted as δ, because it measures how much the vector function $\boldsymbol{v}(x, y, z)$ "spreads out" (i.e., diverges) from a given point. To illustrate, consider the vector function $\boldsymbol{v} = x\boldsymbol{i} + y\boldsymbol{j}$, shown in Figure 9.1A. Notice that all vector arrows spread outward from the point $(x, y) = (0, 0)$. In this case, the divergence is:

$$\nabla \cdot \boldsymbol{v} = \frac{\partial v_x}{\partial x} + \frac{\partial v_y}{\partial y} + \frac{\partial v_z}{\partial z} = \frac{\partial x}{\partial x} + \frac{\partial y}{\partial y} = 2$$

Thus, the function has positive divergence. If we analyze a small region of space centered at the origin, this vector field stretches the region evenly in all directions, as shown in Figure 9.1A.

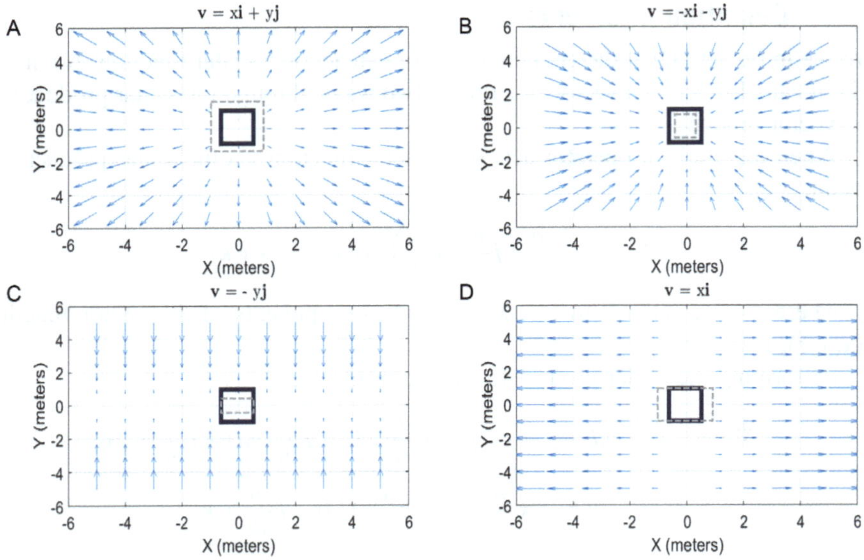

FIGURE. 9.1 Divergence of different vector functions. Four vector functions with divergence are shown here, where a region of space around the origin is highlighted with a black square. (A) The region of space is stretched evenly in all directions (as shown by the dashed gray square). (B) The region of space is compressed evenly in all directions (as shown by the dashed gray square). (C) The region of space is compressed in the *y*-direction (as shown by the dashed gray rectangle). (D) The region of space is stretched in the *x*-direction (as shown by the dashed gray rectangle).

-No Permission Required

Figure 9.1B shows the vector function $\boldsymbol{v} = -(x\boldsymbol{i} + y\boldsymbol{j})$. In this case, the divergence is:

$$\nabla \cdot \boldsymbol{v} = \frac{\partial v_x}{\partial x} + \frac{\partial v_y}{\partial y} + \frac{\partial v_z}{\partial z} = -\left(\frac{\partial x}{\partial x} + \frac{\partial y}{\partial y}\right) = -2$$

When the divergence is negative, we refer to it as **convergence**. Here, all arrows point *toward* the origin, indicating that the small region in space is being compressed evenly in all directions.

It's important to note that a vector field can exhibit convergence even if it compresses a region of space in only one direction. For example, the divergence of the vector function $\boldsymbol{v} = -y\boldsymbol{j}$ (plotted in Figure 9.1C) is $\delta = -1$. In this case, the arrows compress the region at the origin only along the y-direction. Conversely, the divergence of the vector function $\boldsymbol{v} = x\boldsymbol{i}$ (illustrated in Figure 9.1D) is $\delta = +1$, which indicates that the arrows stretch the region of space only along the x-direction.

9.1.2 Identifying divergence on synoptic charts

An examination of synoptic weather charts reveals regions of both divergence and convergence within atmospheric flow. If the vector functions in Figure 9.1 represent wind velocity at specific points on a synoptic chart, there are two general ways in which wind divergence can occur.

1. **Speed Divergence and Convergence**: Divergence can occur when wind speed changes along the path of motion. Specifically, when wind speed increases in the direction of the flow (i.e., downstream), this is called **speed divergence.** An example is given in Figure 9.1D, where the atmosphere is stretched along the flow direction. Conversely, when wind speed decreases downstream, this is known as **speed convergence**, as demonstrated in Figure 9.1C, where the atmosphere is compressed along the flow direction.
2. **Directional Divergence and Convergence**: Divergence can also result from changes in wind direction along the path of motion. When the wind spreads outward from a central point, it is referred to as **directional divergence** or **diffluence**, as illustrated in Figure 9.1A. In contrast, when the wind curves inward toward a central point, it is called **directional convergence** or **confluence**, as shown in Figure 9.1B.

We can apply these concepts of divergence and convergence to synoptic map analysis. Figure 9.2 provides 300-hPa analysis from January 8th, 2024. In the highlighted region "A," wind speed increases downstream, indicating speed divergence.

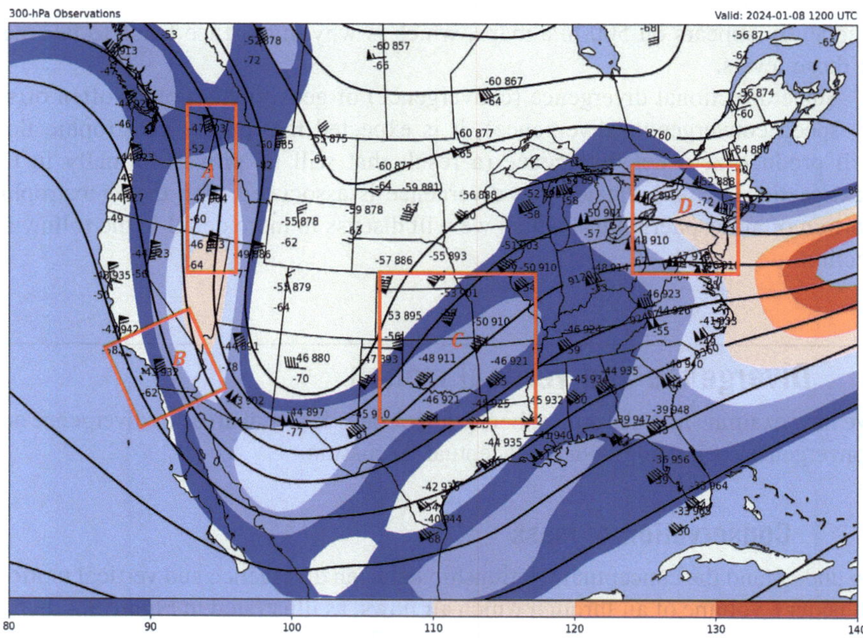

FIGURE. 9.2 Divergence on a 300-hPa map. 300-hPa analysis on January 8, 2024, at 12 UTC. Solid black contours correspond to isohypses. Isotachs greater than 80 knots are contour filled. See the text for a description of the four highlighted regions on the chart.

-No Permission Required

In region "D," wind speed decreases downstream, indicating speed convergence. In region "C," the wind barbs spread apart as the flow moves west to east, indicating diffluence. In contrast, in region "B," the flow curves inward from north to south, indicating confluence.

Divergence is maximized when speed divergence occurs together with diffluence, while convergence is maximized when speed convergence occurs together with confluence. However, it is common to see speed divergence paired with confluence, and speed convergence paired with diffluence. To understand why, recall the geostrophic wind equation from Ch. 7:

$$\boldsymbol{V}_g = \frac{g_0}{f}(\boldsymbol{k} \times \nabla_P Z)$$

When confluence occurs, the winds curve inward, producing closer isohypse spacing and stronger geostrophic flow, as seen in region "B" of Figure 9.2. Conversely, when diffluence occurs, the winds spread outward, leading to wider isohypse spacing and weaker geostrophic flow, as shown in region "C".

There are many instances in the synoptic atmosphere where directional divergence is effectively offset by speed convergence, as shown in the synoptic map in Figure 9.3. Following the atmospheric flow from point A to point B in the highlighted region we can see that diffluence is counteracted by speed convergence. The net effect of these two processes is minimal divergence. This pattern frequently appears on 500-hPa maps, which is why divergence tends to be small at those levels.

Since directional divergence (convergence) of geostrophic flow is often offset by speed convergence (divergence), it is expected that purely geostrophic flow will produce zero net divergence (a result that will be proven formally in the next section). This indicates that divergence is associated with the ageostrophic portion of atmospheric flow, which we will discuss in more detail in the following section.

9.2 Divergence and vertical motion

We now examine how the atmosphere compensates for local areas of divergence and convergence, beginning with a conceptual discussion.

9.2.1 Conservation of mass

To understand the conceptual relationship between divergence and vertical motion, consider a volume of air through which air flows, as illustrated in Figure 9.4. In this case, the rate of inflow exceeds the rate of outflow. Because mass inflow is proportional to velocity, this situation corresponds to speed convergence. Consequently, more air enters the volume than exits, casuing mass to accumulate. Conceptually,

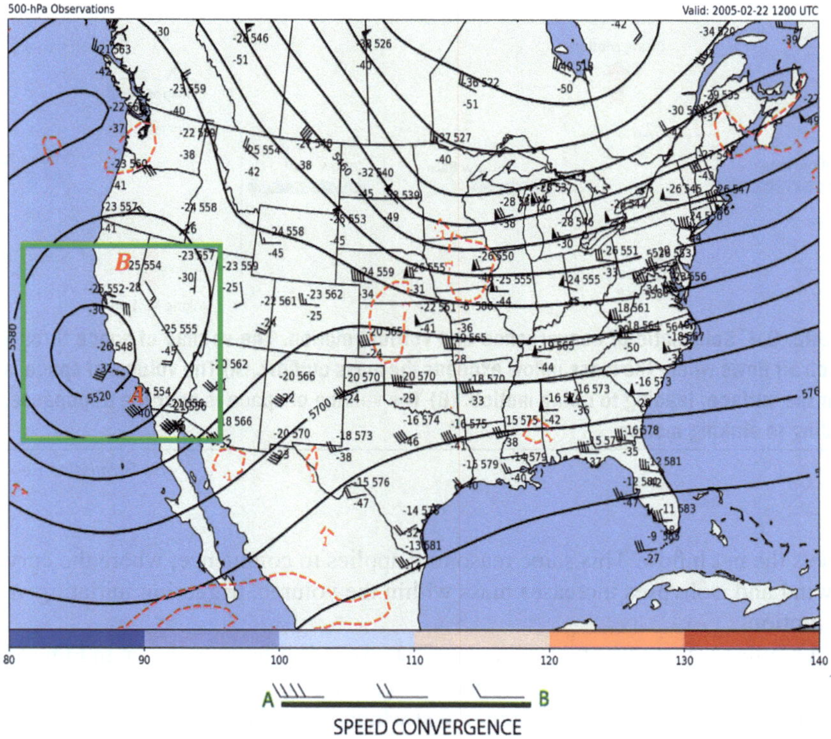

FIGURE. 9.3 Nondivergence for a 500-hPa map. 500-hPa analysis on February 22, 2005, at 12 UTC. Solid black contours correspond to isohypses. Dashed red contours correspond to divergence with units of 10^{-5} s. See the text for a description of the highlighted region.

-No Permission Required

convergence can therefore be viewed as the rate at which mass is added to a volume.

If this process were to continue indefinitely, the density of the air inside the volume would increase without bound. In reality, the atmosphere responds through vertical motion. The nature of this motion depends on the location of the convergence. In Figure 9.4A, convergence occurs near the surface, producing upward motion that removes mass through the top of the volume. In Figure 9.4B, convergence occurs near the tropopause. Because the stratosphere is strongly stable, little air escapes upward, and the tropopause height remains nearly constant. Thus, convergence near the tropopause forces downward motion, ejecting mass through the bottom of the volume.

In both cases, the rate of mass inflow balances the rate of mass outflow, maintaining steady density within the volume. In other words, the induced vertical motion

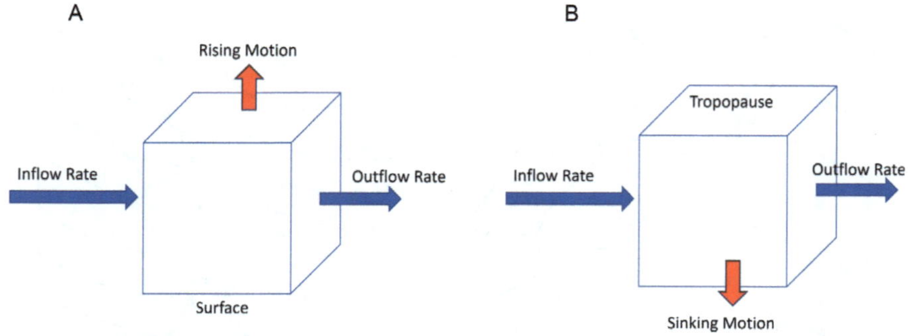

FIGURE. 9.4 Schematic of convergence and vertical motion. The volume of space through which air flows where the mass inflow exceeds the mass outflow. (A) The volume of space is near the surface, leading to rising motion. (B) The volume of space is near the tropopause, leading to sinking motion.

-No Permission Required

offsets the net inflow. This same reasoning applies to confluence, where the curving of wind and isohypses increases mass within the volume, triggering initiating vertical motion.

A similar argument applies when outflow exceeds inflow, as shown in Figure 9.5. Here, the net outflow represents speed divergence, implying a decrease in mass. Divergence therefore represents the rate at which mass is being removed from a volume. If sustained indefinitely, this would lead to decreasing air density.

To restore balance, the atmosphere compensates through vertical motion. As shown in Figure 9.5A, divergence near the surface induces sinking motion aloft. In Figure 9.5B, divergence near the tropopause produces rising motion at the bottom of the volume. In each case, the induced vertical motion is just enough to offset the imbalance. The same reasoning applied to diffluence, which, by spreading out wind and isohypses, removes mass from the volume and initiates vertical motion.

Combining the arguments from Figures 9.4 and 9.5, we can construct a three-dimensional conceptual model of atmospheric motion. Net divergence in the upper-troposphere induces rising motion, which is typically balanced by convergence in the lower troposphere (Figure 9.6A). This process is known as the **chimney effect**. Conversely, net divergence in the upper troposphere induces sinking motion, which produces divergence below (Figure 9.6B). This process is known as the **damper effect**.

These effects help explain how upper-tropospheric divergence and convergence influence surface pressure systems. If upper-tropospheric convergence exceeds lower-tropospheric divergence, sinking motion strengthens, adding mass below the convergence region and raising surface pressure. Thus, strong upper-tropospheric convergence is associated with rising surface pressure.

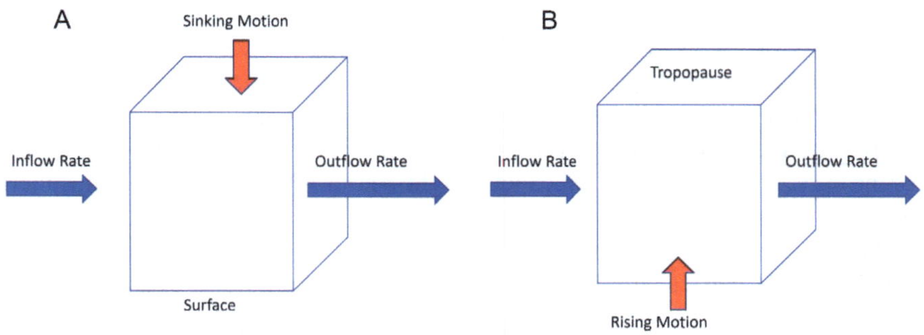

FIGURE. 9.5 Schematic of divergence and vertical motion. The volume of space through which air flows where the mass outflow exceeds the mass inflow. (A) The volume of space is near the surface, leading to sinking motion. (B) The volume of space is near the tropopause, leading to rising motion.

-No Permission Required

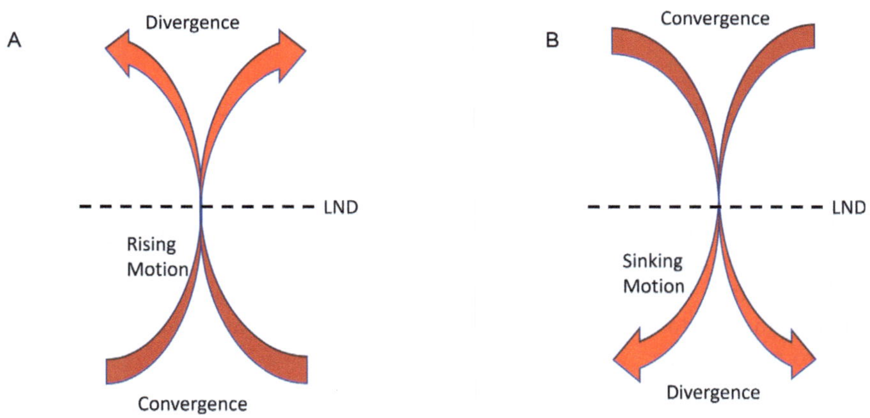

FIGURE. 9.6 Schematic of the Dines' compensation principle. The three-dimensional atmospheric flow in response to divergence. (A) The chimney effect. (B) The damper effect. LND, level of nondivergence.

-No Permission Required

Conversely, if upper-tropospheric divergence exceeds lower-tropospheric convergence, rising motion intensifies, removing mass from the surface and lowering surface pressure. Hence, strong upper-tropospheric divergence is linked to falling surface pressure.

Together, the chimney effect and the damper effect imply that convergence at one level is offset by divergence at another. This leads to the conclusion that there must exist a level in the troposphere where divergence is zero. This is the **Dines' compensation principle**, which states that at least one **level of nondivergence (LND) must exist**, as shown in Figure 9.6. The LND typically lies near 550 hPa, though its exact height varies depending on atmospheric stability.

Having developed a conceptual framework for the relationship between divergence and vertical motion, we now turn to a quantitative treatment. Using mass conservation, we will formally derive the relationship between divergence and vertical motion - known as the continuity equation - in the following section.

9.2.2 The continuity equation

Consider a small region of space with a volume $\Delta V = \Delta x \Delta y \Delta z$ through which air flows. First, we will derive an expression for the mass of this region. Using the hydrostatic equation $\Delta P = -\rho g \Delta z$, we can rewrite ΔV as:

$$\Delta V = \frac{-\Delta x \Delta y \Delta P}{\rho g}$$

Therefore, the mass of the air mass in this volume is given by:

$$\Delta M = \rho \Delta V = -\frac{\Delta x \Delta y \Delta P}{g}$$

The rate at which mass flows at the center of the cube is known as **mass flux** and is defined as $d(\Delta M)/dt$.

To maintain constant density in the cube, the *net* mass flux must be zero, implying that the total mass of this region is constant. We can express the condition for zero mass flux as:

$$\frac{1}{\Delta M}\frac{d(\Delta M)}{dt} = 0$$

$$\Rightarrow -\frac{g}{\Delta x \Delta y \Delta P}\frac{d}{dt}\left(\frac{-\Delta x \Delta y \Delta P}{g}\right) = 0$$

Applying the chain rule to the above expression gives:

$$\frac{1}{\Delta x \Delta y \Delta P}\left[\frac{d(\Delta x)}{dt}\Delta y \Delta P + \frac{d(\Delta y)}{dt}\Delta x \Delta P + \frac{d(\Delta P)}{dt}\Delta x \Delta y\right] = 0$$

We note that:

$$\frac{d(\Delta x)}{dt} = \Delta u, \quad \frac{d(\Delta y)}{dt} = \Delta v, \quad \frac{d(\Delta P)}{dt} = \Delta \omega$$

where Δu is the zonal velocity component, Δv is the meridional velocity component, and $\Delta\omega$ is the omega velocity (i.e., vertical velocity in isobaric coordinates) introduced in Ch. 5. Therefore, we obtain:

$$\frac{1}{\Delta x \Delta y \Delta P}[\Delta u \Delta y \Delta P + \Delta v \Delta x \Delta P + \Delta\omega \Delta x \Delta y] = 0$$

$$\Rightarrow \frac{\Delta u}{\Delta x} + \frac{\Delta v}{\Delta y} + \frac{\Delta\omega}{\Delta P} = 0$$

As $\Delta x, \Delta y, \Delta P \rightarrow 0$, these difference quotients become derivatives, and we arrive at the final form:

$$\frac{\partial u}{\partial x} + \frac{\partial v}{\partial y} + \frac{\partial\omega}{\partial P} = 0 \tag{9.3}$$

Eq. (9.3) represents the formal statement of mass conservation and is known as the **continuity equation**. Notice that the first two terms on the left-hand side of Eq. (9.3) correspond to the horizontal divergence, denoted as δ_H. Therefore, we can rewrite the continuity equation as:

$$\delta_H = -\frac{\partial\omega}{\partial P}$$

This expression shows that horizontal divergence on an isobaric surface is directly related to vertical motion ω, as expected from our conceptual arguments in the previous section.

If we know the vertical distribution of horizontal divergence in an atmospheric column, we can also determine the vertical motion profile using the continuity equation. Consider the situation in Figure 9.6A, where air converges horizontally near the surface ($\delta_H < 0$) and diverges at the top of the column ($\delta_H < 0$). To demonstrate that this must result in rising motion within the column, we integrate Eq. (9.3) with respect to pressure:

$$\int_{p_s}^{p_t} \delta_H dP = -\int_{p_s}^{p_t} d\omega$$

$$\Rightarrow \delta_{H,t} - \delta_{H,s} = -(\omega_t - \omega_s)$$

where $\delta_{H,t}$ and $\delta_{H,s}$ represent the horizontal divergence at the top of the atmosphere and at the surface, respectively. Since there is no vertical motion at ground level, $\omega_s = 0$. Given that $\delta_{H.t} > \delta_{H.s}$, it follows that $\omega_t < 0$, confirming the presence of upward vertical motion at the top of the column.

In general, divergence varies throughout the entire atmospheric column. Integrating Eq. (9.3) more generally with respect to pressure gives:

$$\int_{p_s}^{p} \delta_H dP = -\int_{p_s}^{p} d\omega$$

$$\Rightarrow (\omega - \omega_s) = -\int_{p_s}^{p} \delta_H dP$$

where ω is evaluated at pressure p. Since $\omega_s = 0$ at ground level, this reduces to:

$$\omega(p) = -\int_{p_s}^{p} \delta_H \, dP \tag{9.4}$$

Because δ_H is typically computed from surface or upper-air observations, we evaluate Eq. (9.4) numerically. Box 9.1 illustrates how horizontal divergence can be approximated using meteorological data, and Box 9.2 demonstrates how Eq. (9.4) can be applied in practice.

9.2.3 Divergence and ageostrophic flow

Another valuable physical insight can be gained from the continuity equation by considering horizontal wind as the sum of its balanced geostrophic component and its unbalanced ageostrophic component. Recall from Ch. 7 that the ageostrophic wind vector is defined as $\boldsymbol{V}_{ag} = \boldsymbol{V} - \boldsymbol{V}_g$. Substituting this into Eq. (9.3) gives:

$$\nabla \cdot \boldsymbol{V}_g + \nabla \cdot \boldsymbol{V}_{ag} = -\frac{\partial \omega}{\partial P} \tag{9.5}$$

Here, the first term on the left-hand side of Eq. (9.5) represents the horizontal divergence of the geostrophic wind, whereas the second term corresponds to the horizontal divergence of the ageostrophic wind. It was shown previously that the geostrophic wind on isobaric surfaces can be expressed as:

$$\boldsymbol{V}_g = \frac{g_0}{f}(\boldsymbol{k} \times \nabla_P Z) = \frac{g_0}{f}\left(-\frac{\partial Z}{\partial y}\boldsymbol{i} + \frac{\partial Z}{\partial x}\boldsymbol{j}\right)$$

where Z is the geostrophic height. Taking the divergence of geostrophic wind vector (assuming the Coriolis parameter is constant) yields:

$$\nabla \cdot \boldsymbol{V}_g = \frac{g_0}{f}\left[\frac{\partial}{\partial x}\left(-\frac{\partial Z}{\partial y}\right) + \frac{\partial}{\partial y}\left(\frac{\partial Z}{\partial x}\right)\right] = 0 \tag{9.6}$$

Thus, f the geostrophic wind is nondivergent, as we discussed in Ch. 9.1. Substituting Eq. (9.6) into Eq. (9.5) we find:

$$\nabla \cdot \boldsymbol{V}_{ag} = -\frac{\partial \omega}{\partial P} \tag{9.7}$$

Since wind divergence arises from the ageostrophic component of atmospheric flow, the horizontal divergence of this ageostrophic flow determines the vertical

Box 9.1 Approximating the horizontal divergence using meteorological data

To calculate the divergence at a specific location, we can approximate the derivative using what is known as a finite different approach. Introductory calculus shows that the definition of the derivative is given by:

$$\frac{df}{dx} = \lim_{\Delta x \to 0} \frac{f(x+\Delta x) - f(x)}{\Delta x}$$

For sufficiently small values of Δx, the derivative can be approximated as an algebraic difference:

$$\frac{df}{dx} \approx \frac{\Delta f}{\Delta x}$$

where Δf is the change of the function f over the interval Δx. Applying this approximation to the horizontal divergence gives the following result:

$$\delta_H = \frac{\partial u}{\partial x} + \frac{\partial v}{\partial y} \approx \frac{\Delta u}{\Delta x} + \frac{\Delta v}{\Delta y}$$

where Δu is the change of the zonal component of the wind across the zonal distance Δx and Δv is the change of the meridional component of the wind across the meridional distance Δy. Let's apply this concept to the following example.

At a certain surface observation station, we have obtained the following data: (1) 50 km east of the station, the wind is 10 m/s from 90°; (2) 50 km north of the station, the wind is 4 m/s from 120°; (3) 50 km west of the station, the wind is 8 m/s from 90°; and (3) 50 km west of the station, the wind is 4 m/s from 60°. A sketch of this scenario is given in Figure 9.7. Based on this figure, the horizontal divergence can be approximated as:

$$\delta_H \approx \frac{\Delta u}{\Delta x} + \frac{\Delta v}{\Delta y} = \frac{u_E - u_W}{100 \quad \text{km}} + \frac{v_N - v_S}{100 \quad \text{km}}$$

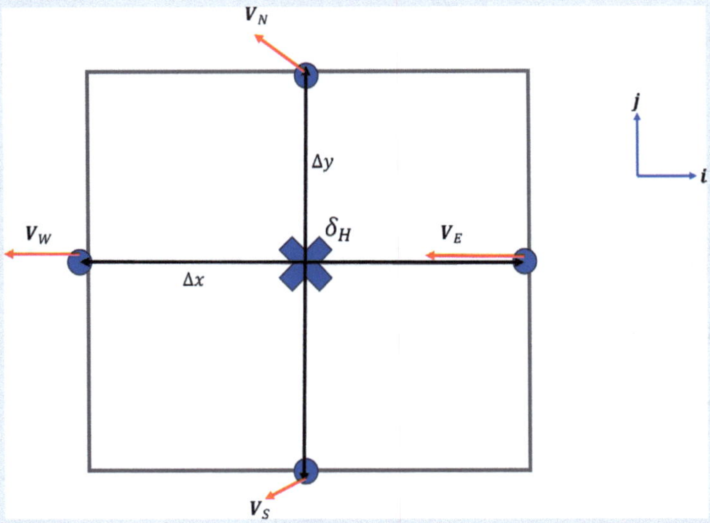

FIGURE. 9.7 Schematic associated with Box 9.1. The wind vectors at each location associated with Box 9.1. The subscripts "E," "N," "S," and "W" correspond to east, north, south, and west of the station, respectively.

-No Permission Required

Computing the components of the wind vectors gives us:

Continued

Box 9.1 Approximating the horizontal divergence using meteorological data—cont'd

$$u_E = -V_E \sin(\alpha) = -\left(10.0\frac{m}{s}\right)\sin(90^\circ) = -10.0\ \frac{m}{s}$$

$$u_W = -V_W \sin(\alpha) = -\left(8.0\frac{m}{s}\right)\sin(90^\circ) = -8.0\ \frac{m}{s}$$

$$v_N = -V_N \cos(\alpha) = -\left(4.0\frac{m}{s}\right)\cos(120^\circ) = 2.0\ \frac{m}{s}$$

$$v_S = -V_S \cos(\alpha) = -\left(4.0\frac{m}{s}\right)\cos(60^\circ) = -2.0\ \frac{m}{s}$$

Therefore, the approximate horizontal divergence is given by:

$$\delta_H \approx = \frac{\left(-10.0\frac{m}{s}\right) - \left(-8.0\frac{m}{s}\right)}{1.00 \times 10^5\ \ m} + \frac{\left(2.0\frac{m}{s}\right) - \left(-2.0\frac{m}{s}\right)}{1.00 \times 10^5\ \ m} = 2 \times 10^{-5}\ \ s^{-1}$$

Box 9.2 Approximating omega velocity using meteorological data

Introductory calculus shows that the integral is defined as

$$\int_a^b f dx = \lim_{n\to\infty} \sum_{i=1}^{n} \overline{f}(x_i)\Delta x_i, \quad \Delta x_i = \frac{b-a}{n}$$

where $\overline{f}(x_i)$ is the average value of the function $f(x)$ over the width Δx_i. For sufficiently small values of Δx_i, the integral can be approximated as a summation:

$$\int_a^b f dx \approx \sum_{i=1}^{n} f(x_i)\Delta x_i$$

Applying this approximation to Eq. (9.4) gives us:

$$\omega \approx -\sum_{i=1}^{N} \overline{\delta}_{H,i}\Delta P_i$$

where $\overline{\delta}_{H,i}$ is the average horizontal divergence within a given layer and ΔP_i is the pressure change over a given layer. Therefore, if we know the horizontal divergence over multiple layers in an atmospheric column, we can evaluate the omega velocity at the top of the atmospheric column.

As an example, suppose that the horizontal divergence at a certain station has been calculated, as shown in Table 9.1. The average horizontal divergence in each layer is given in Table 9.2.

Table 9.1 Horizontal divergence at different pressure levels

Pressure (hPa)	Horizontal divergence $\left(\times 10^{-5}\ \ s^{-1}\right)$
1000	+0.9
850	+0.6
700	+0.3
500	0
300	−0.8
100	−1.0

No Permission Required

Box 9.2 Approximating omega velocity using meteorological data—cont'd

Table 9.2 Average horizontal divergence in each layer

Layer	ΔP_i (hPa)	Average horizontal divergence $\left(\times 10^{-5}\ s^{-1}\right)$
1000–850 hPa	150	+0.75
850–700 hPa	150	+0.45
700–500 hPa	200	+0.15
500–300 hPa	200	−0.40
300–100 hPa	200	−0.90

No Permission Required

Therefore, the omega velocity at 100-hPa is given by:

$$\omega \approx -\sum_{i=1}^{N} \delta_{H,i}\Delta P_i$$

$$= -\left[(0.75\times 10^{-5}\ s)(15000\ Pa) + (0.45\times 10^{-5}\ s)(15000\ Pa) + (0.15\times 10^{-5}\ s)(20000\ Pa) + (-0.40\times 10^{-5}\ s)(20000\ Pa) + (-0.90\times 10^{-5}\ s)(20000\ Pa)\right]$$

$$= 0.0500\frac{Pa}{s}$$

Notice that divergence within the lower troposphere and convergence within the upper troposphere lead to sinking motion.

No Permission Required

motion within the atmosphere, as shown in Eq. (9.7). Consequently, it is the ageostrophic component of the flow that governs significant weather phenomena- such as clouds, precipitation, and frontal systems- which are central to the study of synoptic meteorology.

We now derive an explicit relationship between the ageostrophic wind vector and the horizontal acceleration vector. Recall that the frictionless equations of motion for the horizontal wind can be expressed in vector form as:

$$\frac{d\boldsymbol{V}_H}{dt} = -g_0\nabla_P Z - f\boldsymbol{k}\times\boldsymbol{V}_H$$

Taking the vector product of this expression gives:

$$\frac{\boldsymbol{k}}{f}\times\frac{d\boldsymbol{V}_H}{dt} = -\frac{g_0}{f}\boldsymbol{k}\times\nabla_p Z + \frac{\boldsymbol{k}}{f}\times(-f\boldsymbol{k}\times\boldsymbol{V}_H)$$

Using $\boldsymbol{k} \times (\boldsymbol{k} \times \boldsymbol{V}_H) = -\boldsymbol{V}_H$ and the definition of the geostrophic wind vector, we obtain:

$$\frac{\boldsymbol{k}}{f} \times \frac{d\boldsymbol{V}_H}{dt} = \boldsymbol{V} - \boldsymbol{V}_g$$

$$\Rightarrow \boldsymbol{V}_{ag} = \frac{\boldsymbol{k}}{f} \times \frac{d\boldsymbol{V}_H}{dt} \tag{9.8}$$

From Eq. (9.8), we see that the ageostrophic wind vector has two key properties:

1. The magnitude of the ageostrophic wind is proportional to the magnitude of the horizontal acceleration in the atmosphere.
2. Because of the properties of the cross product, the ageostrophic wind vector is directed 90° to the left of the acceleration vector (i.e., 90° counterclockwise) in the Northern Hemisphere.

Having discussed divergence and vertical motion, we can now turn to the dynamics of upper-tropospheric jets.

9.3 Observations of upper-tropospheric jets

In synoptic meteorology, there are two primary types of upper-tropospheric jets: the polar jet and the subtropical jet. We will begin by examining the structure of the polar jet before turning to the subtropical jet.

9.3.1 The polar jet stream

The **polar jet** is the most prominent upper-tropospheric jet and is readily identified on upper-tropospheric charts, as shown in Figure 9.8. Observations indicate that the polar jet commonly occurs in temperate, midlatitude regions near the −17°C isotherm at 500 hPa, which roughly corresponds to latitudes of 50°-60° in both hemispheres (Vasquez, 2006). This position generally aligns with the 500-hPa front, which results from the strong thermal contrast between cold polar air masses and warmer tropical air masses. The boundary between these two air masses is known as the **polar front**. For this reason, the polar jet is sometimes referred to as the polar front jet.

To first order, the strength and vertical position of the jet stream can be explained by thermal wind balance. As shown in Ch. 7, the thermal wind vector can be written in component form as follows:

$$\frac{\partial u_g}{\partial p} = \frac{R_d}{fp}\frac{\partial T_v}{\partial y}, \quad \frac{\partial v_g}{\partial p} = -\frac{R_d}{fp}\frac{\partial T_v}{\partial x}$$

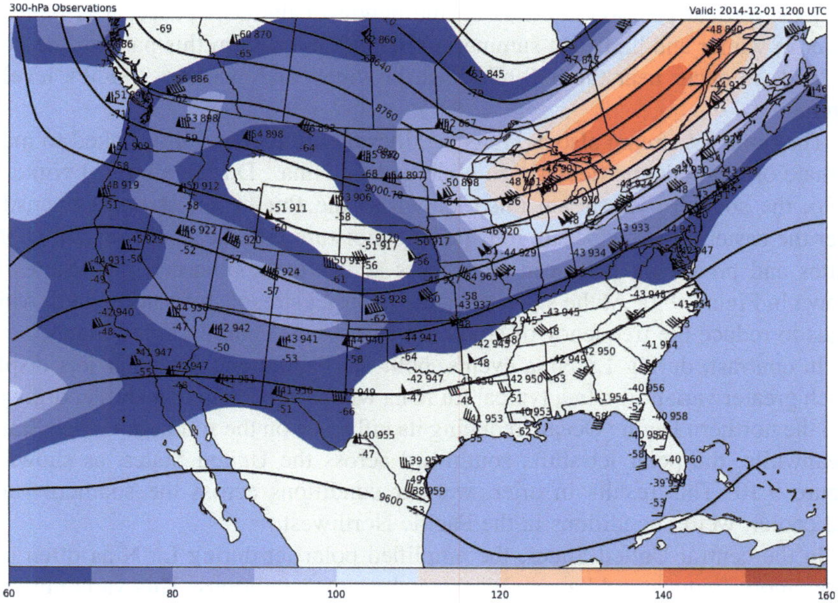

FIGURE. 9.8 300-hPa map analysis of polar jet stream. 300-hPa map analysis from December 1, 2014, at 12 UTC. Solid black contours are isohypses and contour colors are isotachs.

-No Permission Required

In the Northern Hemisphere, the polar front creates a temperature gradient such that $\partial T_v/\partial y < 0$. Since $f > 0$ in the Northern Hemisphere, thermal wind balance implies that $\partial u_g/\partial p < 0$, corresponding to strong westerly flow in the upper troposphere. In the Southern Hemisphere, the polar front produces $\partial T_v/\partial y > 0$. Because $f < 0$ in this hemisphere, thermal wind balance again suggests that $\partial u_g/\partial p < 0$, resulting in strong westerly flow as well. Thus, in both hemispheres, the polar jet is associated with strong upper-tropospheric westerlies concentrated above the polar front.

Thermal wind balance also helps to explain the seasonal variability of the polar jet. During winter, the polar front shifts equatorward, intensifying the temperature gradient and causing the mean position of the polar jet to migrate equatorward and strengthen. In summer, the polar jet shifts poleward and weakens. Because the jet's vertical position depends on the height of the tropopause, it is typically lower in winter and higher in summer. Observations confirm this pattern: the polar jet tends to occur near the 300-hPa level in winter and near the 200-hPa level in summer.

The polar jet stream also exhibits significant natural variability linked to major climate oscillations, particularly El Niño and La Niña.[1] During a typical winter El Niño, the southern jet stream (also known as the **Pacific jet stream**) intensifies over the eastern Pacific Ocean, transporting moisture into the Southwestern United States and producing enhanced storminess across the southern United States (as shown in Figure 9.9). At the same time, the polar jet stream shifts northward, which tends to reduce the frequency of cold outbreaks in the central United States.

In contrast, during La Niña events, both the Pacific and the polar jets display much greater variability. In a typical La Niña winter, the Pacific jet shifts northward into the northern Pacific Ocean, reducing its influence on the southern United States. Meanwhile, the polar jet shifts southward across the United States, as shown in Figure 9.10. This results in drier, warmer conditions across the southern states and cooler, wetter conditions in the Pacific Northwest.

In the central United States, the amplified polar jet during La Niña often produces more frequent cold air outbreaks and wetter conditions. This variability can be seen by comparing the average structure of the polar jet stream during El Niño and La Niña years, as shown in Figure 9.11. During El Niño winters, the polar jet tends to adopt a more zonal pattern, often associated with calmer conditions. In contrast, during La Niña winters, the polar jet exhibits stronger meridional flow. Additionally, the associated trough shifts westward from its normal position toward the central Pacific.

9.3.2 The subtropical jet stream

Forecasters often observe bands of strong winds in subtropical latitudes, such as along the U.S.-Mexico border. These jets, known as the **subtropical jet**, resemble the polar jet but differ in thermal structure and characteristics. The subtropical jet is an important synoptic-scale feature, with wind speeds that can exceed 150 knots.

---

[1] El Niño and La Niña are the atmospheric component of the coupled atmospheric-ocean oscillation that occurs 3–7 years on average known as the El Niño Southern Oscillation (ENSO). Along with the jet stream changes mentioned in the text, there is above average warming of the central and eastern Tropical Pacific Ocean during an El Niño season, and there is below-average cooling of the central and eastern tropical Pacific Ocean during a La Niño season.

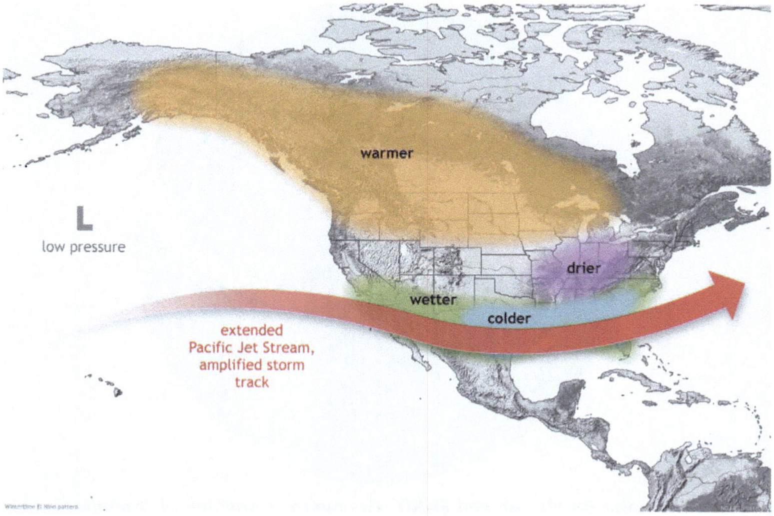

FIGURE. 9.9 The effect of El Niño on winter jet stream pattern. The average position of the jet stream during a typical winter El Niño season. Notice that the Pacific jet stream dominates, and that the southern half of the United States receives enhanced moisture, whereas the central United States receives drier weather.

-No Permission Required

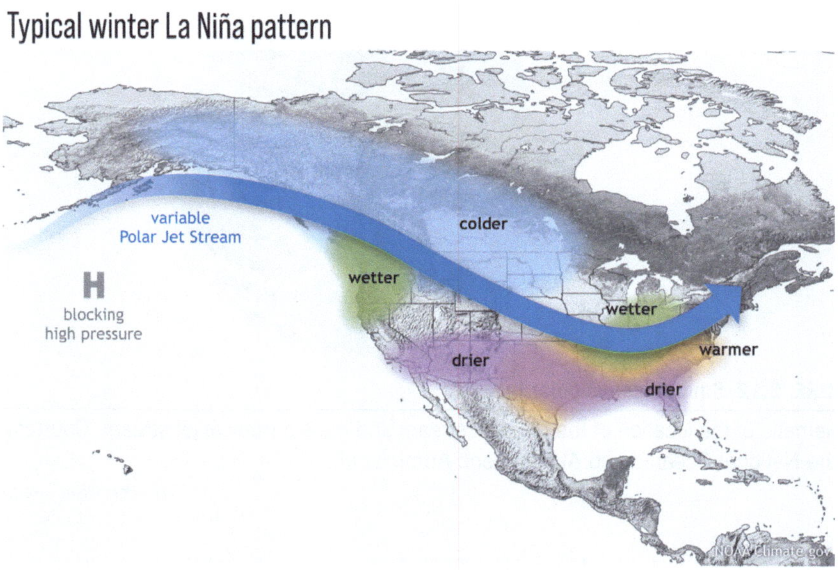

FIGURE. 9.10 The effect of El Niño on winter jet stream pattern. The average position of the jet stream during a typical winter La Niña season. The change in jet stream location leads to various shifts in precipitation across the United States.

-No Permission Required

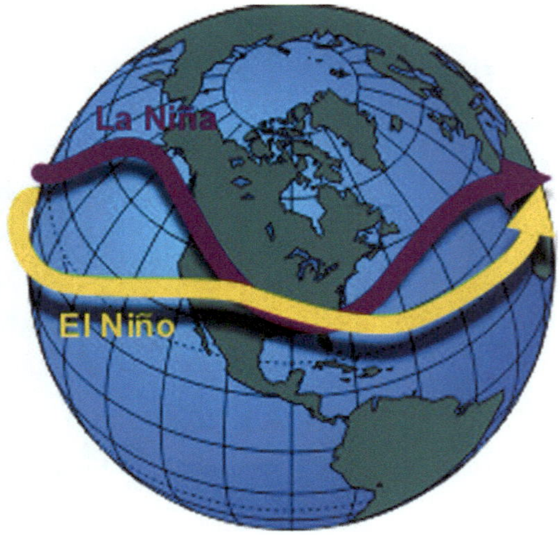

FIGURE. 9.11 The polar jet stream and ENSO. The average position of the polar jet stream during El Niño (in yellow) and La Niña (in purple). Courtesy of the National Oceanic and Atmospheric Administration. ENSO, El Niño Southern Oscillation.

-No Permission Required

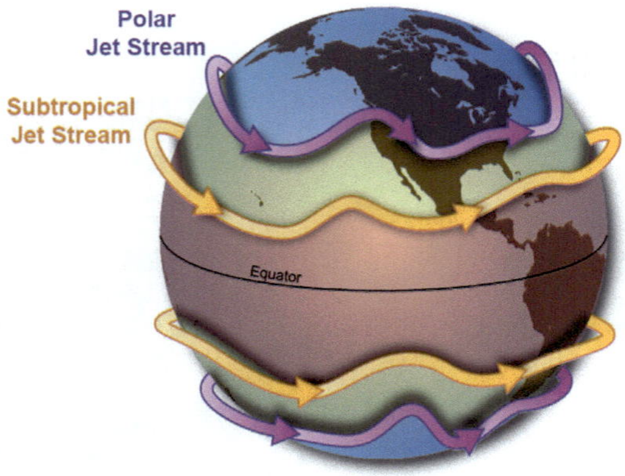

FIGURE. 9.12 Subtropical vs polar jet

Schematic of the location of the polar jet stream and the subtropical jet stream. Courtesy of the National Oceanic and Atmospheric Administration.

-No Permission Required

Observations show that the subtropical jet is typically located near the −11°C isotherm at the 500-hPa level, roughly corresponding to latitudes of 20° and 30° in both hemispheres, as shown in Figure 9.12 (Vasquez, 2006). It is not a permanent feature, but when present, its mean latitude is about 28°. The jet is strongest during winter, as illustrated in Figure 9.13, but often disappears during summer, especially in the Southern Hemisphere, leaving behind weak zonal winds. In winter, the subtropical jet plays a key role in transporting tropical air poleward, which enhances midlatitude weather systems. Unlike the polar jet, the subtropical jet is confined to the upper troposphere and rarely extends into the middle troposphere, as shown in Figure 9.13.

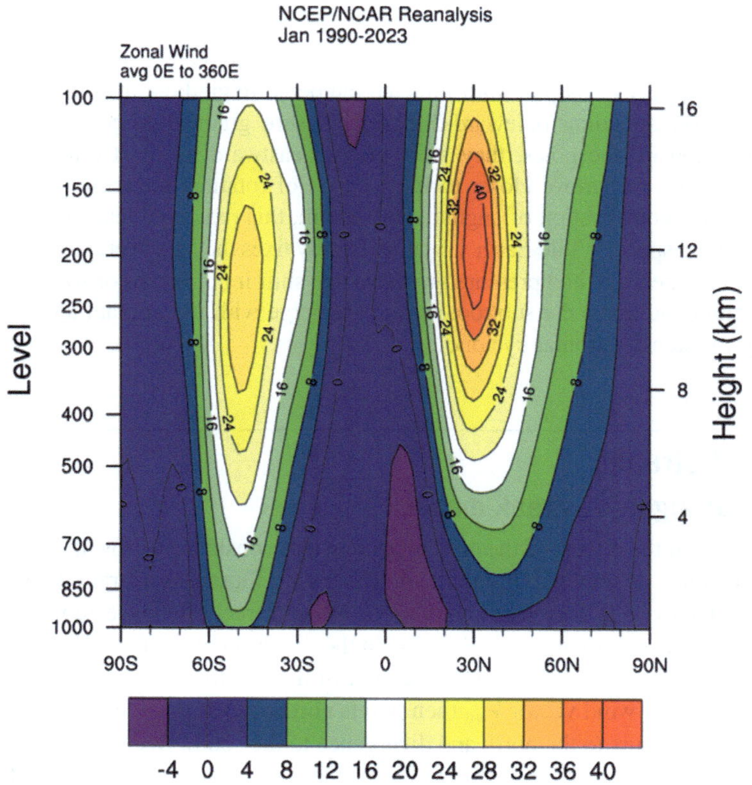

FIGURE. 9.13 Reanalysis of subtropical jet

Vertical cross-section of the longitudinally averaged zonal wind (averaged over all longitudes) from January 1990 through January 2023 from NCEP/NCAR reanalysis. Notice the presence of the subtropical jet in the Northern Hemisphere winter and the absence of the subtropical jet in the Southern Hemisphere summer. *NCEP*, National Centers for Environmental Prediction; *NCAR*, National Center for Atmospheric Research.

Image provided by the National Oceanic and Atmospheric Administration Physical Sciences Laboratory, Boulder, Colorado, from their website at https://psl.noaa.gov/.

During winter, the subtropical jet acts as a *semi-permanent* feature, with air parcels accelerating along its axis as segments of the jet extend into higher latitudes. This behavior can be partly explained by conservation of angular momentum. In introductory physics, angular momentum is given by $L = I\omega$, where ω is the angular speed and I is the moment of inertia (a measure of resistance to changes in rotational motion).[2] The moment of inertia depends on both the total mass and its distribution relative to the axis of rotation: it is smaller when mass is concentrated near the axis, and larger when mass is distributed farther away. In simple terms, objects with mass closer to their axis of rotation are easier to spin, whereas those with mass farther away are harder to rotate.

Applying this principle to the atmosphere, consider air parcels near the equator, as shown in Figure 9.12. A poleward-directed temperature gradient in both hemispheres creates a corresponding poleward-directed pressure gradient force. As air parcels accelerate toward the poles, they are deflected eastward by the Coriolis force. At the same time, they move closer to Earth's axis of rotation, reducing their moment of inertia. By conservation of angular momentum, this reduction leads to an increase in zonal wind speed. While Earth's rotation largely fixes the jet's average position, variations in temperature and pressure gradients can shift it northward or southward. Under certain conditions, the subtropical jet can merge with the midlatitude jet stream, helping to intensify frontal cyclones.

9.4 Jet streaks

9.4.1 Straight jet streak model

To understand the effects of the ageostrophic wind, consider the flow through an isolated jet streak, as shown in Figure 9.14. This figure illustrates a straight jet streak at the 300-hPa level, with isotachs contoured. An air parcel on the western edge of the entrance region experiences acceleration in the direction of the flow. Hence, the vector $d\boldsymbol{V}/dt$ points eastward toward the center of the jet streak. According to Eq. 9.8, the ageostrophic wind vector $\boldsymbol{V}_{ag}$ then points northward at this location. As a result, convergence occurs to the north and divergence to the south. Because this schematic represents airflow at the 300-hPa level, upper-level divergence corresponds to rising motion in the air column, while upper-level convergence corresponds to sinking motion, consistent with the continuity equation.

Similarly, an air parcel on the eastern edge of the exit region experiences deceleration opposite the flow. In this case, the horizontal acceleration vector $d\boldsymbol{V}/dt$ points westward toward the jet streak center, causing $\boldsymbol{V}_{ag}$ to point southward. This distribution of ageostrophic winds in the exit region produces convergence to

[2] This relationship assumes that we are examining a rigid body rotating with a fixed axis of rotation.

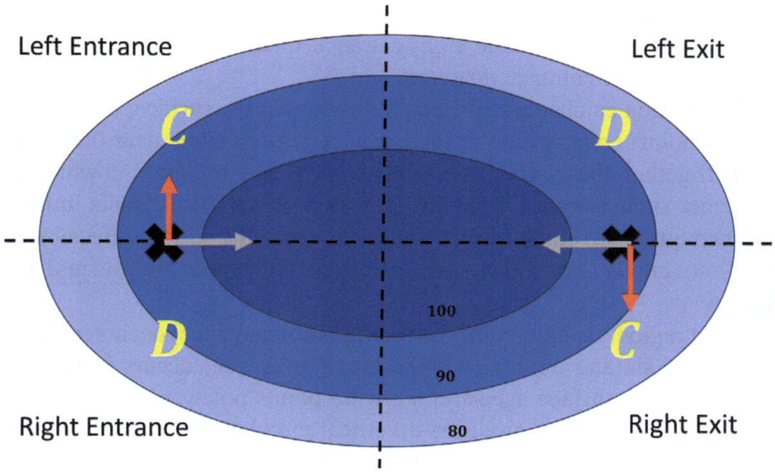

FIGURE. 9.14 Schematic of a straight jet streak

A straight jet streak. Solid black contours are isotachs in m/s. The gray vector arrow corresponds to the horizontal acceleration vector, and the red vector arrow corresponds to the ageostrophic wind vector. "C" corresponds to a region of convergence and "D" corresponds to a region of divergence.

-No Permission Required

the south and divergence to the north, with upward motion beneath the area of maximum divergence.

The flow through a jet streak can also be understood in terms of the pressure gradient force (PGF) and the Coriolis force. Air in geostrophic balance entering the entrance region encounters tighter isohypses. Initially, the PGF acts to the left of the parcel's motion. As the parcel moves into the region of tighter isohypses, the PGF becomes dominant, making the motion ageostrophic and accelerating it leftward. As velocity increases, the Coriolis force strengthens until balance is re-established. The parcel then continues through the jet streak in balanced flow.

Upon exiting the streak, the parcel enters a region of broader isohypses, where the PGF weakens. At this stage, the Coriolis force (acting to the right) dominates, causing the parcel to move ageostrophically to the right and decelerate. As velocity decreases, the Coriolis force decreases, allowing the parcel to return to balance. These imbalances generate convergence and divergence, which in the upper troposphere must be compensated by vertical motion.

As argued earlier, enhanced divergence and convergence around jet streaks produce vertical motions according to the continuity equation. In the entrance region, rising motion occurs in the right-entrance quadrant and sinking motion in the left-entrance quadrant. Dines' compensation principle implies corresponding lower-tropospheric convergence in the right-entrance quadrant and lower-tropospheric divergence in the left-entrance quadrant.

This circulation can be described as follows: near the surface, the ageostrophic wind vector points equatorward in the entrance region, while near the jet it points poleward. Together, these motions establish a vertical circulation in which air rises in the warmer right-entrance quadrant, moves poleward aloft, sinks in the cooler left-entrance quadrant, and returns equatorward near the surface. Because warmer air rises and cooler air sinks, this pattern is known as a **thermally direct circulation**.

In the exit region, rising motion (with surface convergence) occurs in the left-exit quadrant, while sinking motion (with surface divergence) occurs in the right-exit quadrant. Here, the surface ageostrophic wind points poleward, while near the jet it points equatorward. This produces a vertical circulation opposite to that of the entrance region: warm air sinks and cool air rises, which defines a **thermally indirect circulation**.

In summary, a straight jet streak is associated with two vertical circulations: a thermally direct circulation in the entrance region (which transports warm air upward, decreases the environmental lapse rate, and stabilizes the atmosphere) and a thermally indirect circulation in the exit region (which transports cool air upward, increases the lapse rate, and destabilizes the atmosphere). For this reason, the left-exit region of a straight jet streak is typically characterized by enhanced rising motion and weak static stability.

9.4.2 Curved jet streaks

As mentioned earlier, ageostrophic flow produces convergence in the right-rear and left-front quadrants of straight jet streaks, which induces compensating rising motion that transports mass from the lower to the upper troposphere. In contrast, divergence occurs in the left-rear and right-front quadrants, where compensating sinking motion transfers mass from the upper to the lower troposphere.

Now consider a cyclonically curved jet streak, as shown in Figure 9.15A. Cyclonically curved flow enhances divergence and convergence on the poleward side of the jet streak (i.e., the left-exit and left-entrance regions), while suppressing divergence and convergence on the equatorward side (i.e., the right-exit and right-entrance regions).

To understand this behavior, recall that horizontal divergence consists of two components: directional divergence and speed divergence. In the left exit region of the jet streak (see Figure 9.15A), strong synoptic-scale diffluence combines with speed divergence to enhance total divergence. In the left-entrance region, strong synoptic-scale confluence combines with speed convergence to enhance total convergence.

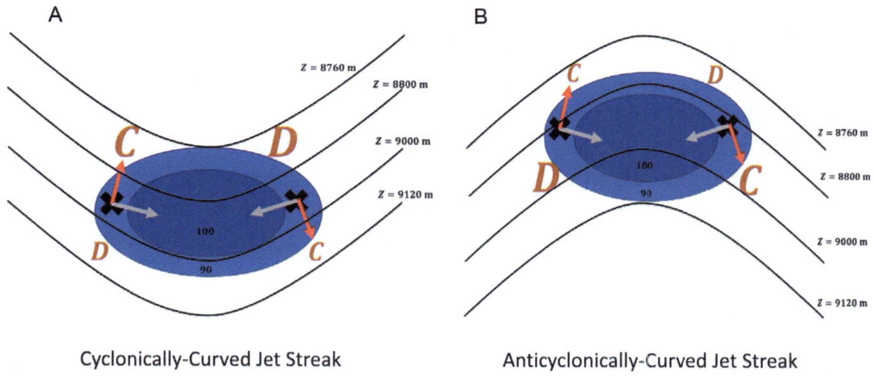

FIGURE. 9.15 Curved jet streaks

(A) Cyclonically curved jet streak. (B) Anticyclonically curved jet streak. The jet core is shaded in blue (with wind speed in m/s), the solid black contours are isohypses associated with 300-hPa flow, the gray vector arrow corresponds to the horizontal acceleration vector, and the red vector corresponds to the ageostrophic wind vector. Schematic of a straight jet streak. Solid black contours are isotachs in m/s. "C" corresponds to a region of convergence and "D" corresponds to a region of divergence.

-No Permission Required

In contrast, in the right-entrance region, speed divergence offsets confluence, and in the right-exit region, speed convergence offset diffluence. These opposing effects weaken the total divergence and convergence equatorward of the jet streak. As a result, rising motion is strongest in the left-exit region and sinking motion in the left-entrance region, with minimal vertical motion changes in the right-entrance and right-exit regions. Thus, enhanced vertical motion on the poleward side of a cyclonically curved jet streak is especially significant.

A similar framework applies to anticyclonic flow associated with a jet streak, as shown in Figure 9.15B. In this case, the directional divergence and convergence generate an enhanced response equatorward and a suppressed response poleward of the jet streak. Consequently, rising motion intensifies in the right-entrance region and sinking motion in the right-exit region. Hence, enhanced vertical motion on the equatorward side of an anticyclonically curved jet streak is of partiular importance.

Questions

1. Under what precise conditions can diffluent flow have zero divergence? Under what precise conditions can confluent flow have zero convergence? Explain your answers. Draw schematic diagrams to illustrate your answers.
2. Explain why the divergence of the ageostrophic wind must be associated with geopotential height decreases.

3. The schematic in Figure 9.16 gives the isohypses associated with a trough-ridge system at 300-hPa level. Suppose that the wind speed is constant and is everywhere parallel to the isohypses.
   a. At the eight points on the figure, use a black pencil to draw the direction of the horizontal acceleration vectors. Defend your answer. If there is no horizontal acceleration at a particular point, explain why.
   b. At the eight points on the figure, use a red pencil to draw the direction of the ageostrophic wind vectors. Defend your answer. If there is no ageostrophic wind vector at a given point, explain why.
4. Repeat Problem 3 if we include the effects of wind speed changes as air parcels move through the trough-ridge system.
5. Ch. 9.2.3 showed that when the Coriolis parameter is constant, the divergence of the geostrophic wind is zero. Would this conclusion still follow if we include change in Coriolis parameter with latitude? Explain your answers. Draw schematic diagrams to illustrate your answers.
6. Consider the 300-hPa contoured chart as shown in Figure 9.17.
   a. Mark the location of troughs, ridges, and short waves using standard conventions.
   b. Use a light blue-colored pencil to shade the regions associated with the presence of the jet stream. Use a purple "X" to mark the location of jet streaks.
   c. Using an orange-colored pencil, shade the regions associated with convergence. Using a yellow-colored pencil, shade the regions of divergence.
7. Consider the 300-hPa contoured chart as shown in Figure 9.18. Identify the states over which rising motion is the strongest. Identify the states over which sinking motion is the strongest. Defend your answer.

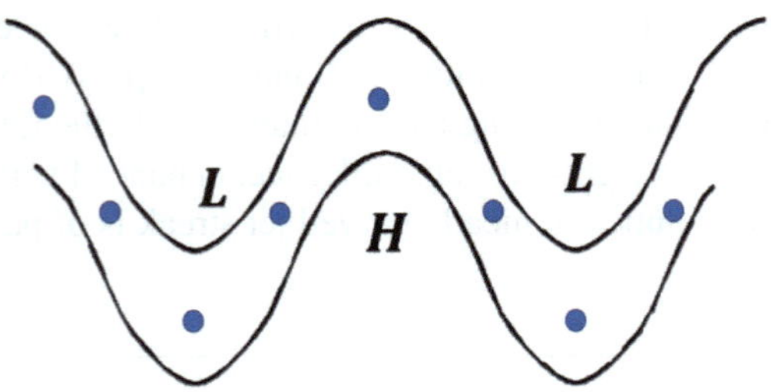

FIGURE. 9.16

Schematic of trough and ridge pattern.

-No Permission Required

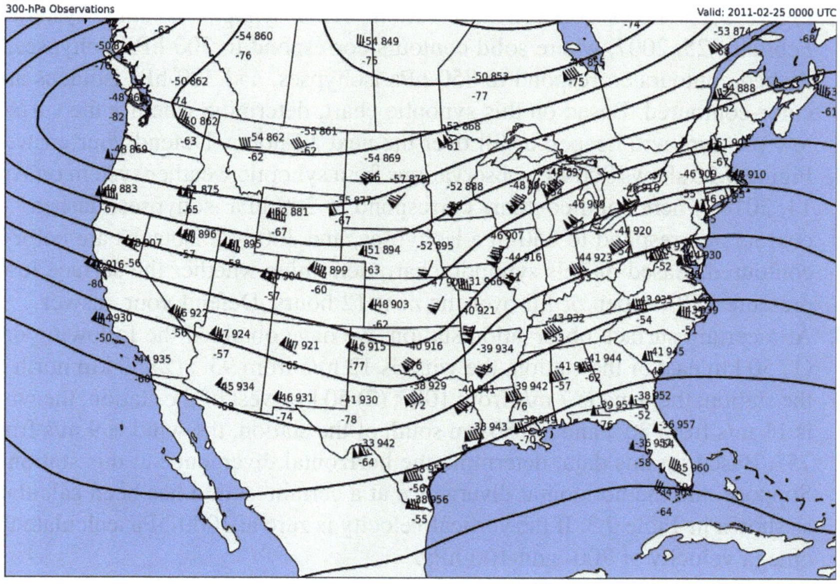

FIGURE. 9.17 300-hPa observations 1

300-hPa observations for February 11, 2005, at 00Z. The solid contours correspond to isohypses.

-No Permission Required

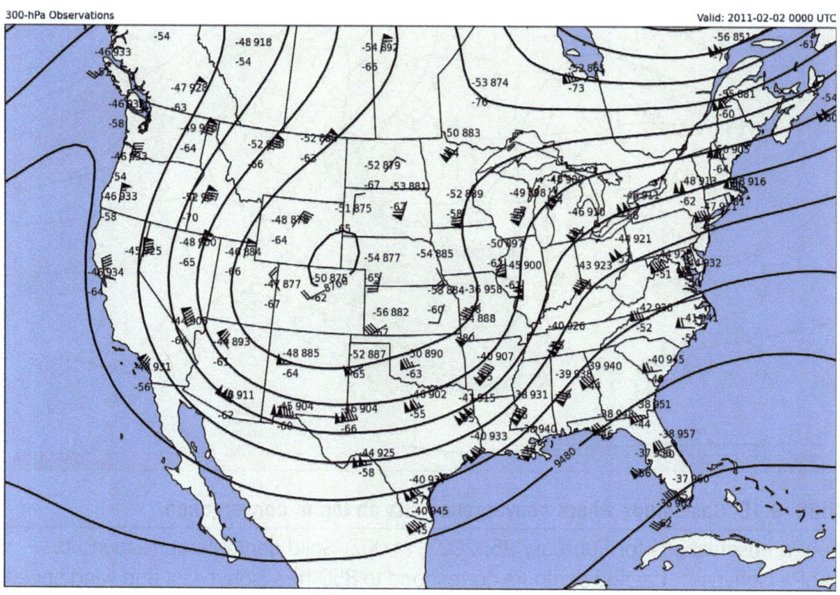

FIGURE. 9.18 300-hPa observations 2

300-hPa observations for February 2, 2011, at 00Z. The solid contours correspond to isohypses.

-No Permission Required

8. Figure 9.19 shows 300-hPa observations for a synoptic weather system on February 25, 2007, where solid contours correspond to 300-hPa isohypses, dashed contours correspond to 850-hPa isohypses, and 300-hPa isotachs are color-contoured. Based on this synoptic chart, determine whether the surface low pressure will deepen or fill over the next 12 hours. Defend your answer.
9. Figure 9.20 shows 300-hPa observations for a synoptic weather system on April 14, 2018, where solid contours correspond to 300-hPa isohypses, dashed contours correspond to 850-hPa isohypses, and 300-hPa isotachs are color-contoured. Based on this synoptic chart, determine whether the surface low pressure will deepen or fill over the next 12 hours. Defend your answer.
10. At a certain surface observation station, we have obtained the following data: (1) 50 km east of the station, the wind is 12 m/s from 95°; (2) 50 km north of the station, the wind is 7 m/s from 100°; (3) 50 km west of the station, the wind is 11 m/s from 20°; and (4) 50 km south of the station, the wind is 9 m/s from 75°. Based on this data, determine the horizontal divergence at this station.
11. Suppose that the horizontal divergence at a certain station has been calculated as shown in Table 9.3. If the vertical velocity is zero at 1000-hPa, calculate the omega velocity at 700- and 100-hPa.

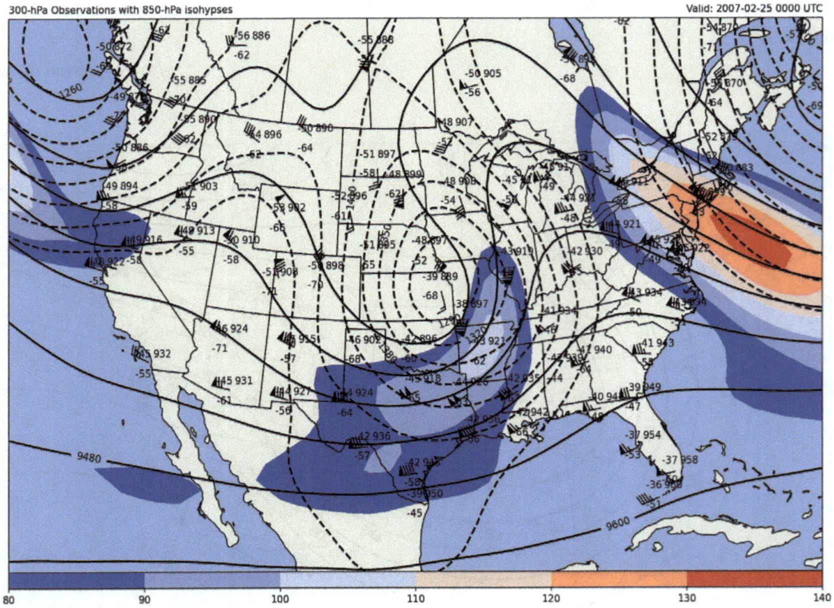

FIGURE. 9.19 Case study where convergence sits on top of convergence

300-hPa observations for February 25, 2007, at 00Z. Solid contours correspond to 300-hPa isohypses, dashed contours correspond to 850-hPa isohypses, and wind speed is given in contour color fill.

-No Permission Required

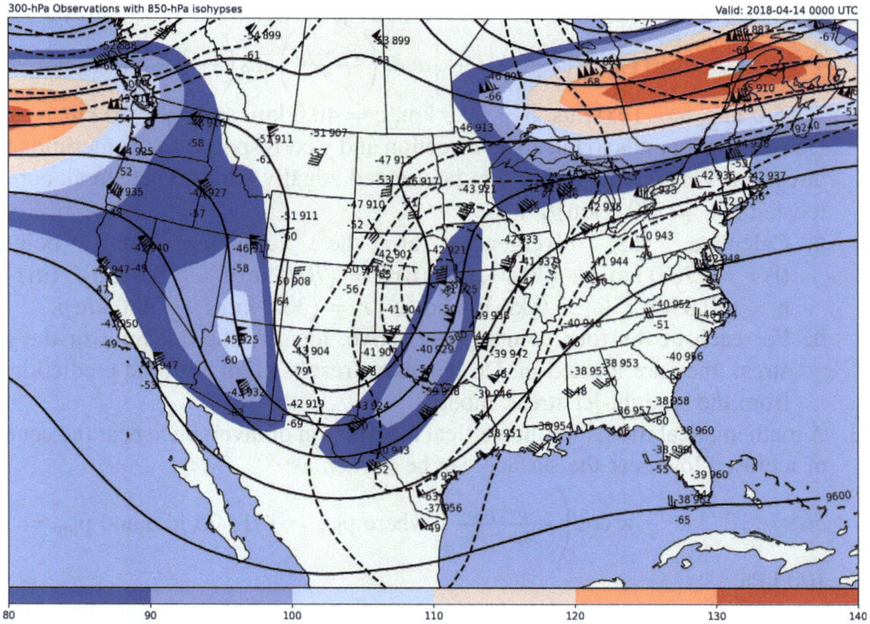

FIGURE. 9.20 Case study where divergence sits on top of convergence

300-hPa observations for April 14, 2018, at 00Z. Solid contours correspond to 300-hPa isohypses, dashed contours correspond to 850-hPa isohypses, and wind speed is given in contour color fill.

-No Permission Required

Table 9.3 Vertical distribution of horizontal divergence for Problem 11

Pressure (hPa)	Horizontal divergence $\left(\times 10^{-5}\ s^{-1}\right)$
1000	−2.9
850	−1.6
700	−0.3
500	1.0
300	1.6
100	2.0

No Permission Required

12. A mathematical model of a 300-hPa jet streak can be written as: $V = \left[U_g + a\sin\left(\frac{y}{b}\right)\cos\left(\frac{x - x_0}{c}\right)\right]\boldsymbol{i} + \left[a\cos\left(\frac{y - y_0/2}{b}\right)\cos\left(\frac{x + x_0/4}{c}\right)\right]\boldsymbol{j}$ where U_g = 55.0 m/s, a = 10.0 m/s, b = 30.0 km, c = 40.0 km, and x_0 = 50 km = y_0. Here, *x* corresponds to the zonal direction and *y* corresponds to the meridional direction. This is a jet streak that is oriented zonally with the jet streak core located at $\boldsymbol{r} = (50.0\text{ km})(\boldsymbol{i} + \boldsymbol{j})$.

a. Calculate the divergence at the following four points: (i) $\boldsymbol{r} = (70.0\text{ km})\boldsymbol{i} + (30.0\text{ km})\boldsymbol{j}$; (ii) $\boldsymbol{r} = (70.0\text{ km})\boldsymbol{i} + (70.0\text{ km})\boldsymbol{j}$; (iii) $\boldsymbol{r} = (30.0\text{ km})\boldsymbol{i} + (30.0\text{ km})\boldsymbol{j}$, and (iv) $\boldsymbol{r} = (30.0\text{ km})\boldsymbol{i} + (70.0\text{ km})\boldsymbol{j}$.

b. For each of the four points given above, identify the quadrant of the jet streak they are located in. Do the above results align with the predictions from the straight-jet streak model?

13. A mathematical model for the vertical distribution of divergence near the center of a closed low near the surface can be written as:

$\delta_H = \left(10^{-4}\text{ s}^{-1}\right) \times \cos\left(\pi\frac{p - p_{top}}{p_{bot} - p_{top}}\right)$ where p_{bot} = 1013.25 hPa and p_{top} = 100 hPa.

a. Assuming that the vertical velocity is zero at p_{top}, determine the omega velocity (in dPa/s) at all mandatory pressure levels. Express your results in tabular form and use your results to make a graph of omega velocity (in dPa/s) vs. pressure (in hPa). What mandatory pressure level has the largest upward velocity?

b. Using the derived expression for omega velocity from part (a), find the pressure level with the largest upward velocity. What is the omega velocity (in dPa/s) and the horizontal divergence (in s^{-1}) at this position?

14. A mathematical model for the vertical distribution of omega velocity near the center of a closed high near the surface can be written as:

$\omega = \left(5.00\ \frac{\text{dPa}}{\text{s}}\right) \times \sin\left(\pi\frac{p - p_{top}}{p_{bot} - p_{top}}\right)$ where p_{bot} = 1013.25 hPa and p_{top} = 100 hPa.

a. Assuming that the vertical velocity is zero at p_{bot}, determine the divergence (in s) at all mandatory pressure levels. Express your results in tabular form and use your results to make a graph of horizontal divergence (in s^{-1}) vs. pressure (in hPa). At what mandatory levels is convergence present? At what mandatory levels is divergence present?

b. Using your derived function from part (a), determine the level of non-divergence. What is the omega velocity at the level of non-divergence?

References

Vasquez, T., 2006. Weather Forecasting Red Book, first ed. Weather Graphics Technologies.

CHAPTER

Troughs and ridges

10

Chapter outline

Upper-tropospheric jet streams, as discussed in Ch. 9, play a central role in directing the movement of air masses, surface fronts, and cyclonic systems. The presence of jet streaks within these jet streams produces localized regions of divergence and convergence, which in turn drive transverse vertical circulation in the synoptic-scale atmosphere. Disturbances in the jet stream, such as those generated when it flows over mountain ranges, can produce large-scale wave patterns. These wave patterns, superimposed on upper-tropospheric jets, are referred to as **Rossby waves**, as introduced in Ch. 4.

As Rossby waves propagate around the globe, they develop a north–south orientation. The equatorward portion of a Rossby wave is identified as a trough, whereas the poleward portion is identified as a ridge. Therefore, troughs and ridges represent the manifestation of wave disturbances within the jet stream. In this chapter, we will examine the dynamics and evolution of Rossby waves and apply this framework to

An Introduction to Synoptic-Dynamic Meteorology. https://doi.org/10.1016/B978-0-443-24858-0.00010-1

understand the development of troughs and ridges. To do so, we will also introduce an important diagnostic tool in atmospheric dynamics: the concept of as vorticity.

10.1 Introduction to vorticity

10.1.1 The curl of a vector field

In Ch. 9, we learned that when the gradient operator acts on a vector function through the scalar product, it yields the divergence. Now, let us consider what happens when the gradient operator acts on a vector function $\boldsymbol{v}\ (x, y, z)$via the vector product. This operation gives:

$$\nabla \times \boldsymbol{v} = \left(\boldsymbol{i}\frac{\partial}{\partial x} + \boldsymbol{j}\frac{\partial}{\partial y} + \boldsymbol{k}\frac{\partial}{\partial z}\right) \times (v_x\boldsymbol{i} + v_y\boldsymbol{j} + v_z\boldsymbol{k})$$

where v_x, v_y, and v_z are the x-, y-, and z-components of the vector function $\boldsymbol{v}\ (x, y, z)$, respectively. Expanding the vector product yields:

$$\nabla \times \boldsymbol{v} = \widehat{\imath} \times \frac{\partial}{\partial x}(v_x\boldsymbol{i}) + \boldsymbol{i} \times \frac{\partial}{\partial x}(v_y\boldsymbol{j}) + \boldsymbol{i} \times \frac{\partial}{\partial x}(v_z\boldsymbol{k}) + \boldsymbol{j} \times \frac{\partial}{\partial y}(v_x\boldsymbol{i}) + \boldsymbol{j} \times \frac{\partial}{\partial y}(v_y\boldsymbol{j}) + \boldsymbol{j} \times \frac{\partial}{\partial y}(v_z\boldsymbol{k})$$
$$+ \boldsymbol{k} \times \frac{\partial}{\partial z}(v_x\boldsymbol{i}) \times \boldsymbol{k} \cdot \frac{\partial}{\partial z}(v_y\boldsymbol{j}) \times \boldsymbol{k} \cdot \frac{\partial}{\partial z}(v_z\boldsymbol{k})$$

Since each unit vector is a constant, we can factor them out of the parentheses, resulting in:

$$\nabla \times \boldsymbol{v} = (\boldsymbol{i} \times \boldsymbol{i})\frac{\partial v_x}{\partial x} + (\boldsymbol{i} \times \boldsymbol{j})\frac{\partial v_y}{\partial x} + (\boldsymbol{i} \times \boldsymbol{k})\frac{\partial v_z}{\partial x} + (\boldsymbol{j} \times \boldsymbol{i})\frac{\partial v_x}{\partial y} + (\boldsymbol{j} \times \boldsymbol{j})\frac{\partial v_y}{\partial y} + (\boldsymbol{j} \times \boldsymbol{k})\frac{\partial v_z}{\partial y}$$
$$+ (\boldsymbol{k} \times \boldsymbol{i})\frac{\partial v_x}{\partial z} + (\boldsymbol{k} \times \boldsymbol{j})\frac{\partial v_y}{\partial z} + (\boldsymbol{k} \times \boldsymbol{k})\frac{\partial v_z}{\partial z}$$

From our discussion on the vector product in Ch. 6, recall the following vector product identities:

$$\boldsymbol{i} \times \boldsymbol{i} = 0, \boldsymbol{j} \times \boldsymbol{j} = 0, \boldsymbol{k} \times \boldsymbol{k} = 0, \boldsymbol{i} \times \boldsymbol{j} = \boldsymbol{k}, \boldsymbol{i} \times \boldsymbol{k} = -\boldsymbol{j}, \boldsymbol{j} \times \boldsymbol{i} = -\boldsymbol{k}, \boldsymbol{j} \times \boldsymbol{k} = \boldsymbol{i}, \boldsymbol{k} \times \boldsymbol{i}$$
$$= \boldsymbol{j}, \boldsymbol{k} \times \boldsymbol{j} = -\boldsymbol{i},$$

Applying these to the above expression gives:

$$\nabla \times \boldsymbol{v} = \boldsymbol{k}\frac{\partial v_y}{\partial x} + (-\boldsymbol{j})\frac{\partial v_z}{\partial x} + (-\boldsymbol{k})\frac{\partial v_x}{\partial y} + (\boldsymbol{i})\frac{\partial v_z}{\partial y} + (\boldsymbol{j})\frac{\partial v_x}{\partial z} + (-\boldsymbol{i})\frac{\partial v_y}{\partial z}$$

This simplifies to:

$$\Rightarrow \nabla \times \boldsymbol{v} = \left(\frac{\partial v_z}{\partial y} - \frac{\partial v_y}{\partial z}\right)\boldsymbol{i} + \left(\frac{\partial v_x}{\partial z} - \frac{\partial v_z}{\partial x}\right)\boldsymbol{j} + \left(\frac{\partial v_y}{\partial x} - \frac{\partial v_x}{\partial y}\right)\boldsymbol{k} = \begin{vmatrix} \boldsymbol{i} & \boldsymbol{j} & \boldsymbol{k} \\ \frac{\partial}{\partial x} & \frac{\partial}{\partial y} & \frac{\partial}{\partial z} \\ v_x & v_y & v_z \end{vmatrix} \quad (10.1)$$

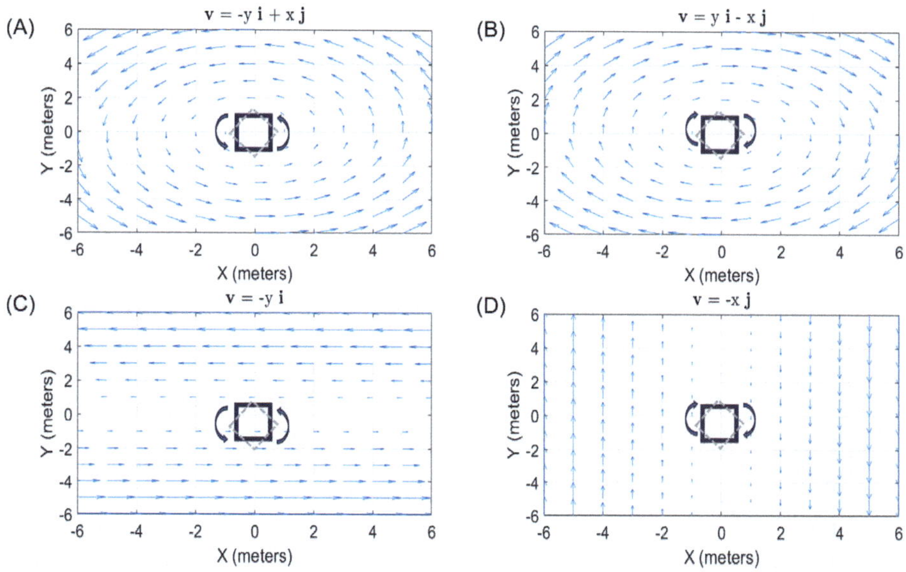

FIGURE 10.1 The curl of different vector functions. Four vector functions with curl are shown here, where a region of space around the origin is highlighted with a black square. (A) and (C) The region of space is rotated counterclockwise (as shown by the dashed gray square). (B) and (D) The region of space is rotated clockwise (as shown by the dashed gray square).

No Permission Required

The vector operation in Eq. (10.1) is called the **curl**. Unlike divergence, which is a scalar, the curl is a vector function. This means that at every point in space (x, y, z), the curl has both a magnitude and a direction.

The curl measures how much the vector function $\boldsymbol{v}$ "wraps" or circulates around a specific point. To illustrate, consider the vector function $\boldsymbol{v} = -y\boldsymbol{i} + x\boldsymbol{j}$, shown in Figure 10.1A. Notice that the vectors circulate counterclockwise around the origin $(x, y) = (0, 0)$.

Let's now calculate its curl:

$$\nabla \times \boldsymbol{v} = \boldsymbol{i}\left(\frac{\partial v_z}{\partial y} - \frac{\partial v_y}{\partial z}\right) + \boldsymbol{j}\left(\frac{\partial v_x}{\partial z} - \frac{\partial v_z}{\partial x}\right) + \boldsymbol{k}\left(\frac{\partial v_y}{\partial x} - \frac{\partial v_x}{\partial y}\right)$$

Substituting the components gives:

$$\nabla \times \boldsymbol{v} = \boldsymbol{i}\left(\frac{\partial(0)}{\partial y} - \frac{\partial(x)}{\partial z}\right) + \boldsymbol{j}\left(\frac{\partial(-y)}{\partial z} - \frac{\partial(0)}{\partial x}\right) + \boldsymbol{k}\left(\frac{\partial x}{\partial x} - \frac{\partial(-y)}{\partial y}\right) = 2\boldsymbol{k}$$

Thus, the curl points in the $\boldsymbol{k}$ direction, representing the axis of rotation. Physically, the magnitude of the curl measures the strength of the local rotation produced by the vector function $\boldsymbol{v}$, while its direction indicates the axis of rotation. The orientation follows the right-hand rule: if you curl the fingers of your right hand in the direction of circulation, your thumb points along the direction of the curl (out of the page, in this case).

Figure 10.1B displays a plot of the vector function $\boldsymbol{v} = y\boldsymbol{i} - x\boldsymbol{j}$. Calculating the curl gives:

$$\nabla \times \boldsymbol{v} = \boldsymbol{i}\left(\frac{\partial v_z}{\partial y} - \frac{\partial v_y}{\partial z}\right) + \boldsymbol{j}\left(\frac{\partial v_x}{\partial z} - \frac{\partial v_z}{\partial x}\right) + \boldsymbol{k}\left(\frac{\partial v_y}{\partial x} - \frac{\partial v_x}{\partial y}\right)$$
$$= \boldsymbol{i}\left(\frac{\partial(0)}{\partial y} - \frac{\partial(-x)}{\partial z}\right) + \boldsymbol{j}\left(\frac{\partial(y)}{\partial z} - \frac{\partial(0)}{\partial x}\right) + \boldsymbol{k}\left(\frac{\partial(-x)}{\partial x} - \frac{\partial y}{\partial y}\right) = -2\boldsymbol{k}$$

Here, the vectors circulate clockwise around the origin. Applying the right-hand rule shows the curl points in the $-\boldsymbol{k}$ direction.

Importantly, a vector field can have vorticity even if the vectors do not appear to rotate around a specific point. For example, for $\boldsymbol{v} = -y\boldsymbol{i}$ (as plotted in Figure 10.1C), the curl is $\nabla \times \boldsymbol{v} = \boldsymbol{k}$. Here, the vector arrows above $x = 0$ displace the top of a small region leftward, while those below $x = 0$ displace the bottom rightward. Together, these effects induce a counterclockwise rotation.

Similarly, for $\boldsymbol{v} = x\boldsymbol{j}$ (shown in Figure 10.1D), the curl is $\nabla \times \boldsymbol{v} = -\boldsymbol{k}$. In this case, the vector arrows to the right of $y = 0$ displace the right side of the region downward, while those to the left push the left side upward $y = 0$. The result is a clockwise rotation.

10.1.2 Identifying vorticity

When the vector function v represents the wind, the curl is referred to as **relative vorticity**. If u, v, and w are the zonal, meridional, and vertical components of the wind, respectively, then the relative vorticity is defined as:

$$\nabla \times \boldsymbol{V} = \begin{vmatrix} \boldsymbol{i} & \boldsymbol{j} & \boldsymbol{k} \\ \frac{\partial}{\partial x} & \frac{\partial}{\partial y} & \frac{\partial}{\partial z} \\ u & v & w \end{vmatrix} = \left(\frac{\partial w}{\partial y} - \frac{\partial v}{\partial z}\right)\boldsymbol{i} + \left(\frac{\partial u}{\partial z} - \frac{\partial w}{\partial x}\right)\boldsymbol{j} + \left(\frac{\partial v}{\partial x} - \frac{\partial u}{\partial y}\right)\boldsymbol{k}$$

In synoptic meteorology, we focus on horizontal winds and their zonal and meridional variations. For this reason, our primary interest is in the vertical component of relative vorticity, commonly referred to as **vertical vorticity** (ζ). The vertical component can be obtained by taking the scalar product of the unit vector $\boldsymbol{k}$ with the relative vorticity vector:

$$\zeta = \boldsymbol{k} \cdot (\nabla \times \boldsymbol{V}) = \boldsymbol{k} \cdot \left[\left(\frac{\partial w}{\partial y} - \frac{\partial v}{\partial z}\right)\boldsymbol{i} + \left(\frac{\partial u}{\partial z} - \frac{\partial w}{\partial x}\right)\boldsymbol{j} + \left(\frac{\partial v}{\partial x} - \frac{\partial u}{\partial y}\right)\boldsymbol{k}\right] = \frac{\partial v}{\partial x} - \frac{\partial u}{\partial y} \quad (10.2)$$

As previously earlier, $\zeta > 0$ corresponds to counterclockwise rotation, whereas $\zeta < 0$ corresponds to clockwise rotation. In the Northern Hemisphere, surface cyclones rotate counterclockwise, so we typically refer to $\zeta > 0$ as **cyclonic vertical vorticity** (or **positive vertical vorticity**) and $\zeta < 0$ as **anticyclonic vertical vorticity** (or **negative vertical vorticity**).

Vertical vorticity arises in two main ways when analyzing wind velocity patterns, such as those illustrated in Figure 10.1:

1. **Curvature Vorticity:** This type results from changes in wind direction along its path while maintaining constant speed. Positive curvature vorticity (or cyclonic curvature vorticity) develops when the wind curves counterclockwise, as shown in Figure 10.1A. Negative curvature vorticity (or anticyclonic curvature vorticity) develops when the wind curves clockwise, as shown in Figure 10.1B.
2. **Shear vorticity:** This type results from changes in wind speed across the flow. Positive shear vorticity (or cyclonic shear vorticity) occurs when wind speed increases as one moves away from a reference point, as illustrated in Figure 10.1C (flow along the x-direction with wind speed increasing in the y-direction). Negative shear vorticity (or anticyclonic shear vorticity) occurs when wind speed decreases away from a reference point, as shown in Figure 10.1D (flow along the x-direction with wind speed decreasing in the y-direction).

We can now apply this framework to synoptic map analysis. Figure 10.2 shows a 500-hPa chart from February 24, 2022. In the highlighted region "A," winds rotate counterclockwise as they moves through the trough, indicating cyclonic curvature vorticity. In contrast, near the ridge, the wind rotates clockwise, indicating anticyclonic curvature vorticity.

Figure 10.2 also shows regions of shear vorticity. In highlighted region "C," meridional flow weakens eastward, indicating cyclonic shear vorticity. In region "D," zonal flow strengthens northward, producing anticyclonic shear vorticity.

From these patterns, we can make two general observations: First, the curvature of the flow is strongest at the base of a trough, making it a local maximum for cyclonic curvature vorticity. Conversely, the peak of a ridge typically corresponds to a local maximum in anticyclonic curvature vorticity. Thus, air typically gains cyclonic vorticity as it enters a trough and loses cyclonic vorticity when it enters a ridge.

Second, decreasing wind speed away from the trough axis produces anticyclonic shear vorticity, while increasing wind speed away from the axis produces cyclonic shear vorticity. As a result, troughs often exhibit strong positive vorticity, where cyclonic shear vorticity poleward of the trough combines with cyclonic curvature vorticity near the base of the trough.

10.2 Vorticity tendency

As we will see in Ch. 12, the intensification of synoptic-scale systems is often accompanied by changes in vertical vorticity. The time evolution of vorticity is called the **vorticity tendency**. In this section, we explore the dominant processes that cause changes in vertical vorticity. Within an Eulerian reference frame (ie, a frame fixed relative to Earth), the two main synoptic-scale processes that modify

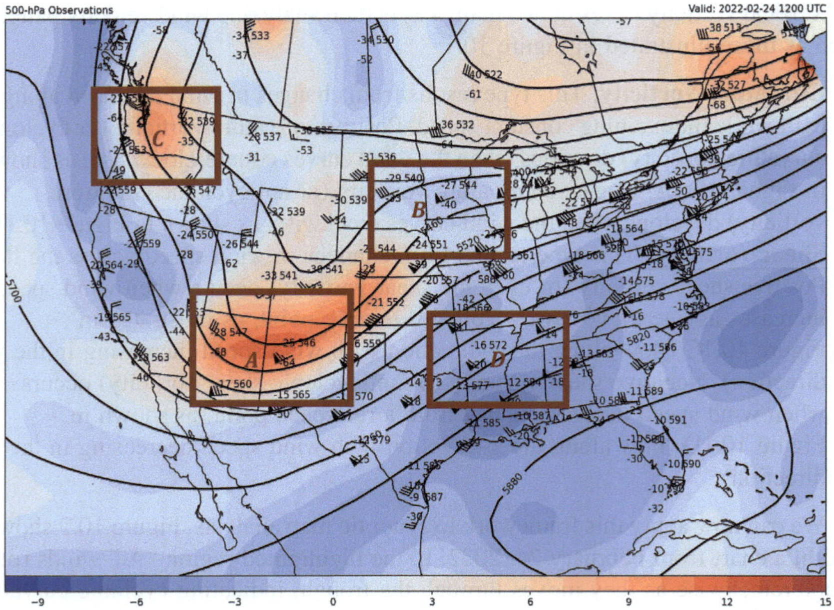

FIGURE 10.2 Vertical vorticity on a 500-hPa map. 500-hPa analysis on February 24, 2022, at 12 UTC. Solid black contours correspond to isohypses. Contour color fill corresponds to vertical vorticity in units of 10^{-5} s^{-1}. See the text for a description of the four highlighted regions on the chart.

No Permission Required

vorticity are vorticity advection and divergence. We will discuss each of these in turn.

10.2.1 Vorticity advection

Recall from Ch. 6 that any scalar property can be advected by the wind, and vertical vorticity is no exception. Mathematically, advection of vertical vorticity by the horizontal wind can be expressed as:

$$\left(\frac{\partial \zeta}{\partial t}\right)_{adv} = -\boldsymbol{V}_H \cdot \nabla \zeta \qquad (10.3)$$

Eq. (10.3) shows that vertical vorticity advection depends on three factors:

1. The gradient of vertical vorticity in a region (represented by $\nabla \zeta$).
2. The wind speed across the vorticity gradient (represented by $\boldsymbol{V}_{\mathrm{H}}$).
3. The angle at which the wind crosses the gradient (from the scalar product).

Thus, the strongest vorticity advection occurs when strong winds blow parallel to a strong vorticity gradient.

To understand how this appears on a 500-hPa map, consider the schematic trough-ridge couplet given in Figure 10.3. Upstream of the 500-hPa trough, the geostrophic wind flows from the vorticity minimum at the ridge crest toward the vorticity maximum at the trough base. In this case, negative vorticity is carried into the trough base, a process known as **negative vorticity advection (NVA). NVA reduces** cyclonic vertical vorticity upstream of the trough.

Conversely, when the wind flows from the trough toward the ridge, positive vorticity is transported into the ridge crest. This is called **positive vorticity advection (PVA)**, which increases cyclonic vorticity downstream of the trough. When PVA and NVA act together in a trough-ridge system with uniform wind speeds on both sides, the net effect is an eastward (downstream) progression of the system.

In reality, winds are rarely uniform on either side of a trough. Suppose, as shown in Figure 10.4A, that winds downstream of the trough axis are weaker than those upstream. In this case, the combined effects of shear and curvature vorticity shift the maximum vertical vorticity upstream of the trough axis. As a result, the maximum PVA occurs near the base of the trough rather than east of it, producing an equatorward component of motion for the trough.

Conversely, if the winds downstream of the trough are stronger than those upstream, as shown in Figure 10.4B, the maximum vorticity shifts downstream, toward the region of strongest winds. In this case, the PVA maximum lies poleward and downstream of the trough axis, producing a poleward component of motion. In summary, vorticity advection provides a useful diagnostic for predicting trough movement. When the strongest winds are upstream of the trough, the trough tends to dig equatorward. Conversely, when the strongest winds are downstream of the trough, the trough tends to lift poleward.

10.2.2 Vorticity and divergence

In Ch. 9, we used the principle of angular momentum conservation to explain the wind speeds associated with the subtropical jet. The same principle helps explain the relationship between vertical vorticity and divergence. Consider an analogy of a rotating figure skater: when the skater pulls her arms inward, her moment of inertia decreases (because more mass is concentrated near the rotation axis). Conservation of angular momentum then requires her angular speed to increase. Conversely, when she extends her arms outward, her moment of inertia increases, forcing her angular speed to decrease. Importantly, changes in the skater's effective rotation radius must be accompanied by corresponding changes in angular speed.

In a rotating fluid, the conserved variable analogous to angular momentum is **circulation, which can be thought of** as the product of the surface area of the air mass and its vertical vorticity. Just as angular momentum is conserved in the absence of torque, **Kelvin's circulation theorem** states that the circulation of a barotropic fluid is conserved when no net torque acts on it.

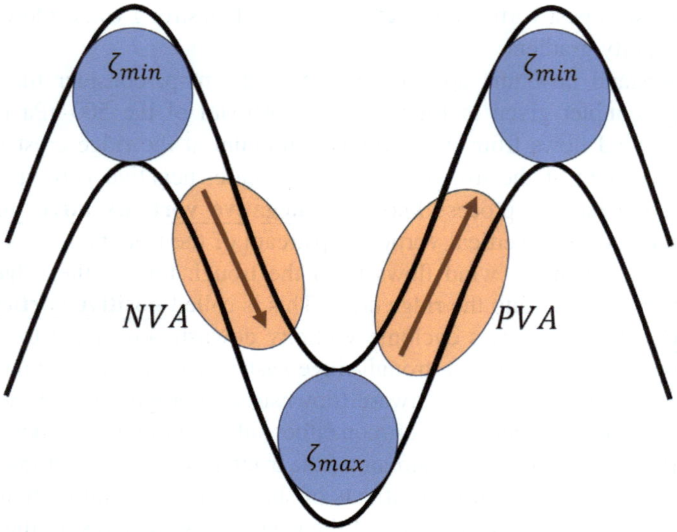

FIGURE 10.3 Schematic of vorticity advection

Schematic of air moving through a trough-ridge couplet at 500-hPa. There is a cyclonic vertical vorticity maximum at the base of the trough, and there are anticyclonic vertical vorticity maxima at the crest of both ridges. Notice that negative vorticity advection (NVA) is found upstream of the trough, while positive vorticity advection (PVA) is found downstream of the trough.

No Permission Required

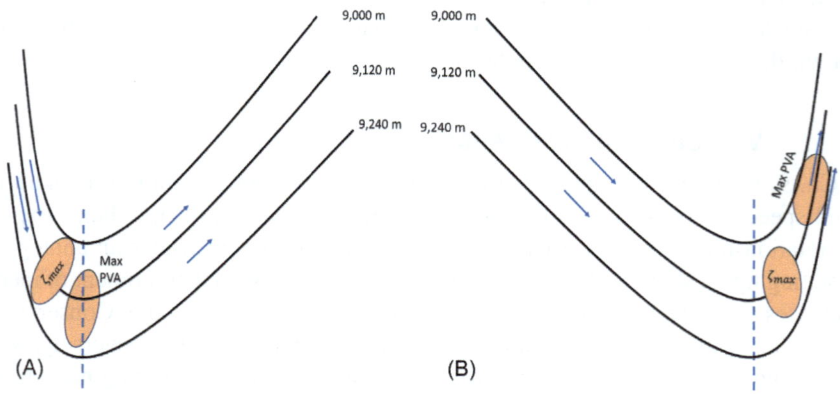

FIGURE 10.4 The effect of downstream winds on trough motion

Schematic of a 300-hPa trough in which (A) the downstream winds are weaker than the upstream winds and (B) the downstream winds are stronger than the upstream winds. The blue arrows correspond to the geostrophic flow. ζ_{max} corresponds to the location of maximum cyclonic vorticity and the location of the maximum PVA is marked. *PVA*, positive vorticity advection.

No Permission Required

According to this theorem, any change in the surface area of a fluid must be accompanied by a compensating change in vertical vorticity. To illustrate, consider an air mass with initial positive vertical vorticity, as shown in Figure 10.5. If the air mass is in a region of divergence (see Figure 10.5A), its volume increases over time. As divergence grows, the air mass expands, and its surface area increases. To conserve circulation, cyclonic vorticity must decrease. Equivalently, this can be described as the generation of anticyclonic vorticity. Thus, **divergence generates anticyclonic vertical vorticity** in an air mass with existing vorticity. Conversely, if convergence occurs within the air mass, as shown in Figure 10.5B, its volume decreases, leading to a reduction in surface area. To conserve circulation, cyclonic vorticity must increase. Therefore, **convergence generates cyclonic vertical vorticity**.

10.2.3 Vorticity advection and vertical motion

So far, we have seen that the primary processes affecting vertical vorticity are divergence/convergence and vorticity advection. We now turn to the important connection between vorticity advection and vertical motion.

Consider geostrophic flow through a trough-ridge couplet at 300-hPa level, as shown in Figure 10.6. For simplicity, assume wind speed remains constant, so any horizontal acceleration arises only from changes in wind direction. As air passes through the ridge, the required acceleration points toward the ridge center, turning the wind from southwesterly upstream to northwesterly downstream. Similarly, as air passes through the trough, the required acceleration points toward the trough center, turning the wind from northwesterly upstream to southwesterly downstream.

From Ch. 9, we know that the ageostrophic wind vector is directed 90° to the left (ie, counterclockwise) of the horizontal acceleration vector in the Northern Hemisphere. This means that ageostrophic winds are directed downstream of ridges and upstream of troughs, as shown in Figure 10.6. The resulting flow pattern produces convergence upstream of the trough and divergence downstream. Because vertical motion is driven by divergence and convergence of the ageostrophic flow, and because this system lies above the level of nondivergence, we expect sinking motion upstream of the trough and rising motion downstream. This explains why clouds and precipitation are often observed downstream of upper-level troughs.

Comparing Figure 10.3 with Figure 10.6, notice that the region of upper-level convergence coincides with NVA, while the region of divergence coincides with PVA. This is not a coincidence but a direct consequence of the vorticity-divergence relationship discussed earlier. As an air mass enters a trough, it encounters higher positive vorticity. To maintain equilibrium with its environment, the air mass must increase its vorticity, which requires horizontal convergence. Similarly, in the region of PVA downstream of the trough, the air must divergence. This leads to the following conclusion: ***PVA implies divergence and NVA implies convergence at any level in the atmosphere***. Additionally, if this PVA and NVA occur above the

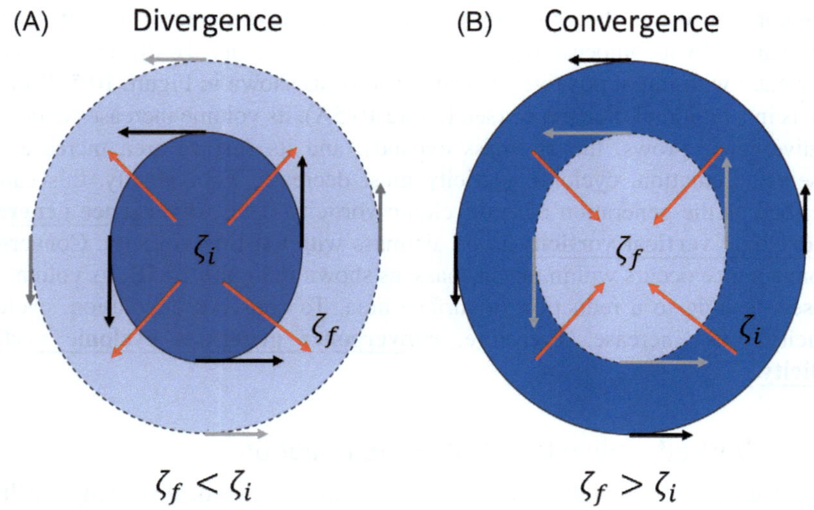

FIGURE 10.5 Schematic of the relationship between vorticity and divergence

(A) An air mass with positive vorticity under the effects of divergence shown by the red vector arrows. (B) An air mass with positive vorticity under the effects of convergence shown by the red vector arrows. For both parts of the figure, the solid circle corresponds to the initial state of the air mass where the initial velocity vectors are given in black arrows, and the dashed circle corresponds to the final state of the air mass where the final velocity vectors are given in gray arrows.

No Permission Required

level of nondivergence, this suggests that PVA will be linked to rising motion, whereas NVA will be linked to sinking motion.

So far, our discussion on troughs and ridges has assumed neutral orientation. In practice, troughs often tilt because of the influence of jet streaks. As discussed in Ch. 4, a trough axis oriented from southwest to northeast (as shown in Figure 10.7A) is a positively tilted trough, while a trough axis oriented from northwest to southeast (as shown in Figure 10.7B) is a negatively tilted trough. Both positively and negatively tilted troughs are associated with clouds and precipitation downstream. However, their thermal advection characteristics differ. In a positively tilted trough, as shown in Figure 10.7A, momentum is advected northward away from the trough base. To sustain the low geopotential heights associated with the trough, cold advection must occur at its base. In contrast, in a negatively tilted trough, momentum is advected toward the trough base, as shown in Figure 10.7B. As discussed in Ch. 8, cold advection aloft increases lapse rates and destabilizes the atmosphere. Furthermore, negatively tilted troughs exhibit rapid acceleration as flow transitions from subgeostrophic at the trough base to supergeostrophic downstream. This enhanced speed divergence promotes stronger rising motion downstream compared with positively tilted troughs.

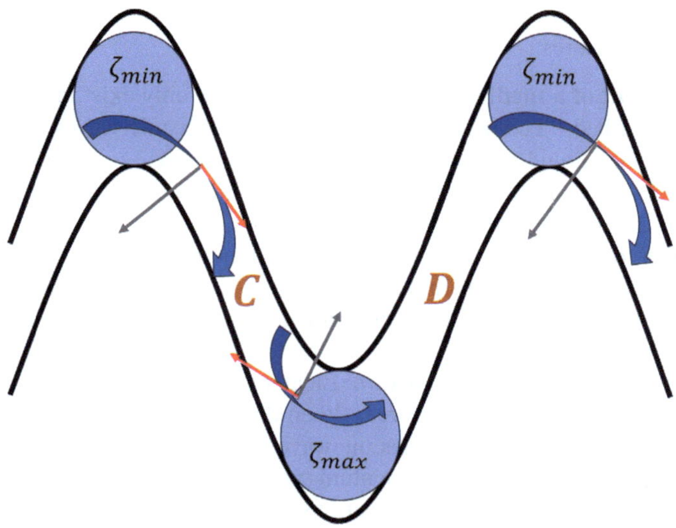

FIGURE 10.6 Schematic of vorticity advection and divergence

Schematic of geostrophic flow through a trough-ridge couplet at 300-hPa where the solid black contours represent isohypses. The blue arrows denote the direction of the flow, the gray arrow denotes the direction of the horizontal acceleration vector, and the red arrow denotes the direction of the ageostrophic wind vector. There is a cyclonic vertical vorticity maximum at the base of the trough, and there are anticyclonic vertical vorticity maxima at the crest of both ridges. Note that there is convergence (indicated with "C") upstream of the trough and divergence (indicated with "D") downstream of the trough.

No Permission Required

10.3 Introduction to mechanical waves

Our earlier discussion of vorticity provides the foundation for analyzing the motion and the dynamics of troughs and ridges. As mentioned in the introduction to this chapter, troughs and ridges can be understood as wave disturbances superimposed on the jet stream. These disturbance are known as **Rossby waves** or **planetary waves**. To begin our analysis, we first introduce the general properties of mechanical waves. This section will describe what mechanical waves are and outline the basic features of their propagation.

10.3.1 Description of mechanical waves

Mechanical waves are oscillations within a physical medium that propagate through space. Common examples include ocean waves, sound waves, and seismic waves. In each case, mechanical waves transfer energy and momentum through the medium without transferring mass. This definition implies that all mechanical waves share three essential features:

1. **Requirement of a medium**: Mechanical waves can only exist in a medium that can be disturbed.[1] The medium may be solid (eg, seismic waves traveling through rock), liquid (eg, sound waves in water), or gas (eg, atmospheric waves). As we will show later, the properties of the medium determine both the speed of the wave and the amount of energy it transports.
2. **Source of disturbance**: Mechanical waves require a source to generate the disturbance. This can be a single pulse (e.g., a rock thrown into a pond producing ripples) or a continuous generator (e.g., a vibrating tuning fork producing sound waves).
3. **Physical mechanism for interaction**: Elements within the medium must be able to interact through a restoring force. When disturbed, the medium exerts an elastic restoring force that enables the wave to propagate. Therefore, mechanical waves carry energy and momentum while the particles of the medium oscillate about their equilibrium positions. For example, as sound travels through air, molecules oscillating locally. The resulting variations in pressure and velocity allow us to track the wave's progress.

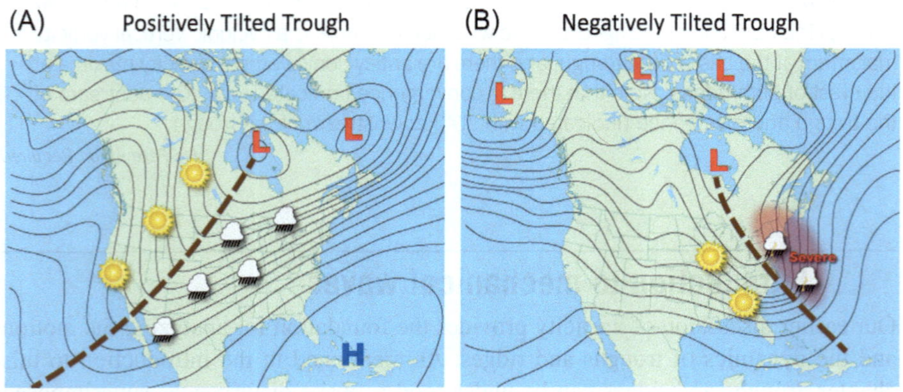

FIGURE 10.7 Schematic of tilted troughs

(A) A positively tilted trough. (B) A negatively tilted trough.

Courtesy of the National Oceanic and Atmospheric Administration.
No Permission Required

[1] This property distinguishes mechanical waves from electromagnetic waves, which do not require a medium to propagate.

Mechanical waves, like all waves, are characterized by amplitude, wavelength, period or frequency, and energy. As illustrated in Figure 10.8, the **amplitude** of a wave is the maximum displacement from equilibrium (the undisturbed position of the medium). The crest corresponds to $y = +A$, while the trough corresponds to $y = -A$. The **wavelength** is the distance between two adjacent crests, representing one complete cycle of the wave. The **period** is the time it takes for one wavelength to pass a fixed point. Thus, longer wavelengths correspond to longer periods. Conversely, the **frequency** is the number of wavelengths passing a point per unit time, and is therefore inversely related to wavelength.

10.3.2 Propagation of mechanical waves

Mechanical waves cause mass to oscillate within a medium, allowing us to analyze the motion of particles in the simplest case: waves generated by a linear oscillator. In introductory mechanics, a simple harmonic oscillator displaced from equilibrium experiences a restoring force proportional to its displacement. Common examples include a mass attached to a spring or a pendulum. Neglecting all other forces, the system oscillates with two key properties: (1) the amplitude and frequency of oscillation remain constant and (2) the amplitude is independent of the frequency of oscillation.

Now consider a simple harmonic oscillator creating a disturbance in a medium. The motion of this oscillator models the displacement of particles in the medium, allowing us to describe the properties of a mechanical wave as it passes propagates. The wave function associated with this disturbance is:

$$y(x,t) = A\cos(kx - \omega t + \varphi), \quad k = \frac{2\pi}{\lambda}, \quad \omega = \frac{2\pi}{T} \tag{10.4}$$

Here, A is the amplitude of the wave, the maximum displacement of the medium's elements as the wave passes. The quantity ω is the **angular frequency (rad/s)**, describing how many oscillations occur per second.

The variable k is the **wavenumber** (rad/m), the spatial analogy of angular frequency. Just as angular frequency gives the number of oscillations per unit time, wavenumber gives the number of oscillations per unit distance. For this reason, wavenumber is often called *spatial frequency*. Lastly, φ is the **phase constant**, which shifts the wave in space or time. We can deduce an important property of harmonic waves using Eq. (10.4). Since amplitude remains constant, each point on the moving wave retains its displacement as the wave propagates. In Eq. (10.4), this means that:

$$y(x,t) = A\cos(kx - \omega t + \varphi) = constant$$

$$\Rightarrow kx - \omega t + \varphi = constant$$

Differentiating with respect to time and defining $c = dx/dt$:

$$c = \frac{\omega}{k} \tag{10.5}$$

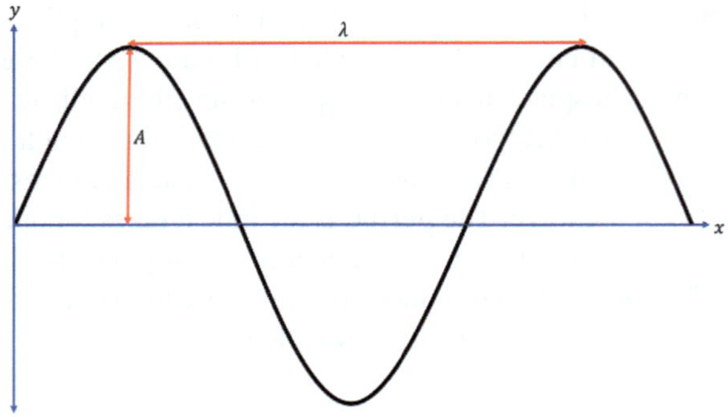

FIGURE 10.8 Schematic of a mechanical wave

A simple mechanical wave with amplitude A and wavelength λ.

No Permission Required

This quantity c is the **phase velocity** of the wave. For simple harmonic waves, where k and ω are constant, the phase velocity represents the actual wave speed. Importantly, in this ideal case, the speed is independent of the properties of the medium.

In reality, propagating waves face two additional complications. First, most waveforms are complex, containing multiple frequencies. However, it is possible to show that any reasonably well-behaved function can be represented as a combination of sinusoidal waves.[2] Thus, propagating waves can be thought of as a collection of sinusoidal waves of different frequencies, each with its own phase velocity determined by Eq. (10.5).

Second, in real media, the angular frequency generally depends on the wavenumber of the oscillator generating the wave and the physical properties of the medium. As a result, phase velocity varies with wavenumber. Waves for which phase speed depends on wavenumber are known as **dispersive waves**. Unlike simple harmonic waves, dispersive waves do not preserve their shape; instead, the wave packet spreads apart as energy disperses. In this case, the velocity at which energy propagates is generally different from the phase velocity. This energy transport speed is the **group velocity**, defined as:

$$c_g = \frac{\partial \omega}{\partial k} \tag{10.6}$$

Applying Eq. (10.6) requires knowledge of the **disperson relation** $\omega(k)$. Two important conclusions follow:

[2] This is a statement of Fourier's theorem whose proof can be found in many textbooks in differential equations or applied mathematics.

1. If ω is linearly proportional to k (i.e., $\omega = ck$), the group velocity is the same for all wavenumbers and equals the phase velocity. In this case, the wave shape remains unchanged as it propagates - just like simple harmonic waves.
2. If ω is not linearly proportional to k, then group velocity depends on wavenumber. Because real waves consist of multiple sinusoidal components, each component travels at a different speed. As a result, the overall wave packet becomes distorted as it travels.

10.4 Rossby waves

With this background on mechanical waves, we are ready to examine the properties of Rossby waves and conclude this chapter with a discussion of their formation and motion.

10.4.1 Rossby wave formation

As discussed in Ch. 10.3, all mechanical waves arise from a restoring force. By analogy, Rossby waves must also form in response to a restoring force within the atmosphere. We will begin with the simpler case of a barotropic atmosphere, which is characterized by a constant depth.

Consider tropospheric air of constant depth within the jet stream at latitude φ_0, with a purely westerly flow at the initial time t_0, as shown in Figure 10.9. Because the jet stream has no shear or curvature, its initial relative vorticity will be zero. Therefore, the absolute vorticity is simply the Coriolis parameter: $f_0 = 2\Omega \sin \varphi_0$.

Now, suppose a disturbance displaces the jet stream poleward by an amount dy at time t_1. By Kelvin's circulation theorem, the circulation of the flow must remain constant over time. However, since the disturbance does not change the volume of the flow, this implies conservation of absolute vorticity:

$$(f+\zeta)_i = (f+\zeta)_f$$

$$\Rightarrow f_0 = f_1 + \zeta_1$$

Expanding f_1 in terms of latitude gives:

$$f_1 \approx f_0 + \beta\, dy, \quad \beta = \left(\frac{df}{dy}\right)_{\varphi=\varphi_0}$$

where β is the planetary vorticity gradient. Substituting yields:

$$\zeta_1 = f_0 - f_1 = -\beta\, dy$$

Thus, a poleward displacement increases the Coriolis parameter, requiring a decrease in relative vorticity to conserve absolute vorticity. Since the initial velocity was zero, this produces anticyclonic vorticity, bending the jet stream equatorward.

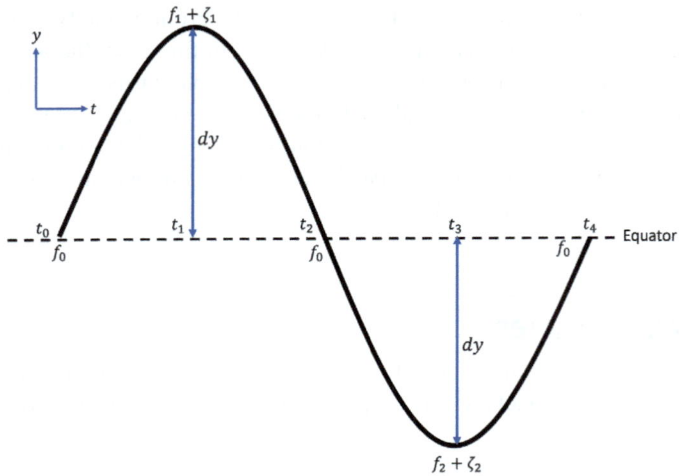

FIGURE 10.9 Schematic of barotropic Rossby wave formation

The formation of a barotropic Rossby wave within the Northern Hemisphere. The dashed line shows the original position of the jet stream, and the solid line shows the oscillation of the jet stream when given a poleward displacement *dy*. See the text for more details.

No Permission Required

When the jet returns to its original latitude at time t_2, the Coriolis parameter recovers its initial value and the relative vorticity returns to zero. However, because of inertia, the jet overshoots its equilibrium position and continues equatorward. For an equatorward displacement dy at time t_3:

$$(f+\zeta)_{t_2} = (f+\zeta)_{t_3}$$

$$\Rightarrow f_0 = f_2 + \zeta_2$$

$$\Rightarrow f_0 \approx f_0 + \beta\, dy + \zeta_2$$

$$\Rightarrow \zeta_2 \approx \zeta_0 - \beta\, dy$$

Since $dy < 0$, the Coriolis parameter decreases, requiring an increase in vertical vorticity to conserve absolute vorticity. This positive curvature bends the jet stream poleward, completing the oscillation. This conceptual model shows how a meridional oscillation, a Rossby wave, forms in the westerly jet stream. The restoring force driving this oscillation is provided by the planetary vorticity gradient β. In general, the meridional gradient of absolute vorticity serves as the restoring mechanism for Rossby waves.

So far, we have assumed a barotropic atmosphere. However, the real atmosphere is baroclinic with variable thickness. To expand our argument, we introduce **potential vorticity (PV)**, defined as the ratio of absolute vorticity to static stability.

To conceptualize PV, consider an air column of initial depth z_i. If static stability increases, vertical motion is suppressed, leading to sinking motion. By continuity, this sinking motion produces horizontal divergence, which reduces the column's absolute vorticity. Conversely, decreasing static stability promotes ascent and convergence, thereby increasing absolute vorticity. This discussion indicates that potential vorticity remains conserved unless a diabatic process changes the thickness of the column or friction reduces its vorticity.

Now consider Rossby wave formation in a baroclinic atmosphere, as shown in Figure 10.10. The poleward temperature gradient implies a poleward thickness gradient so that the thickness decreases from the equator to the poles. Based on the definition given above, we can express potential vorticity as:

$$PV = \frac{f+\zeta}{\Delta Z}$$

Here, ΔZ represents the thickness of the air column at a particular latitude. We can analyze the jet stream at each point indicated in Figure 10.10. As the air moves poleward at time t_1, planetary vorticity increases while thickness decreases. Both effects raise PV, meaning relative vorticity must decrease even more strongly than in the barotropic case. This causes a sharper equatorward turning of the jet stream.

Conversely, when the air moves toward the equator, planetary vorticity decreases while thickness increases. Both effects lower PV, requiring greater increases in relative vorticity than in the barotropic case. This sharpens the poleward turning of the jet stream. The net effect is that conservation of PV strengthens the restoring force in a baroclinic atmosphere. As a result, Rossby waves in baroclinic conditions have shorter wavelengths than those in barotropic conditions.

10.4.2 Rossby wave motion

As mentioned in Ch. 10.3.2, the propagation of a mechanical wave can be analyzed by examining the motion of particles within the medium as the wave passes. For a Rossby wave, this oscillation of the atmosphere can be represented with a simple sinusoidal function:

$$y' \approx A\cos\left[k\left(x' - ct\right)\right], \quad k = \frac{2\pi}{\lambda} \tag{10.7}$$

Here, y' is the meridional displacement of the wave from its starting latitude, x' is the eastward distance from the wave origin, c is the phase speed (the speed of the wave crests relative to Earth), A is the amplitude of the wave, and λ is the wavelength.

Barotropic Rossby waves in the jet stream typically have wavelength of $\lambda \approx 6000$ km and amplitude of $A \approx 1665$ km (Stull, 2017). To determine the phase speed, we differentiate Eq. (10.7) with respect to time to obtain the meridional speed:

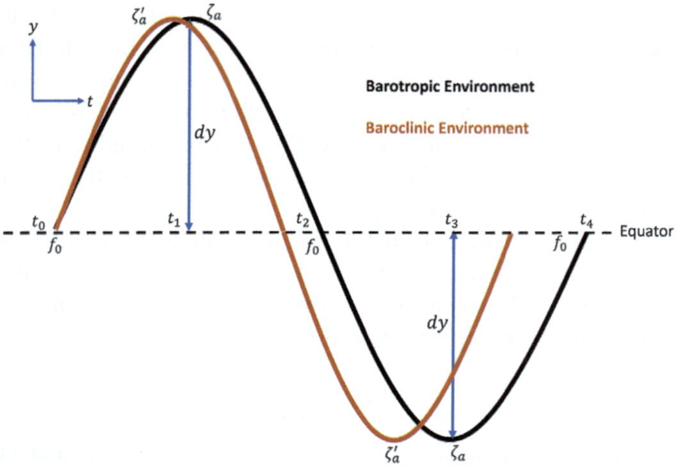

FIGURE 10.10 Schematic of baroclinic Rossby wave formation

The formation of a baroclinic Rossby wave within the Northern Hemisphere. The dashed line shows the original position of the jet stream. The solid black line shows the formation of a barotropic Rossby wave, and the solid brown line shows the formation of a baroclinic Rossby wave. See the text for more details.

No Permission Required

$$v = \frac{dy'}{dt} = \frac{d}{dt}\{A\cos[k(x' - ct)]\} = kcA\sin[k(x' - ct)]$$

The associated relative vorticity is:

$$\zeta = \frac{\partial v}{\partial x} = \frac{\partial}{\partial x}\{kcA\sin[k(x' - ct)]\} = k^2cA\cos[k(x' - ct)]$$

From the previous section, we know that $\zeta \approx -\beta y$. Substituting gives:

$$k^2cA\cos[k(x' - ct)] = -\beta A\cos[k(x' - ct)]$$

$$\Rightarrow c = -\frac{\beta}{k^2} = -\beta\left(\frac{\lambda}{2\pi}\right)^2$$

This expression gives the **intrinsic phase speed** of Rossby waves. The negative sign indicates that, relative to the air, the wave crests propagate westward. However, because the air itself (the jet stream) flows eastward with mean speed U_0, the ground-relative phase speed is:

$$c = U_0 - \beta\left(\frac{\lambda}{2\pi}\right)^2 \tag{10.8}$$

For typical jet stream speeds exceeding 100 knots, the total phase speed of Rossby waves is positive. This means Rossby waves generally drift eastward with the flow, but the intrinsic westward propagation reduces their effective speed.

Importantly, the intrinsic speed decreases with increasing wavelength. Therefore, short Rossby waves (short waves) have slow intrinsic westward speeds, so the jet advects them rapidly eastward. In contrast, long Rossby waves (ie, planetary waves such as troughs and ridges) have larger intrinsic phase speed, so they move more slowly eastward relative to the jet stream. This explains why short waves often propagate through through planetary-scale waves, as seen on upper-tropospheric weather charts.

Moreover, Eq. (10.8) shows that phase speed is a nonlinear function of wavenumber, which means barotropic Rossby waves are dispersive. Since $\omega = ck$, the angular frequency is:

$$\omega = ck = \left(-\frac{\beta}{k^2}\right)k = -\frac{\beta}{k}$$

The group velocity is then:

$$\Rightarrow c_g = \frac{\partial \omega}{\partial k} = \frac{\beta}{k^2} = \beta\left(\frac{\lambda}{2\pi}\right)^2 \tag{10.9}$$

Notice that the group velocity is positive in Eq. (10.9), which means Rossby waves transport energy eastward even though the crests themselves propagate westward relative to the air. In other words, Rossby waves move more slowly than the background flow, but their energy propagates faster than the background flow. This explains why new disturbances often develop downstream of existing ones - an important concept for weather forecasting.

Questions

1. Answer the following conceptual questions regarding vorticity.
   a. Under what precise conditions can curved, geostrophic flow have zero relative vorticity? Explain your answer. Draw schematic diagrams to illustrate your answers.
   b. We have shown that the presence of horizontal wind shear leads to vertical vorticity. Use the definition of curl to demonstrate that vertical wind shear must generate horizontal relative vorticity.
2. A common method for diagnosing rising motion in the synoptic atmosphere is to examine whether positive vorticity advection increases with height. Explain why this makes sense conceptually.
3. Consider a trough-ridge couplet superimposed on westerly geostrophic flow. Explain why there cannot be advection of geostrophic vorticity along the trough and ridge axes.
4. Consider the 500-hPa contoured weather chart shown in Figure 10.11. In this chart, solid black contours represent isohypses, dashed red contours represent isotherms, and vertical vorticity is depicted with color contoured.

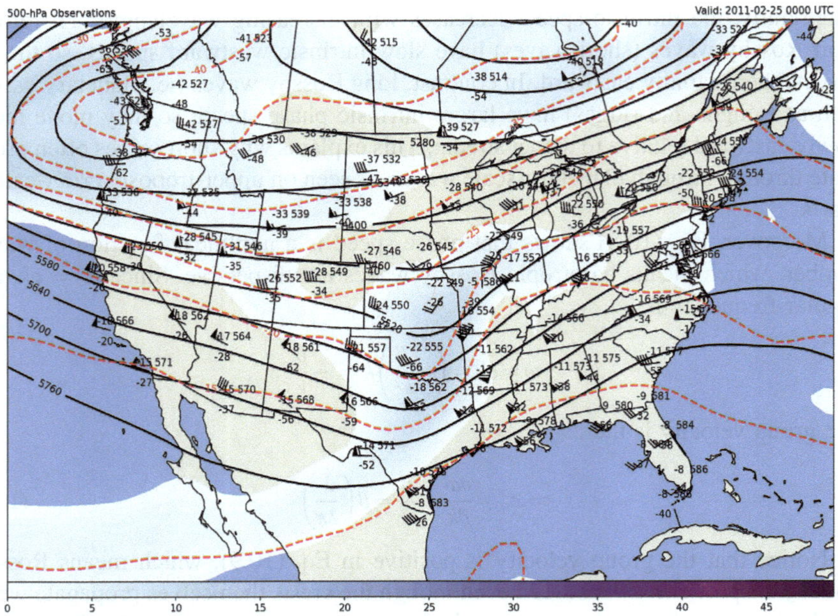

FIGURE 10.11 Longwave trough with short waves

500-hPa observations for February 25, 2011, at 00Z. Solid black contours correspond to isohypses, dashed red contours correspond to isotherms, and vertical vorticity is color contoured.

No Permission Required

- **a.** Identify and mark the locations of long-wave troughs, long-wave ridges, and short-wave troughs using standard convention.
- **b.** Use a purple-colored pencil to shade the areas associated with negative vorticity advection, and use a red-colored pencil to shade the regions associated with positive vorticity advection.

5. Figure 10.12 shows the 500-hPa observations for a synoptic weather system on April 14, 2018. In this figure, solid contours represent the 500-hPa isohypses, dashed contours represent the 850-hPa isohypses, and 500-hPa vertical vorticity is shown using color contours.

- **a.** Identify and mark the locations of long-wave troughs, long-wave ridges, and short-wave troughs using standard convention.
- **b.** Use a purple-colored pencil to shade the areas associated with negative vorticity advection. Use a red-colored pencil to shade the regions associated with positive vorticity advection.
- **c.** Based on this synoptic chart, assess whether the surface low pressure will deepen or fill over the next 12 hours. Defend your answer.

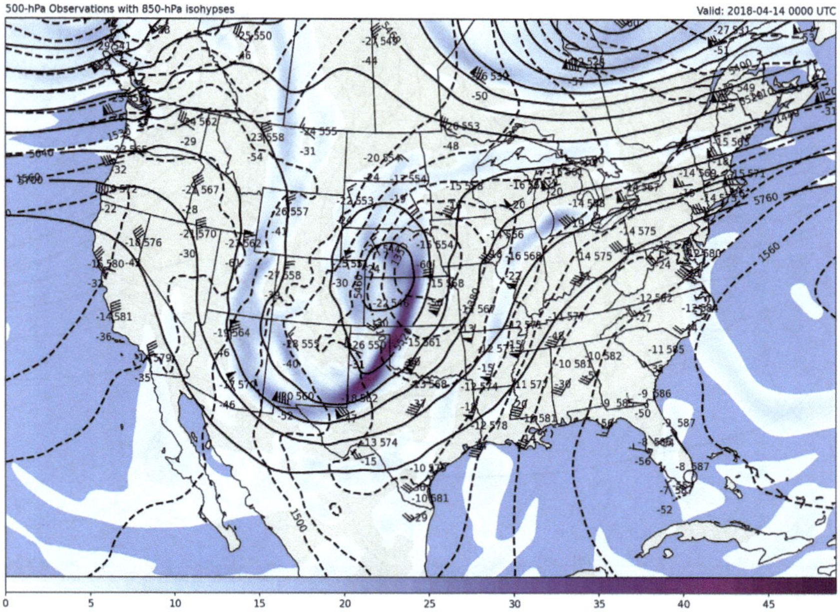

FIGURE 10.12 Case study of surface low pressure below the downstream location of a trough

500-hPa observations for April 14, 2018, at 00Z. Solid contours correspond to 500-hPa isohypses, dashed contours correspond to 850-hPa isohypses, and 500-hPa vertical vorticity is color contoured.

No Permission Required

6. Figure 10.13 presents 500-hPa observations for a synoptic weather system on February 25, 2007. In this figure, solid contours show the 500-hPa isohypses, dashed contours represent the 850-hPa isohypses, and 500-hPa vertical vorticity is depicted using color contours.
   a. Identify and mark the locations of long-wave troughs, long-wave ridges, and short-wave troughs using standard conventions.
   b. Use a purple-colored pencil to shade the regions associated with negative vorticity advection. Use a red-colored pencil to shade the areas associated with positive vorticity advection.
   c. Based on this synoptic chart, evaluate whether the surface low pressure will deepen or fill over the next 12 hours. Defend your answer.
7. Figure 10.14 shows 500-hPa observations of a long-wave trough on April 18, 2013. In this figure, black solid contours correspond to 500-hPa isohypses, dashed red contours correspond to 500-hPa isotherms, and 300-hPa isotachs are color contoured.
   a. Place an "X" at the location of maximum vertical vorticity, place a "PVA" at the location of maximum positive vorticity advection, and place a "NVA" at the location of maximum negative vorticity advection. Defend your answer.

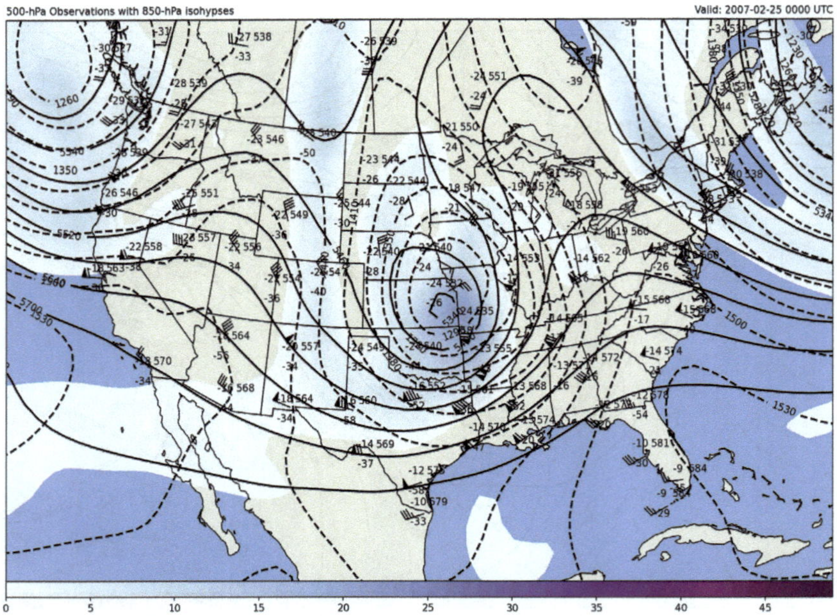

FIGURE 10.13 Case study of surface low pressure below the upstream location of a trough

500-hPa observations for February 25, 2007, at 00Z. Solid contours correspond to 500-hPa isohypses, dashed contours correspond to 850-hPa isohypses, and 500-hPa vertical vorticity is color contoured.

No Permission Required

- **b.** What direction should the trough travel over the next 6 hours? Defend your answer.

8. Assume that the dispersion relation for a wave can be written as $\omega = ak + b$. Calculate the phase velocity and group velocity. Describe the motion of this wave. In this problem, you will estimate the speed associated with various trough systems.

- **a.** Consider the 500-hPa trough shown in Figure 10.11. If we approximate the size of the Rossby wave as 5000 km, provide an estimate of the group velocity of the Rossby wave.
- **b.** Consider the 500-hPa trough shown in Figure 10.12. If we estimate the size of the Rossby wave as 6000 km, provide an estimate of the ground-relative phase speed of the Rossby wave.

9. Consider the vector function $\boldsymbol{F} = (x^2y)\boldsymbol{i} + (xyz)\boldsymbol{j} + (-x^2y^2)\boldsymbol{k}$

- **a.** Perform the following vector operation: $\boldsymbol{F}\cdot\nabla[\boldsymbol{k}\cdot(\nabla\times\boldsymbol{F})]$. Evaluate this function at $(x,y,z) = (1,2,1)$.
- **b.** If $\boldsymbol{F}$ represents the wind velocity, provide a physical interpretation of the vector operation performed above.

10. Let's consider a simplified version of the 300-hPa jet streak model from Ch. 9:

$$\boldsymbol{V} = \left[U_g + a\sin\left(\frac{y}{b}\right)\cos\left(\frac{x-x_0}{c}\right)\right]\boldsymbol{i}$$

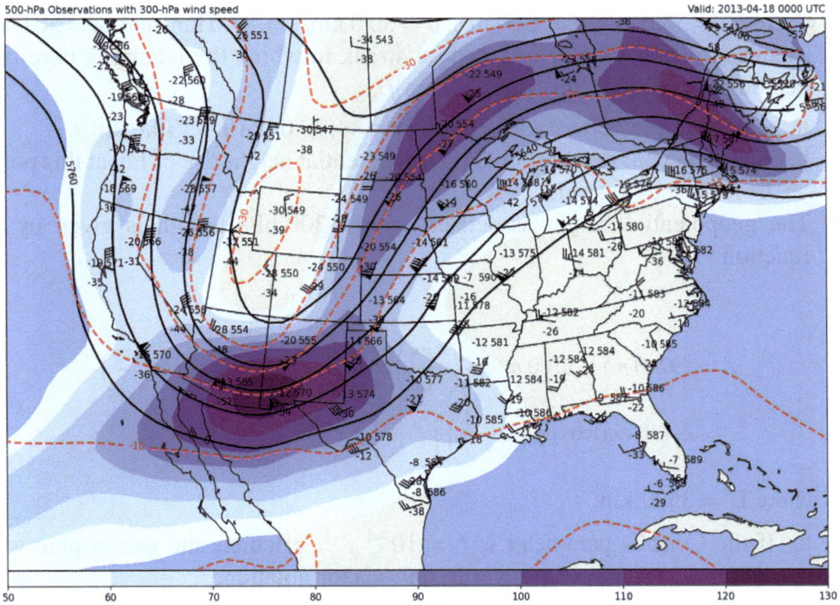

FIGURE 10.14 Case study of trough motion and intensification due to jet streaks

500-hPa observations on April 18, 2013, at 00Z. Solid black contours correspond to 500-hPa isohypses, dashed red contours correspond to 500-hPa isotherms, and 300-hPa isotachs are color-filled.

No Permission Required

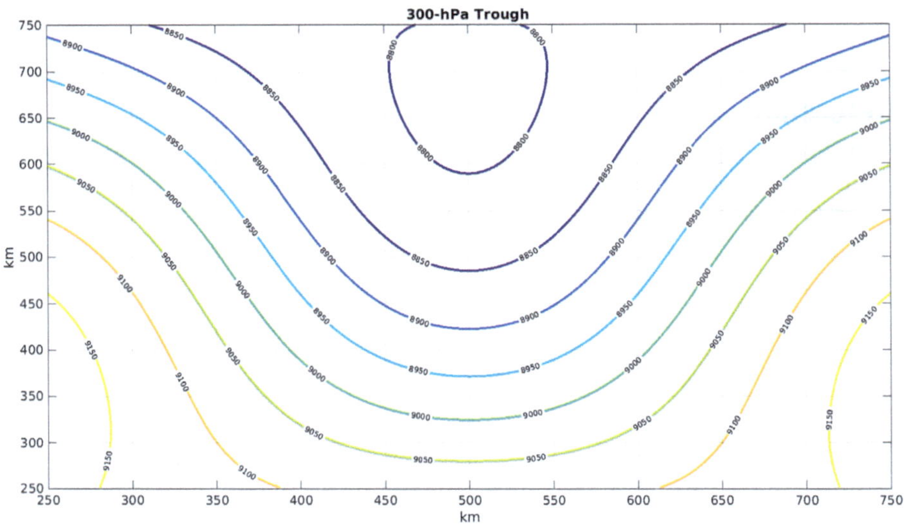

FIGURE 10.15 Schematic of trough from Problem #11

Schematic of trough from Problem 11. Contoured label corresponds to geopotential heights.

No Permission Required

where $U_g = 55.0$ m/s, $a = 10.0$ m/s, $b = 30.0$ km, $c = 40.0$ km, and $x_0 = 50.0$ km. This describes a zonally oriented jet streak in which the jet streak core is located at $\boldsymbol{r} = (50.0 \text{ km})\boldsymbol{i} + (50.0 \text{ km})\boldsymbol{j}$.

a. Calculate the vorticity advection at $\boldsymbol{r} = (70.0 \text{ km})\boldsymbol{i} + (30.0 \text{ km})\boldsymbol{j}$.

b. Based on your result, should the geopotential height rise or fall at this point? Defend your answer.

11. The geopotential height associated with a 300-hPa trough is given by the function:

$$Z = (9240\ m) + (100\ m)\cos\left(\frac{4\pi x}{L}\right)\cos\left(\frac{2\pi y}{L}\right) + (200\ m)\cos\left(\frac{\pi y}{L}\right) + (300\ m)\cos\left[\frac{\pi}{L}\left(x + \frac{L}{2}\right)\right]$$

where L = 3500 km.

a. If the Coriolis parameter is $f = 10^{-4}\ s^{-1}$, calculate the geostrophic wind vector. Express your answer in unit-vector notation.

b. Using your answer from part (a), calculate the vertical vorticity associated with the geostrophic wind vector.

c. Using your answers from parts (a) and (b), calculate the vorticity advection by the geostrophic wind vector.

d. The 300-hPa trough is shown in Figure 10.15. Use your answer from part (c) to calculate the vorticity advection at $\boldsymbol{r} = (L/4)\boldsymbol{i} + (L/2)\boldsymbol{j}$. Is this region characterized by PVA or NVA?

Reference

Stull, Roland, 2017. Practical Meteorology: An Algebra-Based Survey of Atmospheric Science. University of British Columbia, British Columbia. Available from: https://www.eoas.ubc.ca/books/Practical_Meteorology/.

CHAPTER

Surface fronts

11

Chapter outline

As discussed in Ch. 4, the *Glossary of Meteorology* defines a front as an "interface or transition zone between two air masses of different density." In that chapter, we examined surface fronts in relation to the air masses that form them and discussed methods for identifying surface fronts on synoptic charts. In this chapter, we apply atmospheric dynamics and thermodynamics to examine the dynamics, structure, and evolution of surface fronts. We begin by discussing how surface fronts form from air masses.

11.1 Formation of surface fronts

As noted in Ch. 4, surface fronts form when different air masses move and interact. This raises the following question: How do air masses form? In this section, we examine the formation of air masses and how their evolution leads to the development of frontal boundaries.

An Introduction to Synoptic-Dynamic Meteorology. https://doi.org/10.1016/B978-0-443-24858-0.00011-3

11.1.1 Creation of air masses

An air mass is a large body of air, typically at least 1000 km wide, with relatively uniform properties, such as temperature and moisture. Air masses can form in two ways: (1) when air remains stagnant over a surface long enough to adopt that surface's characteristics, or (2) when moving air is advected across a surface with uniform characteristics over a large area. The first mechanism implies that surface high-pressure centers promote the formation of air masses.

To understand why, we consider the flow associated with a surface anticyclone, as shown in Figure 11.1A. In regions with curved isobars within the planetary boundary layer (PBL), the Coriolis force, pressure gradient force, and turbulent drag force are not in balance. This imbalance causes winds to cross isobars and diverge outward from the high-pressure center. According to the continuity equation, this low-level divergence produces sinking motion above the center of the high-pressure area, a process known as subsidence, as shown in Figure 11.1B.

Subsidence brings dry air from the upper troposphere down toward the surface, creating drier conditions above the PBL. Because sinking air undergoes adiabatic compression, the PBL warms and develops higher potential temperature, which produces a temperature inversion and strong static stability above the PBL. This enhanced stability forms a capping inversion that suppresses vertical exchange between the free atmosphere and the PBL. At the same time, subsidence compresses the PBL, reducing its depth, as shown in Figure 11.1B.

This process shows how stable boundary layers develop into air masses. Once established, air masses are modified by both kinematic and thermodynamic processes as they interact with the surface below. For example, when cold air moves over a warmer surface, heat flux from the surface generates a convective mixed layer. As discussed in Ch. 8, thermal turbulence within the PBL depends on the difference between surface potential temperature (θ_{sfc}) and the mixed-layer potential temperature (θ_{ML}). This temperature difference drives turbulent mixing, raising θ_{ML} toward θ_{sfc} and deepening the mixed layer. If the air mass formed under a surface anticyclone, subsidence from above may oppose this deepening. When θ_{sfc} and θ_{ML} are approximately equal, the mixed layer approaches a steady-state depth. Observations show that this equilibrium depth is usually reached within 1 and 4 days (Stull, 2017).

In contrast, cold air masses often develop when air moves over colder surfaces, such as Arctic ice. In this case, the lowest portion of the air mass cools through thermal conduction, thermal radiation, and turbulent exchange with the surface. This cooling increases static stability, which suppresses turbulence and reduces further heat transfer to the surface. As cooling continues, the PBL temperature drops until saturation occurs, producing clouds. Radiative cooling from cloud tops then drives localized sinking, generating turbulence that helps mix cold air downward. Entrainment from above the cloud layer deepens the air mass, which may reach thicknesses of up to 4 km. This process explains the formation of continental polar (cP) and continental arctic (cA) air masses.

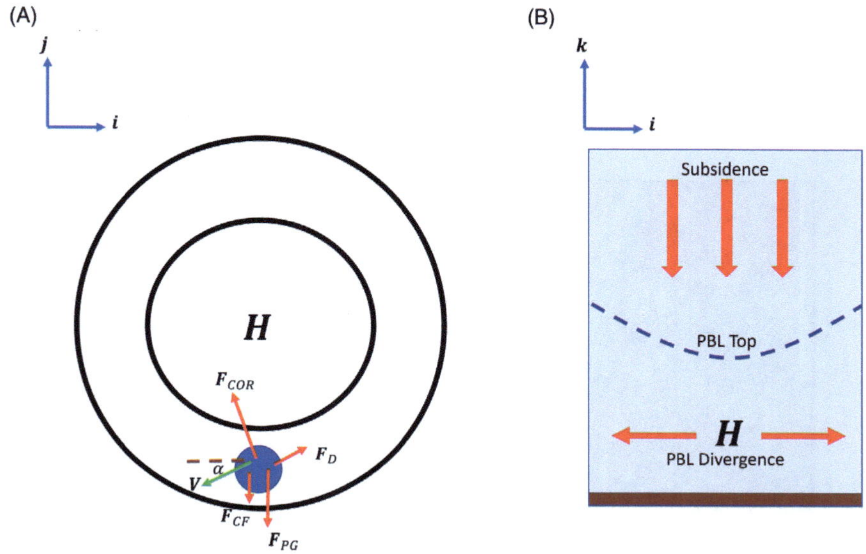

FIGURE 11.1 Schematic of surface high pressure

(A) Horizontal cross-section of a surface high pressure. Red arrows correspond to the forces acting on the air parcel, which are the Coriolis force F_{COR}, the turbulent drag force F_D, the pressure gradient force F_{PG}, and the centrifugal force due to the parcel's motion F_{CF}. The green arrow corresponds to the velocity vector and α is the cross-isobar angle. (B) Vertical cross-section of the surface high pressure given in part A.

No Permission Required

The formation of both warm and cold air masses, depends primarily on the presence of surface anticyclones. For synoptic-scale processes over the United States, two mechanisms contribute to their development:

1. **Atmospheric Global Circulation**: Large-scale circulation produces semi-permanent high-pressure zones, as shown in Figure 11.2A. In particular, upper-level convergence associated with the Hadley and Ferrel cells generates subtropical high-pressure systems, typically over the oceans. The two dominant Northern Hemisphere anticyclones are the North Pacific High and the North Atlantic High, often referred to as the Bermuda-Azores High (see Figure 11.2B).
2. **Rossby Wave Propagation**: As discussed in Ch. 10, surface anticyclones in the midlatitudes often form downstream of ridges or upstream of troughs. Thus, the propagation of Rossby waves - especially in blocking patterns - can promote the development of air masses.

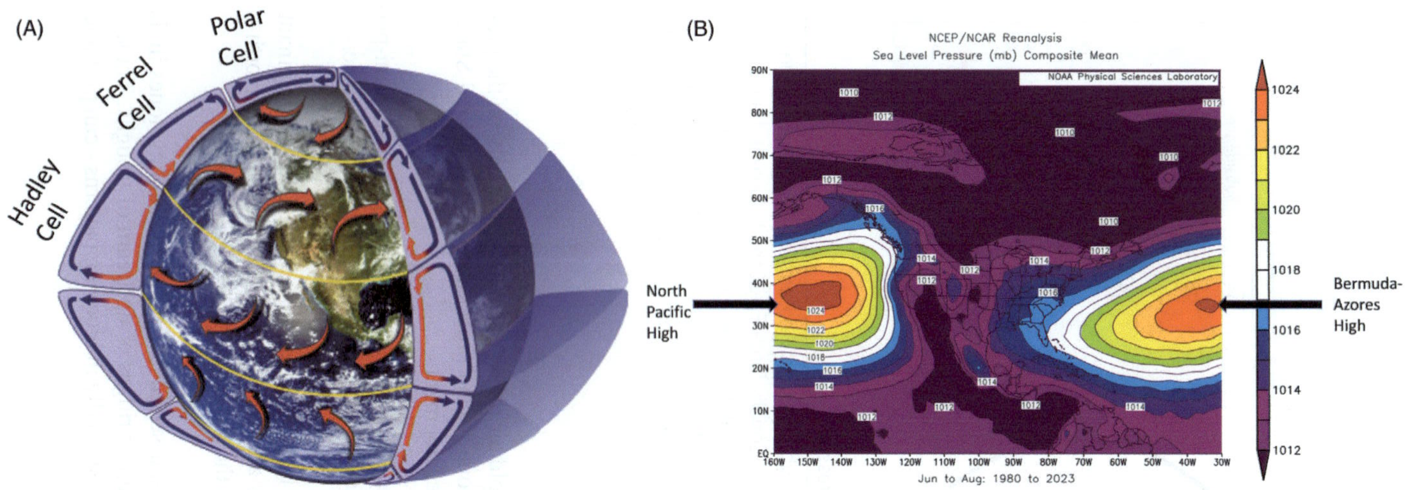

FIGURE 11.2 Global circulation and surface high pressure

(A) The atmospheric global circulation, highlighting the Hadley cell, the Ferrel cell, and the polar cell. (B) NCEP/NCAR reanalysis of the sea-level pressure composite mean during Northern Hemisphere summer from 1980 to 1983. The North Pacific high and the Bermuda-Azores high are highlighted. *NCEP*, National Centers for Environmental Prediction; *NCAR*, National Center for Atmospheric Research.

(A) Courtesy of the National Oceanic and Atmospheric Administration; (B) Image provided by the NOAA Physical Sciences Laboratory, Boulder, Colorado, from their website at https://psl.noaa.gov/)

Once formed, air masses rarely remain stationary; instead, they are advected by synoptic-scale weather patterns. As an air mass moves over a surface with different properties, it is modified by heat and moisture transfer. To quantify this modification, we can construct a simple mathematical model. Assuming that mechanical turbulence creates a well-mixed air mass of depth z_i, the change in its potential temperature is given by:

$$d\theta_{ML} = \left[\frac{C_H\left(\theta_{sfc} - \theta_{ML}\right)}{z_i}\right] dx$$

Here, C_H is the **bulk-transfer coefficient** for heat (a measure of turbulence-driven heat exchange), θ_{ML} is the air mass potential temperature, θ_{sfc} is the surface potential temperature, and dx is the horizontal distance traveled. Rearranging gives:

$$\frac{d\theta_{ML}}{\left(\theta_{ML} - \theta_{sfc}\right)} = -\frac{C_H}{z_i} dx$$

Integrating from the initial state at $x = 0$ (where the initial potential temperature of the air mass is $\Theta_{ML,0}$) to the distance x (where the potential temperature of the air mass is θ_{ML}) yields:

$$\int_{\theta_{ML,0}}^{\theta_{ML}} \frac{d\theta'_{ML}}{\left(\theta'_{ML} - \theta_{sfc}\right)} = -\frac{C_H}{z_i}\int_0^x dx'$$

$$\Rightarrow \ln\left(\frac{\theta_{ML} - \theta_{sfc}}{\theta_{ML,0} - \theta_{sfc}}\right) = -\frac{C_H}{z_i} x$$

$$\Rightarrow \frac{\theta_{ML} - \theta_{sfc}}{\theta_{ML,0} - \theta_{sfc}} = \exp\left(-\frac{C_H}{z_i} x\right)$$

$$\Rightarrow \theta_{ML} = \theta_{sfc} + \left(\theta_{ML,0} - \theta_{sfc}\right)\exp\left(-\frac{C_H}{z_i} x\right) \quad (11.1)$$

Figure 11.3 shows the evolution of θ_{ML} when an air mass with $\theta_{ML,0} = -20°C$ moves across a surface with $\theta_{sfc} = 0°C$. Two key results emerge. First, the rate of change of potential temperature decreases with distance. Initially, the strong temperature contrast drives a large upward heat flux, so θ_{ML} increases rapidly. As the contrast diminishes, the heat flux weakens, and the rate of change decreases accordingly.

Second, wind speed does not affect the modification rate. While stronger winds enhance surface heat transfer, they also increase the distance traveled by the air mass. Because the modification depends on distance, not time, the effects cancel out. Thus, Eq. (11.1) is independent of wind speed.

11.1.2 Geostrophic adjustment and frontal structure

We have shown that when an air mass moves, it can be modified as it travels across a new surface. Additionally, a moving air mass can encounter another air mass.

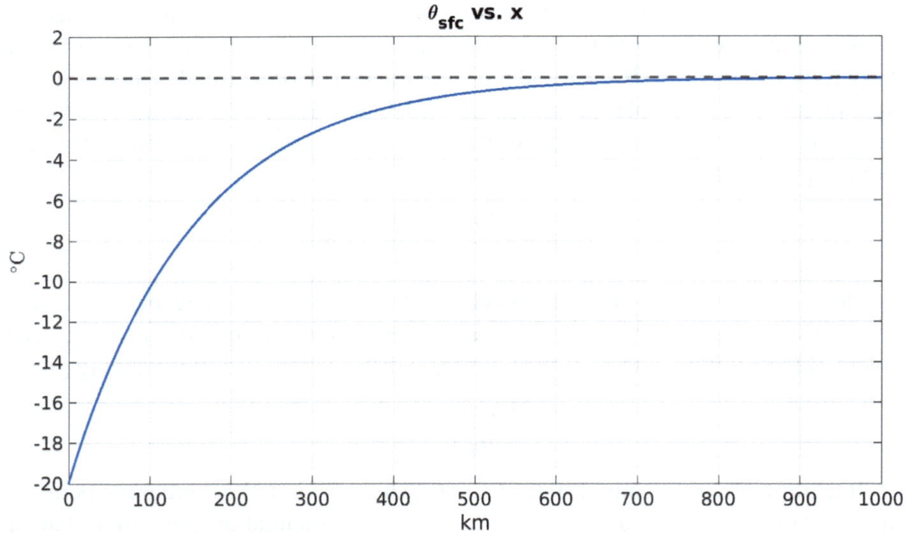

FIGURE 11.3 Graph of mixed layer potential temperature vs. distance traveled

The change in mixed layer potential temperature as a function of the distance traveled by the air mass where $C_H = 0.01$, $\theta_{sfc} = 0°C$, and $\theta_{ML,O} = -20°C$. The black dashed line indicates the surface potential temperature of the surface.

No Permission Required

Figure 11.4 provides an overview of the common air masses present across North America. Notice that the boundary between cA and cP air corresponds to the Arctic front, whereas the boundary between cP and cT air corresponds to the polar front.

This raises an important question: Why doesn't the cA air mass displace the cP air mass at it moves south? Similarly, why doesn't the cP air mass displace the cT air mass when it moves south? As we will show, the locations of the polar and Arctic fronts are largely the result of geostrophic adjustment, introduced in Ch. 8.

Consider two adjacent air masses. In the absence of Earth's rotation, the warmer air would rise above the colder air due to buoyancy, while the cold air would spread across the surface. In this case, no surface front would form because the entire surface would be covered with uniform cold air.

However, on a rotating planet, the southward motion of the cold air is deflected westward by the Coriolis force. Before the flow reaches full balance, the cold air moves a certain distance south while also being pushed westward, eventually aligning parallel to the warm air. In this state, the pressure-gradient force associated with the sloping cold-air interface balances the Coriolis force, creating easterly geostrophic flow. This outcome represents geostrophic adjustment.

From this reasoning, a cold air mass originating near the poles will continue moving south until buoyancy effects are balanced by Earth's rotation acting on

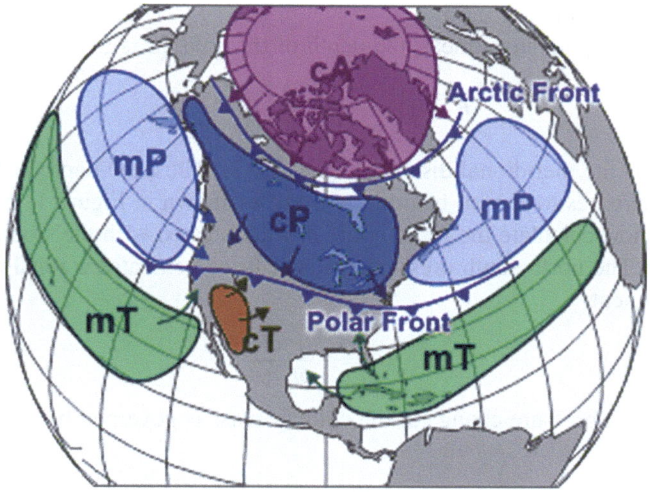

FIGURE 11.4 Air masses across North America. Here, cP corresponds to continental polar, cA corresponds to continental Arctic, cT corresponds to continental tropical, mP corresponds to maritime polar, and mT corresponds to maritime tropical.

Courtesy the National Oceanic and Atmospheric Administration. No Permission Required

the air mass. The distance at which this balance occurs is called the **Rossby deformation radius, which defines the horizontal scale at which rotation becomes important relative to buoyancy**.

In the absence of Earth's rotation, flow would always move toward lower pressure (warmer air). However, on a rotating planet, the Coriolis force deflects the flow. Because the Coriolis force ultimately aligns the cold air parallel to the warm air, the length scale of the frontal boundary should be on the order of the Rossby deformation radius.

The Rossby deformation radius depends strongly upon the static stability of the air mass (ie, how density changes with height) and the Coriolis parameter. For this case, it is defined as:

$$\lambda_R = \frac{1}{f}\sqrt{gH\frac{\Delta\theta_v}{\overline{T}_v}} \tag{11.2}$$

where H is the initial depth of the cold air mass, $\overline{T}_v$ is the average virtual temperature of both air masses, and $\Delta\theta_v$ is the difference in the virtual potential temperature between the two air masses. Virtual potential temperature is defined as the temperature an air parcel would have if brought adiabatically to a standard reference pressure, given by:

$$\theta_v = T_v\left(\frac{P_0}{P}\right)^{2/7} \approx \theta(1+0.608r)$$

where r is the mixing ratio.

After geostrophic adjustment, the depth of the cold air is given by:

$$h = H\left[1 - \exp\left(-\frac{y + \lambda_R}{\lambda_R}\right)\right] \tag{11.3}$$

where y is the meridional distance (Stull, 2017). Thus, the depth of the cold air smoothly increases to H well behind the front, as shown in Figure 11.5.

We can now calculate the geostrophic flow associated with the sloping depth of the cold air. The north-south pressure gradient force generates a zonal geostrophic wind component:

$$U_g = -\frac{1}{\rho f}\frac{\partial P}{\partial y}$$

The vertical pressure change in the cold air mass is governed by the hydrostatic equation:

$$\frac{\partial P}{\partial z} = -\rho g$$

Integrating from $z = 0$ to $z = h$ (assuming uniform density) gives:

$$\int_0^h \frac{\partial P}{\partial z} dz = -\int_0^h \rho g \, dz$$

$$\Rightarrow \Delta P_{cold} = -\rho_{cold} g h$$

Similarly, for the warm air mass, the vertical pressure change is given by:

$$\Delta P_{warm} = -\rho_{warm} g h.$$

Therefore, the surface pressure difference between the air masses is:

$$\Delta P = -(\rho_{cold} - \rho_{warm}) g h = -g h \Delta\rho$$

$$\Rightarrow \Delta P = -\left[\frac{\Delta\rho}{\overline{\rho}}\right]\overline{\rho} g h \tag{11.4}$$

where $\overline{\rho}$ is the average density of the two layers. We can express Eq. (11.4) in terms of virtual potential temperature by applying the ideal gas law.

$$\rho = \frac{P}{R_d T_v}$$

$$\Rightarrow \ln(\rho) = \ln(P) - \ln(R_d) - \ln(T_v)$$

Taking the differential of both sides where it is assumed that pressure is continuous across both air masses and that $\Delta T_v = \Delta\theta_v$ gives:

$$\frac{d\rho}{\rho} = -\frac{dT_v}{T_v} = -\frac{d\theta_v}{T_v}$$

$$\Rightarrow \frac{\Delta\rho}{\rho} \approx -\frac{\Delta\theta_v}{T_v}$$

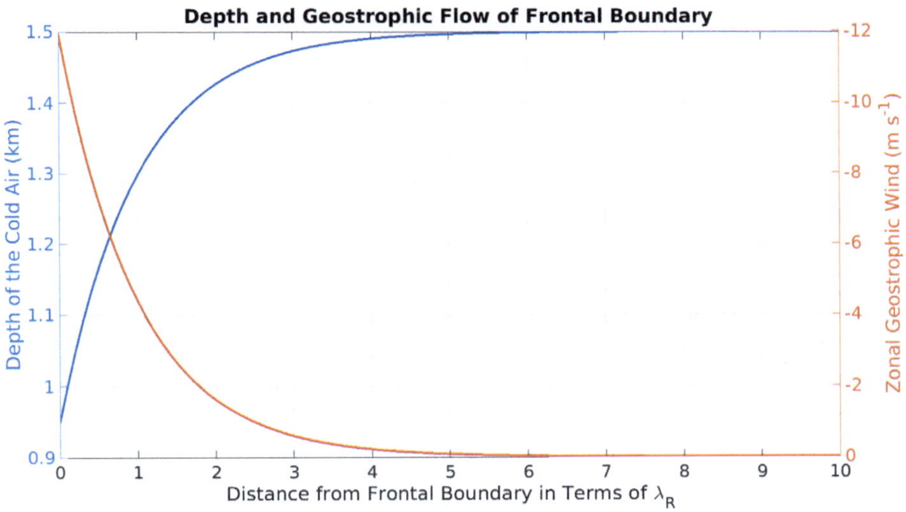

FIGURE 11.5 Depth and geostrophic flow associated with frontal boundary

Graph of the cold air depth (blue) and the zonal geostrophic wind (orange) as a function of the distance from the frontal boundary in terms of the Rossby deformation radius. In this figure, the initial depth of the cold air is 1.50 km, the average virtual temperature of both air masses is 280 K, and the difference in virtual potential temperature is 20 K.

No Permission Required

Therefore, Eq. (11.4) can be expressed as:

$$\Delta P \approx -\left[\frac{\Delta\theta_v}{\overline{T}_v}\right]\overline{\rho} g h \tag{11.5}$$

Substituting Eq. (11.5) into the geostrophic wind equation gives:

$$U_g = -\frac{1}{\rho f}\frac{\partial}{\partial y}\left[\frac{\Delta\theta_v}{\overline{T}_v}\overline{\rho} g h\right] \approx -\frac{g}{f}\frac{\Delta\theta_v}{\overline{T}_v}\frac{\partial h}{\partial y}$$

Substituting Eq. (11.3) into the above expression gives:

$$U_g \approx -\frac{g}{f}\frac{\Delta\theta_v}{\overline{T}_v}\frac{\partial}{\partial y}\left\{H\left[1-\exp\left(-\frac{y+\lambda_R}{\lambda_R}\right)\right]\right\}$$

$$\Rightarrow U_g \approx -\frac{gH}{f\lambda_R}\frac{\Delta\theta_v}{\overline{T}_v}\exp\left(-\frac{y+\lambda_R}{\lambda_R}\right)$$

Recalling that:

$$\lambda_R^2 = \frac{gH}{f^2}\frac{\Delta\theta_v}{\overline{T}_v}$$

from Eq. (11.2), the zonal geostrophic wind becomes:

$$U_g \approx -\frac{1}{f\lambda_R}(f\lambda_R)^2 \exp\left(-\frac{y+\lambda_R}{\lambda_R}\right)$$

$$\Rightarrow U_g \approx -f\lambda_R \exp\left(-\frac{y+\lambda_R}{\lambda_R}\right) \tag{11.6}$$

Thus, as cold air moves southward, the Coriolis force ensures geostrophic balance, producing the geostrophic wind described in Eq. (11.6). Comparing Eq. (11.3) with Eq. (11.6) shows that the cold-air depth increases smoothly behind the front, while the geostrophic wind decreases smoothly, as illustrated in Figure 11.5. Because the Rossby deformation radius depends strongly on static stability, Eq. (11.6) implies that a stronger thermal contrast between the two air masses generates stronger geostrophic flow across the frontal boundary.

As a result of geostrophic adjustment, the horizontal temperature gradient is sharpest near the surface and weakens with height. By thermal wind balance, this implies vertical shear in the geostrophic wind across the frontal zone. Consequently, strong and well-defined fronts are necessarily accompanied by strong upper-level winds. This explains the association between jet streams and frontal zones - for example, the polar jet and the polar front discussed in Ch. 9. The strongest winds are typically on the warm side, and if the cold air advances as a cold front, the associated jet is referred to as a **prefrontal jet**.

Frontal zones also support strong vertical circulations. Air parcels moving into or out of the front encounter sharp changes in horizontal temperature gradients, which will be discussed in more detail in Ch. 11.4. These vertical circulations help explain why fronts are often marked by bands of clouds and precipitation.

Finally, when combining winds from both the cold and warm air masses, strong horizontal shear develops across the frontal boundary. This horizontal shear corresponds to cyclonic vorticity. According to Eq. (11.6), larger density differences between air masses result in greater positive vertical vorticity. This explains why cold fronts, which generally feature large temperature contrasts, are typically associated with stronger wind shifts compared with warm fronts.

11.2 Frontal structure and motion

So far, we have provided a general overview of the formation and two-dimensional structure of fronts. In Ch. 11.1, we examined a stationary front, where neither the cold air mass nor the warm air mass advances relative to each other. Typically, rising motion above a stationary frontal zone is associated with warm advection. This rising motion, often referred to as overrunning, occurs as warm air glides above the cooler air mass. The resulting precipitation is usually stratiform and light, owing to the high static stability in the frontal zone. In this section, we will examine what happens when either the cold or warm air mass advances relative to the other, leading to the development of cold fronts and warm fronts. We begin with cold fronts.

11.2.1 Cold fronts

As discussed in Ch. 4, a cold front occurs when a cold air mass advances equatorward, equatorward and eastward, or eastward relative to the warm air mass. If the cold air advances westward, or both equatorward and westward, the front is referred to as a **back-door cold front** (as shown in Figure 11.6). This name comes from the fact that synoptic-scale weather systems in the midlatitudes are usually steered by westerly flow, so cold fronts typically move westerly. A back-door cold front advances from the east or northeast, appearing to enter the region "through the back door".

Typical cold fronts extend in an equatorward direction from a surface cyclone (discussed in Ch. 12) and are often followed by a surface anticyclone. Back-door cold fronts are common in the northeastern United States during spring and summer, occur east of the Rocky Mountains year-round, and sometimes affect the Great Plains from fall through spring. In New England, back-door cold fronts are strongly influenced by the relatively cold ocean upstream.

Observations show that cold front movement is best correlated with the wind component that is perpendicular to the frontal boundary, known as the **front-normal wind speed**. This behavior arises because the movement of a surface cold front depends on surface pressure tendencies on both sides of the front. Frontal zones are typically associated with a local pressure minimum, which tends to shift toward the region where surface pressure is falling. Behind the frontal boundary, cold advection dominates. This cold advection is associated with sinking motion and increasing surface pressure, which explains why the movement of cold fronts is closely tied to the front-normal wind speed in the cold air mass.

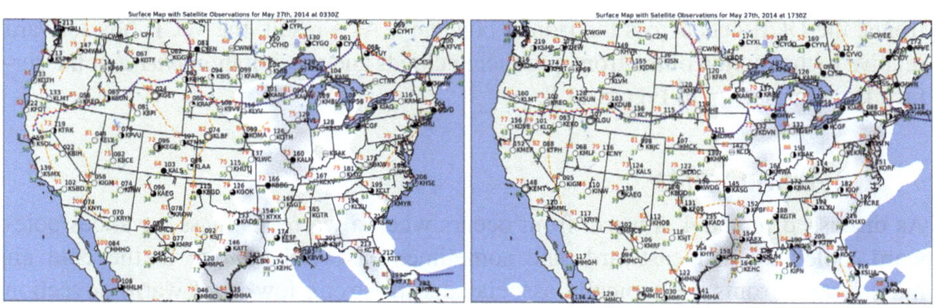

FIGURE 11.6 Example of a back-door cold front

Observations of a back-door front from May 27, 2014. Visual satellite and base radar reflectivity are displayed along with frontal boundaries.

Courtesy the University Corporation of Atmospheric Research.
No Permission Required

The three-dimensional airflow associated with a frontal cyclone can be decomposed into three types of air streams: the **warm conveyor belt (WCB)**, the **dry conveyor belt (DCB)**, and the **cold conveyor belt (CCB)**. These airstreams transport moisture and energy both vertically and horizontally. Because many cold fronts are associated with surface cyclones, it is useful to analyze their structure as part of the low-pressure system, as shown in Figure 11.7.

For cold fronts, the most important conveyor belts are the WCB and the DCB. The WCB originates near the surface ahead of the cold frontal zone, typically in a maritime tropical (mT) air mass. As it transports warm, moist air northward, the WCB ascends steadily, eventually overrunning the colder air to the north. This ascent generates stratiform clouds and precipitation. Owing to thermal wind balance, the poleward-directed temperature gradient causes the winds to veer, producing an eastward turning of the flow with height. Additionally, southeasterly winds near the surface advect the westward portion of the WCB westward, forming the classic S-shaped band of high clouds shown in Figure 11.7.

The DCB originates near the tropopause within the cold sector of the frontal cyclone. As it advects colder, dry air southward, the DCB descends toward the surface. During this descent, adiabatic compression warms the air, making the DCB warmer at the surface than at higher altitudes. As the DCB continues downward, it undergoes diffluence and stretching deformation, processes that will be discussed in more detail in Ch. 11.3.

The interaction between the WCB and the DCB produces two distinct types of cold fronts: anafront cold fronts (or **anafronts**) and katabatic fronts (or **katafronts**). In anafronts, cold air advances strongly against warm air, forcing convergence within the frontal zone and rising motion along the frontal surface. As a result, the primary clouds and precipitation form *behind* the surface front as shown in Figure 11.8. In the conveyor belt framework, an anafront is characterized by a WCB with a rearward component relative to the motion of the front, while the DCB remains largely parallel to the WCB and positioned behind the cold front. This alignment produces a sharp rear cloud edge, marking the transition from the WCB to the DCB. In contrast, in katafronts, the DCB cuts across the frontal boundary. As a result, the WCB tilts forward relative to the motion of the front, producing clouds and precipitation *ahead* of the surface front, as shown in Figure 11.8.

11.2.2 Warm fronts

As discussed in Ch. 4, a warm front occurs when a warm air mass advances poleward, poleward and eastward, and sometimes eastward relative to the cold air mass. Warm fronts are typically associated with strong low-level warm advection ahead of an intensifying surface cyclone. However, not all surface cyclones contain warm fronts, and warm fronts are generally weaker than cold fronts. This leads to two important conclusions.

First, as shown in Eq. (11.3), the depth of a frontal boundary depends on the temperature contrast across it. Because warm fronts usually feature weaker temperature

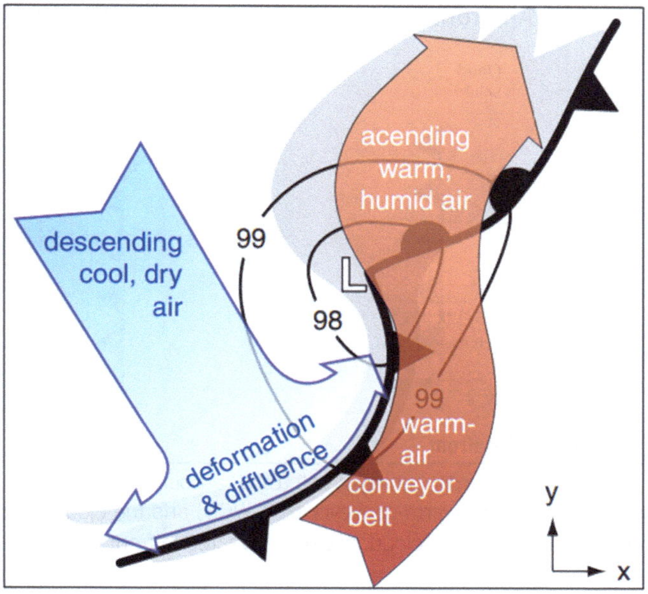

FIGURE 11.7 Schematic of 3D atmospheric flow around midlatitude cyclone

The ascending and descending atmospheric flow in a midlatitude cyclone. Thin black lines with numbers are isobars in kPa. Thick black lines are fronts.

From Stull, R., 2017: Practical Meteorology: An Algebra-Based Survey of Atmospheric Science—*version 1.02b. University of British Columbia. 940 pages.*

contrasts than cold fronts, they are less steeply sloped. Second, as shown in Eq. (11.4), the geostrophic wind near a front depends on the temperature contrast. Consequently, the wind shift along warm fronts is weaker and often less well-defined than along cold fronts.

The three-dimensional structure of a warm front can also be described using the conveyor belt model (see Figure 11.7). The WCB originates behind the warm front in the lower troposphere, crosses the warm frontal boundary, and ascends into the upper troposphere. As it rises, the WCB rotates clockwise (veering eastward) and stops rising once it aligns parallel to the warm front boundary. If the atmosphere contains sufficient humidity, this ascent produces condensation and increased cloudiness.

The CCB originates north of the surface cyclone, typically in a cP or cA air mass. As the CCB approaches the warm front, it turns parallel to the boundary and remains beneath the WCB, since it is denser and less buoyant. The CCB advects cool, moist air north of the warm front, and due to the cyclone's circulation, transports cold air westward on the poleward side of the surface low. Often, the CCB wraps around the low-pressure system, producing the cloud shield that forms the "comma-head"

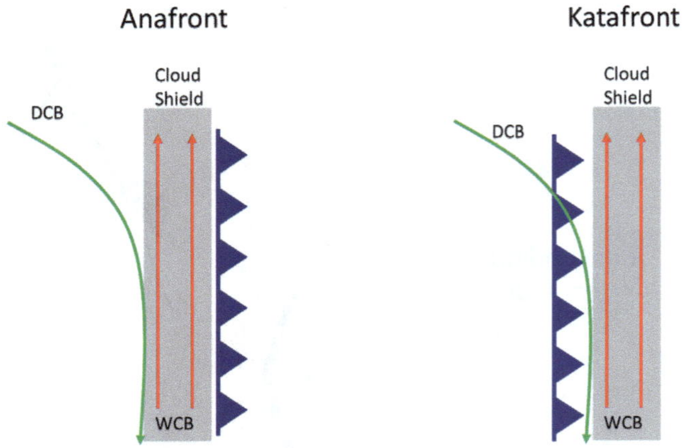

FIGURE 11.8 Anafront vs. katafront

Difference in the position of the dry conveyor belt (DCB) associated with an anafront vs. a katafront. For the anafront, the warm conveyor belt (WCB) and the cloud shield remain behind the cold front, whereas the cloud shield is advected in front of the katafront.

No Permission Required

precipitation pattern common in midlatitude cyclones. As the WCB evaporates precipitation within the CCB, the CCB itself often become saturated, merging with the WCB to produce extensive nimbostratus clouds.

In summary, the conveyor belt model provides a good explanation of the three-dimensional structure of both cold fronts and warm fronts. The WCB and CCB define the warm front: the WCB rises and veers with height across the front, while the CCB flows beneath it, turning cyclonically westward behind the cold front. The DCB, meanwhile, originates near the tropopause, descends behind the cold front, and transports dry, cool air below it. Together, the WCB and CCB shape the wind and precipitation patterns typical of warm fronts, while the interaction of the CCB and DCB controls the structure of cold fronts.

11.2.3 Occluded fronts

Cold and warm fronts are often embedded within surface cyclones. The speed of a front depends on the temperature contrast across the frontal zone, so cold fronts typically move faster than warm fronts. This difference often allows a cold front to overtake the warm front, producing an **occlusion**. This new boundary formed between the cold and warm fronts is called an **occluded front**. During an occlusion, the warm air mass (originally located ahead of the cold front) is lifted aloft, sandwiched between two colder air masses: the advancing cold air behind the cold front and the retreating cold air ahead of the warm front. This lifting forces the warm air upward, where it condenses and enhances cloud development.

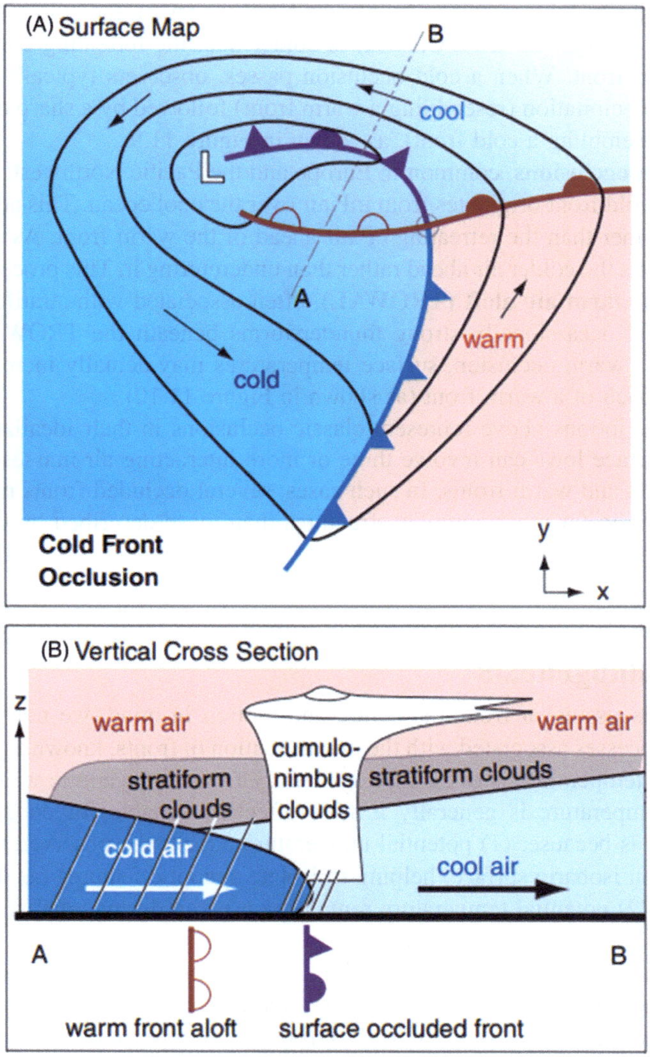

FIGURE 11.9 Schematic of a cold occlusion

(A) A cold occlusion as seen on a surface weather map. The surface cold front is shown with blue triangles, and the surface warm front is illustrated with red semicircles. The occluded front is shown with purple triangles and semicircles. The trailing edge of cool air aloft marks the warm front aloft, which is shown in white semicircles. (B) A vertical cross-section alongside A-B from part A.

From Stull, R., 2017: Practical Meteorology: An Algebra-Based Survey of Atmospheric Science—*version 1.02b. University of British Columbia. 940 pages.*

In **cold occlusions**, common in interior North America, the cold air behind the cold front (often cP air from Canada) is colder than the retreating cold air ahead of the warm front. When a cold occlusion passes, observers typically experience stratiform precipitation (resembling a warm front) followed by a sharp drop in temperature (resembling a cold front), as shown in Figure 11.9.

In **warm occlusions**, common in Europe and the Pacific Northwest, the cold air behind the cold front originates from mP air over the cool ocean. This advancing air mass is warmer than the retreating cP air ahead of the warm front. As a result, the occlusion lifts the colder air ahead rather than undercutting it. This process produces a **trough of warm air aloft (TROWAL)**, often associated with stratiform clouds, showers, and occassionally strong thunderstorms beneath the TROWAL. During passage of a warm occlusion, surface temperatures may actually increase, similar to the approach of a warm front (as shown in Figure 11.10).

The descriptions above represent classic occlusions in their idealized form. In practice, surface lows can involve three or more interacting air masses, producing multiple cold and warm fronts. In such cases, several occluded fronts may develop aloft, leading to far more complex structures than those described by the standard cold and warm occlusion models.

11.3 Frontogenesis

With our discussion of frontal structure and motion in mind, we now turn to the physical processes associated with the intensification of fronts, known as **frontogenesis**. While temperature is often used to identify fronts on synoptic weather charts, potential temperature is generally a more effective variable for defining surface fronts. This is because: (1) potential temperature adjusts the observed temperature to a common isobaric surface, helping to isolate synoptic features (as discussed in Ch. 8) and (2) potential temperature remains conserved during adiabatic processes (also discussed in Ch. 8). Accordingly, frontogenesis corresponds to an increase in the potential temperature gradient over time, whereas **frontolysis** corresponds to a decrease in the potential temperature gradient over time. In this section, we will examine the mechanisms that lead to frontogenesis, beginning with a kinematic process that plays a central role: deformation.

11.3.1 Introduction to deformation

So far in this textbook, we have discussed three primary patterns of atmospheric motion:

(1) translation - straight-line motion associated with constant-speed winds (e.g., uniform geostrophic flow). Translational flow also appears in the context of advection, such as thermal, moisture, and vorticity advection. In all such cases, the air parcel moves without changing its size or shape;

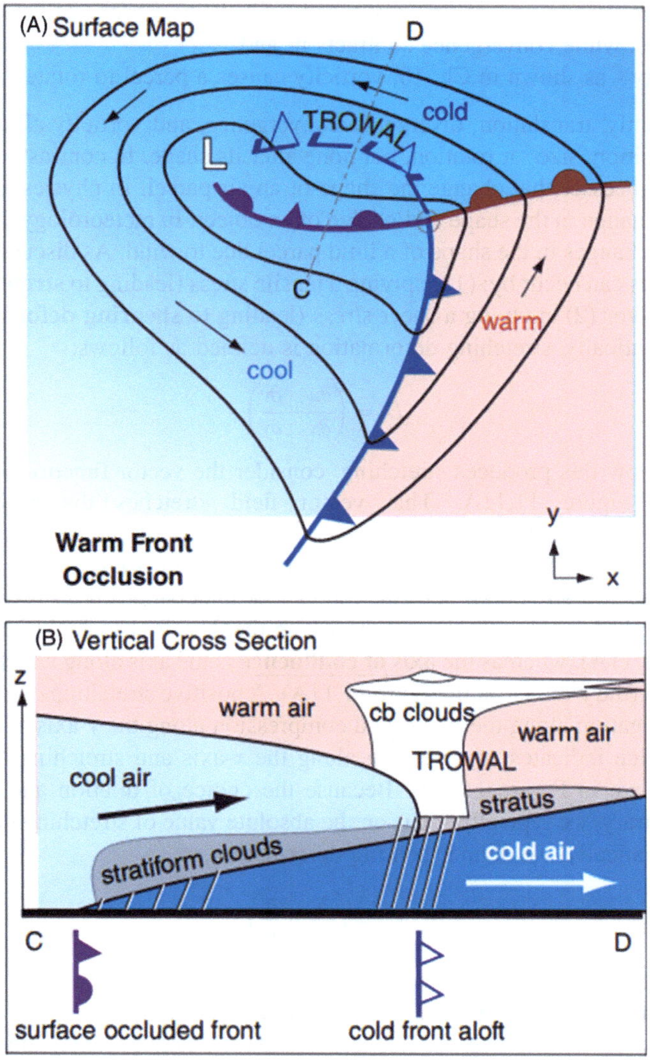

FIGURE 11.10 Schematic of a warm occlusion

(A) A warm occlusion as seen on a surface weather map. The surface cold front is shown with blue triangles, and the surface warm front is illustrated with red semicircles. The occluded front is shown with purple triangles and semicircles. The white triangles denote the cold front aloft, which is shown in white semicircles. (B) A vertical cross section alongside C-D from part A where the trough of warm air aloft (TROWAL) has been displayed.

From Stull, R., 2017: Practical Meteorology: An Algebra-Based Survey of Atmospheric Science—*version 1.02b. University of British Columbia. 940 pages.*

(2) divergence/convergence - as introduced in Ch. 9, divergence enlarges the size of a parcel, while convergence contracts it; and
(3) vorticity - as shown in Ch. 10, vorticity causes a parcel to rotate.

Importantly, translation, divergence/convergence, and vorticity all affect an air parcel's position, size, or rotation, but none alter its shape. In contrast, deformation describes processes that change the shape of an air parcel. In physics, deformation refers to a change in the shape and/or size of an object. In meteorology, it refers specifically to changes in the shape of a fluid parcel due to wind. As discussed in Ch. 1, such changes can occur by: (1) applying a tensile stress (leading to **stretching deformation**) and/or (2) applying a shear stress (leading to **shearing deformation**).

Mathematically, stretching deformation is defined as follows:

$$D_1 = \left(\frac{\partial u}{\partial x} - \frac{\partial v}{\partial y}\right)$$

To see how this produces stretching, consider the vector function $v = xi - yj$, plotted in Figure 11.11A. This vector field stretches the region around $(x, y) = (0, 0)$ along the x-axis, while compressing it along the y-axis, thereby altering its shape. The calculated stretching deformation is $D_1 = 2$.

The effects of stretching deformation can be understood in terms of two axes. The **axis of diffluence** is the axis along which the flow is stretched (the x-direction in Figure 11.11A), whereas the **axis of confluence** is the axis along which the flow is compressed (the y-direction in Figure 11.11A). A positive stretching deformation indicates elongation along the x-axis and compression along the y-axis. Conversely, a negative value indicates compression along the x-axis and stretching along the y-axis, as shown in Figure 11.11B. Because the choice of dilation and contraction axes is arbitrary, we typically focus on the absolute value of stretching deformation.

Mathematically, we define shearing deformation as:

$$D_2 = \left(\frac{\partial u}{\partial y} + \frac{\partial v}{\partial x}\right)$$

To see how this produces shearing, consider the vector function $v = yi + xj$, as plotted in Figure 11.11C. Here, the flow stretches space around $(x, y) = (0, 0)$ along the northeast-southwest diagonal (ie, +45° from the positive x-axis) and compresses it along the northwest-southeast diagonal (ie, −45° from the positive x-axis). As a result, the region surrounding the origin experiences shearing along the x-axis. The resulting shearing deformation has a value of $D_2 = 2$.

Relative to Figure 11.11A, the axes of diffluence and confluence are rotated 45° counterclockwise in Figure 11.11C. A negative shearing deformation would instead rotate the axes 45° clockwise, as shown in Figure 11.11D. As with stretching, we are primarily concerned with the absolute value of the shearing deformation.

The total deformation is the combined effect of stretching and shearing:

$$D = \sqrt{D_1^2 + D_2^2} = \left[\left(\frac{\partial u}{\partial x} - \frac{\partial v}{\partial y}\right)^2 + \left(\frac{\partial u}{\partial y} + \frac{\partial v}{\partial x}\right)^2\right]^{1/2} \tag{11.7}$$

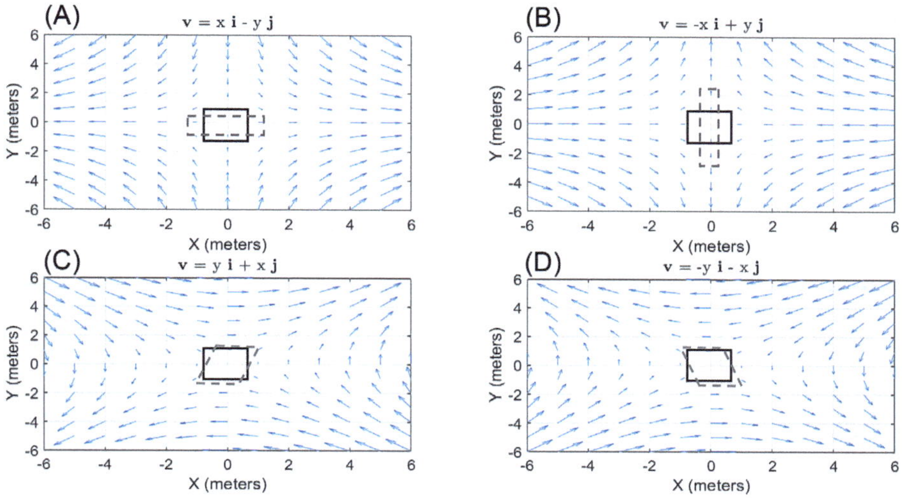

FIGURE 11.11 Schematic of stretching and shearing deformation

Stretching and shearing deformations. (A) The region of space is stretched along the *x*-axis. (B) The region of space is stretched along the *y*-axis. (C) Stretching occurs along the line +45 degrees from the *x*-axis. (D) Stretching occurs along the line –45 degrees from the *x*-axis.

No Permission Required

This measure allows us to quantify the overall distortion of an air parcel due to both stretching and shearing. Within this framework, we are now prepared to examine the physical mechanisms by which deformation contributes to frontogenesis.

11.3.2 Kinematics of frontogenesis

Frontogenesis is strongly influenced by the kinematic properties of the wind field, particularly the geometry of the horizontal flow. For synoptic-scale processes, the two primary contributors are: (1) divergence/convergence and (2) deformation.

As illustrated in Figure 11.12, divergence leads to frontolysis, whereas convergence leads to frontogenesis, regardless of the orientation of the isotherms. Unlike divergence and convergence, deformation of the wind can produce either frontogenesis or frontolysis. As shown in Figure 11.11, stretching and shearing deformation cause stretching along the axis of diffluence and contraction along the axis of confluence, with these axes oriented at right angles to each other. This geometry means that deformation simultaneously causes some regions of the air parcel to spread apart (leading to frontolysis) and others to contract together (leading to frontogenesis).

To illustrate, consider Figure 11.13A, where deformation concentrates the isotherms along the axis of diffluence. In contrast, Figure 11.13B shows deformation leading to greater isotherm spacing along the same axis. The key factor distinguishing frontogenesis from frontolysis is the angle between the isotherms and the axis of diffluence, known as the **angle of dilation**. In Figure 11.13A, the isotherms are parallel to the axis of diffluence, which means that the angle of dilation is 0°.

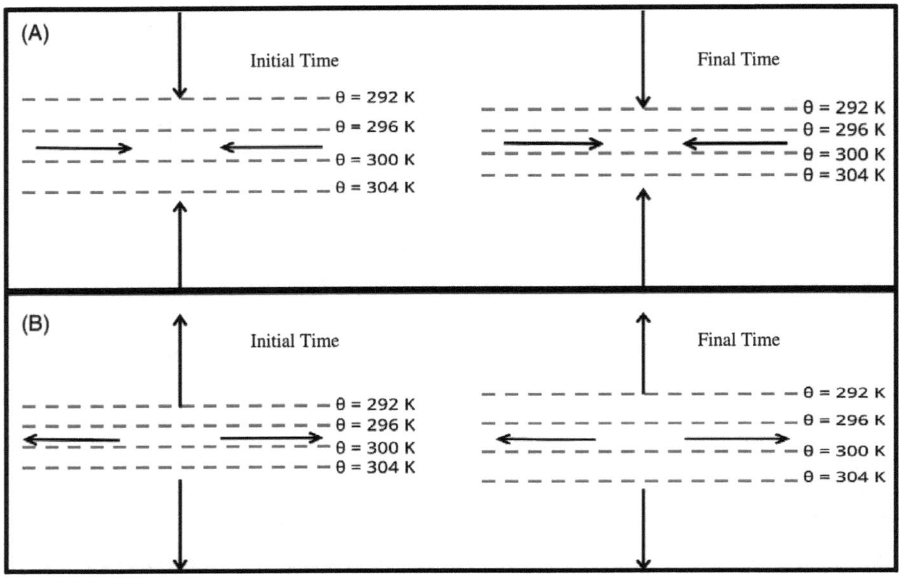

FIGURE 11.12 Schematic of divergence with frontogenesis

(A) The effect of convergence on frontogenesis. (B) The effect of divergence on frontogenesis. Notice that convergence generates frontogenesis and divergence generates frontolysis, regardless of the orientation of the isotherms.

No Permission Required

Conversely, in Figure 11.13B, the isotherms are perpendicular to the axis of diffluence, which means that the angle of dilation is 90°. It can be shown that if the angle of dilation is less than 45°, deformation will lead to frontogenesis, whereas if the angle of dilation is greater than 45°, deformation will lead to frontolysis (Markowski and Richardson, 2010).

Although deformation and divergence are the primary drivers of frontogenesis, shear vorticity can also play a role. As shown in Figure 11.14, horizontal shear affects the horizontal temperature gradient. In cases of shearing frontogenesis, the temperature gradient strengthens when cold advection occurs within the cold air and warm advection occurs within the warm air, both of which enhance frontal contrast, as shown in the cold front depicted in Figure 11.14. In contrast, shearing frontolysis occurs when warm advection occurs within cold air or cold advection occurs within warm air, both of which weaken the frontal contrast, as shown in the warm front depicted in Figure 11.14.

In addition to kinematic processes, thermodynamic factors can also modify frontal strength. Figure 11.15 shows how horizontal variations in diabatic heating influence the horizontal temperature gradient for an anafront. During the daytime, solar radiation heats the region ahead of the cold front, while cloud cover reduces heating behind it. The resulting increase in the temperature gradient strengthens the front, leading to frontogenesis. At night, longwave radiation cools the region ahead of the front, while re-radiated longwave radiation warms the surface behind it. This reduces the temperature gradient, resulting in frontolysis. Because these effects arise

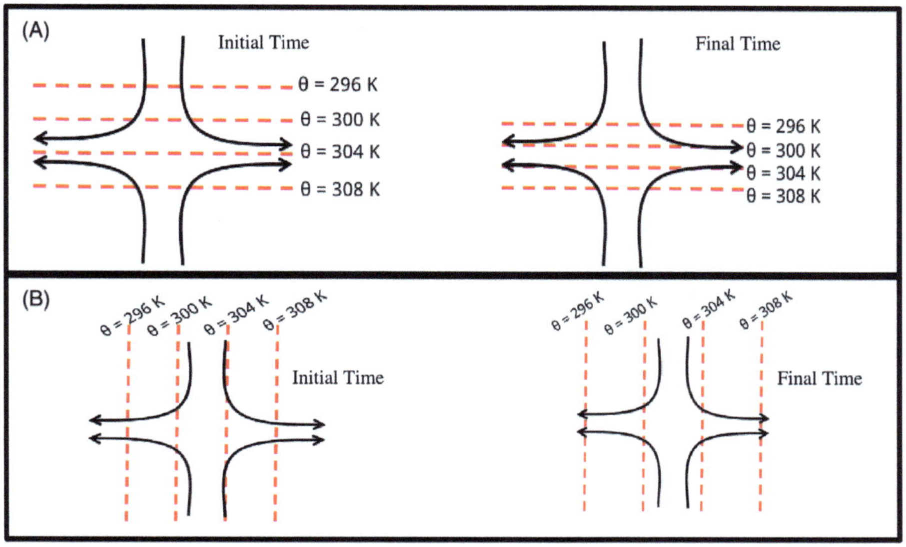

FIGURE 11.13 Schematic of deformation with frontogenesis

(A) The effect of deformation on frontogenesis when the axis of diffluence is parallel to the isotherms. (B) Schematic of the effect of deformation on frontogenesis when the axis of diffluence is perpendicular to the isotherms. Notice that frontogenesis is produced in part A, whereas frontolysis occurs in Part B.

No Permission Required

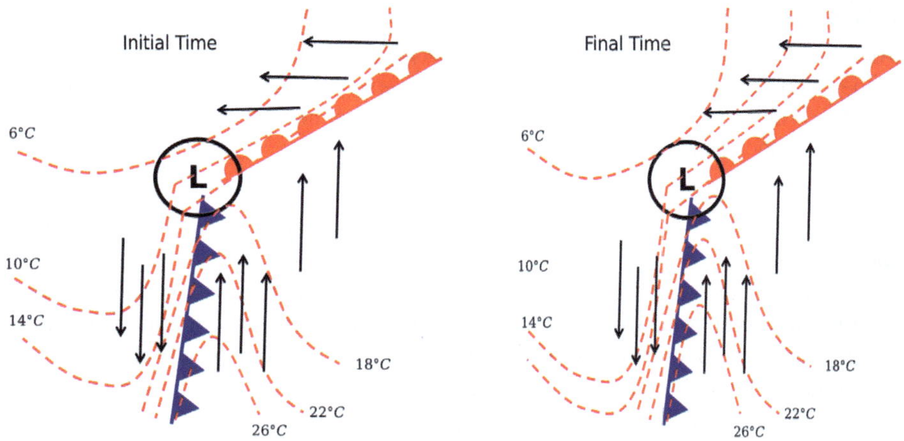

FIGURE 11.14 Schematic of shear with frontogenesis

The effect of horizontal shear on frontogenesis. The figure to the left shows the frontal system at an initial time, and the figure to the right shows the frontal system at a later time. Notice that horizontal shear generates frontogenesis for the cold front, whereas frontogenesis generates frontolysis on the warm front.

No Permission Required

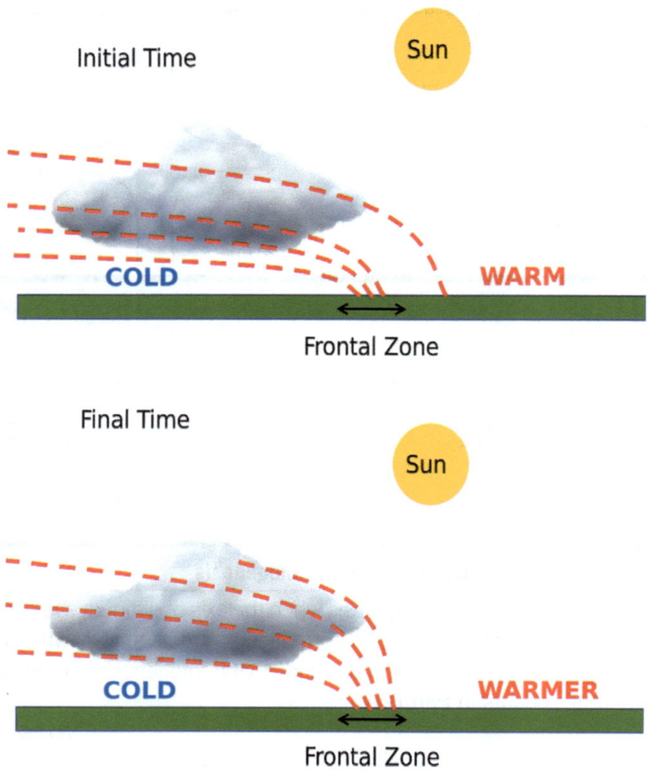

FIGURE 11.15 Schematic of differential diabatic heating on frontogenesis

The effect of differential diabatic heating on frontogenesis. In the top figure, the presence of the cloud deck behind the frontal boundary prevents this region from receiving solar radiation. Therefore, as shown in the bottom figure, the temperature gradient across the frontal zone increases.

No Permission Required

from horizontal variations in heating, they are often referred to as **differential diabatic heating**.

Similar processes can occur in other settings. For example, stronger heating over bare ground relative to snow cover can sharpen the temperature gradient and drive frontogenesis. Evaporation from rain or snow cools the air beneath clouds. Thus, precipitation behind a cold front enhances frontogenesis, while precipitation ahead of a cold front weakens it a front, leading to frontolysis.

11.4 Frontal circulation

Although the geometry of the wind field and diabatic processes can contribute to frontogenesis, they are not sufficient to explain the observed rates of frontal

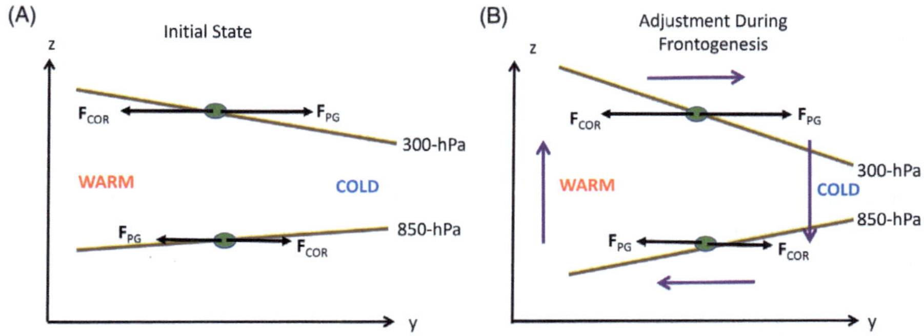

FIGURE 11.16 Effect of ageostrophic flow on frontogenesis. (A) The initial state of the atmosphere in thermal wind balance in which warm air is located south of the cold air. (B) The adjustment in atmospheric flow during a process of frontogenesis. Black arrows indicate the atmospheric forces in which F_{PG} corresponds to the pressure gradient force and F_{COR} corresponds to the Coriolis force. Purple arrows indicate the ageostrophic vertical circulation during the adjustment process.

intensification. To achieve realistic frontogenesis rates, we must account for the influence of ageostrophic flow.

To see this, imagine the atmosphere initially in thermal wind balance, as shown in Figure 11.16A. Because of the thickness gradient associated with the frontal boundary, the geostrophic wind increases with height, producing a stronger pressure gradient force aloft. Near the surface, the geostrophic wind flows eastward (out of the page in Figure 11.16A), while aloft it flows westward (into the page in Figure 11.16A). In other words, the geostrophic flow is parallel to the frontal boundary.

Now, consider what happens as the front strengthens. According to the thickness equation, an increase in horizontal temperature gradient must correspond to a larger geopotential height gradient. This increases the slope of the isobaric surfaces, as shown in Figure 11.16B. A steeper slope implies a stronger pressure gradient force, which pulls the system out of thermal wind balance. As a result, the wind accelerates toward lower geopotential height.

As shown in Figure 11.16B, this imbalance produces enhanced surface convergence into the warm air mass, intensifying cold advection there. Aloft, the stronger pressure gradient force produces convergence toward the cold air mass. By continuity, this imbalance drives a direct transverse circulation: warm air rises on the warm side of the front while cold air sinks on the cold side, as shown in Figure 11.16B.

This transverse circulation introduces a positive feedback mechanism. Near the surface, the ageostrophic inflow enhances the convergence into the warm air, which increases cold advection and further sharpens the thickness gradient. This, in turn, strengthens the frontal circulation. However, the process does not continue indefinitely for two reasons.

First, rising warm air expands adiabatically and cools, while sinking cold air compresses and warms, as discussed in Ch. 8. This vertical circulation acts to weaken the horizontal temperature gradient.

Second, as with other all adjustment processes, the ageostrophic flow redistributes mass until the wind and pressure gradient are once again consistent, as discussed in Ch. 8. Eventually, the horizontal flow returns to geostrophic balance - but with a stronger geostrophic wind than before.

This discussion highlights two important points about frontal circulation. First, frontal circulation develops as a mechanism to restore thermal wind balance. While geostrophically forced thermal advection drives the system away from balance, the direct thermal circulation is the dynamic response that works to bring the system back.

Second, the ageostrophic component of the circulation enhances temperature advection within the frontal zone. In doing so, it intensifies the thermal contrast and sharpens the frontal boundary.

Questions

1. Answer the following questions about the properties of surface fronts:
   a. Explain why there must be large static stability within the frontal zone.
   b. Explain why there must be cyclonic shear vorticity across a front.
   c. Explain why shear vorticity can lead to frontogenesis, whereas curvature vorticity will not lead to frontogenesis.
2. Answer the following questions regarding the properties of anabatic and katabatic fronts:
   a. Explain why anabatic fronts usually have strong temperature gradients across the frontal boundary, whereas katabatic fronts usually have strong dew point gradients across the frontal boundary.
   b. Explain why fronts tend to undergo frontolysis when they transition from an anabatic structure to a katabatic structure.
3. In Ch. 11.1, it was shown that shear vorticity is generated along a cold front boundary due to geostrophic adjustment. Explain how turbulent drag can increase the cyclonic shear vorticity associated with cold fronts.
4. Explain why thermal wind balance should break down near the frontal boundary. How does this lead to the vertical motion observed along fronts?
5. How should the static stability near the frontal boundary affect the structure of the frontal circulation? How would the overturning frontal circulation in a region of strong static stability compare to a region of weak static stability?
6. The vertical circulation of jet streaks was discussed in Ch. 9.
   a. Suppose that the vertical circulation associated with the exit region of an upper-level jet streak is directly above the vertical circulation associated

with a surface front. Will this synoptic setup destabilize or stabilize the environment? Defend your answer.

b. Suppose that the vertical circulation associated with the jet streak is displaced east of the cold front. Will this synoptic setup destabilize or stabilize the environment? Defend your answer.

7. Suppose that an air mass has a mixed layer potential temperature of 0°C, and it is advected across a surface with $\theta_{sfc} = -10°C$. How far must the air mass travel until it is completely modified by the new surface? Assume $C_H = 0.01$.

8. A cold air mass of depth 1.5 km and virtual potential temperature 0°C is embedded in warm air of virtual potential temperature 20°C. Determine the Rossby deformation radius, the maximum geostrophic wind speed, and the equilibrium depth of the cold air mass at the frontal boundary. You may assume that $f = 10^{-4}$ s^{-1}.

9. Figure 11.17 depicts a synoptic-scale system on December 22, 2013, at 00Z and at 12Z.

a. Draw the frontal boundaries associated with each surface map.

b. According to the synoptic maps, determine whether the cold fronts are likely anabatic or katabatic. Defend your answer.

10. Figure 11.18 provides a surface map (with satellite observations) and an 850-hPa map for September 16, 2009, at 12Z. Determine whether the cold front in the Midwest will intensify or weaken. Defend your answer.

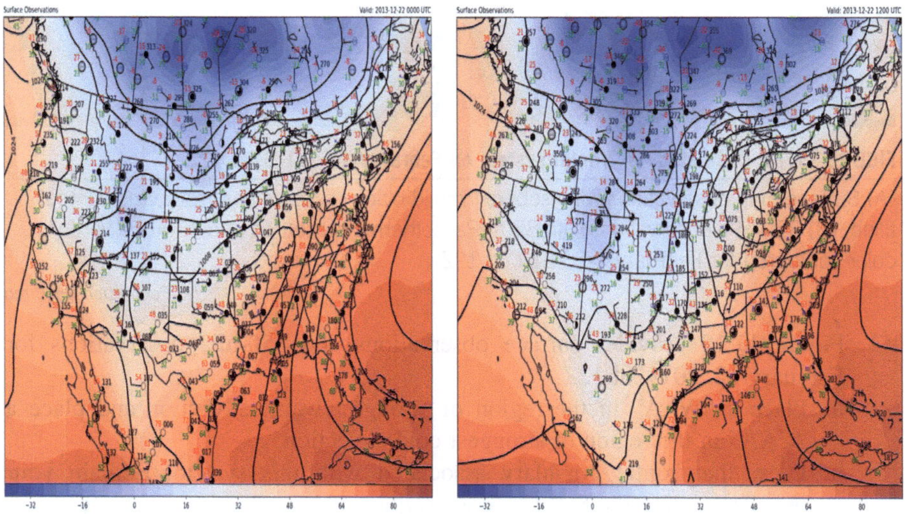

FIGURE 11.17 Case study of frontal system on December 22, 2013

Surface observations of December 22, 2013, at 00Z and at 12Z. Solid black contours are isobars and isotherms are in contour color fill.

No Permission Required

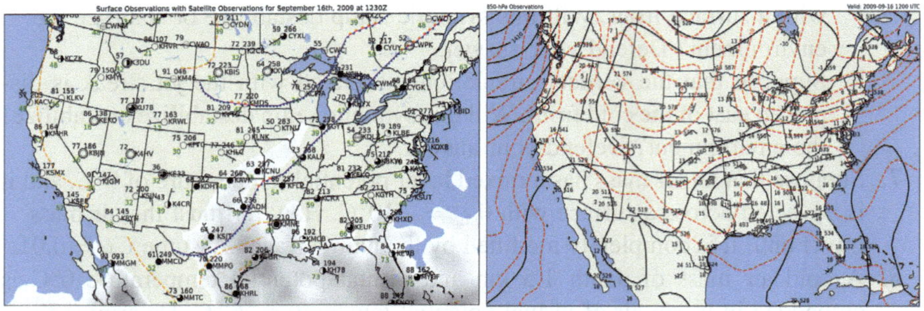

FIGURE 11.18 Case study of cold front on September 16, 2009, at 12Z

(Left) Surface map with satellite observations and (Right) 850-hPa observations for September 16, 2009, with isohypses in solid black contours and isotherms in dashed red contours.

No Permission Required

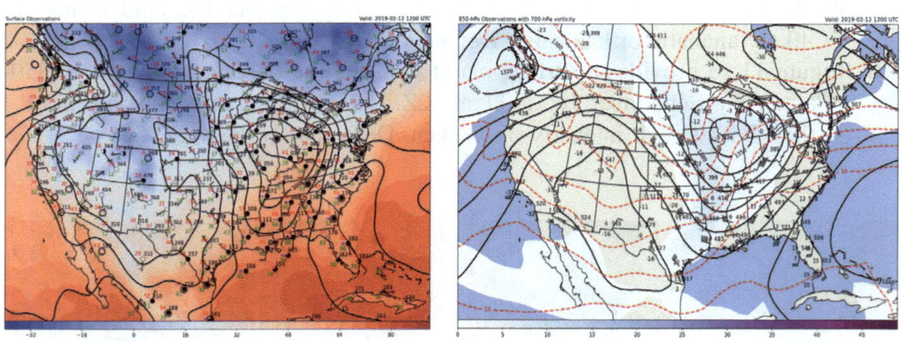

FIGURE 11.19 Case study of frontal system on February 12, 2019, at 12Z

(A) Surface map observations with isobars in solid black contours and isotherms contour filled. (B) 850-hPa observations with 850-hPa isotherms contoured and 700-hPa vorticity color-contoured for February 12, 2019, at 12Z.

No Permission Required

11. Figure 11.19 provides surface observations and 850-hPa observations for February 12, 2019, at 12Z.

- **a.** Place a "WAA" in the region of strongest warm advection and place a "CAA" in the region of strongest cold advection.
- **b.** Draw the frontal boundary associated with this system. Explain your reasoning using the 850-hPa observations and the 700-hPa vorticity.
- **c.** Based on your map, what physical mechanism is most responsible for frontogenesis? Defend your answer.

12. Figure 11.20 provides surface observations for a frontal system on April 23 and 24, 2010. Draw the cold front and warm front boundary associated with each map and determine whether frontogenesis is occurring along each front.

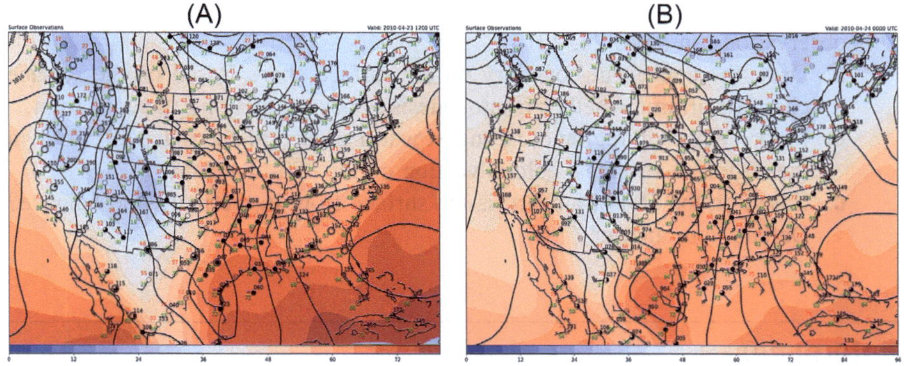

FIGURE 11.20 Case study of frontal system on April 23–24, 2010

Surface map observations with isobars contoured for (A) April 23, 2010, at 12Z and (B) April 24, 2011, at 00Z. For both maps, isobars are in solid black contours and isotherms are contour filled.

No Permission Required

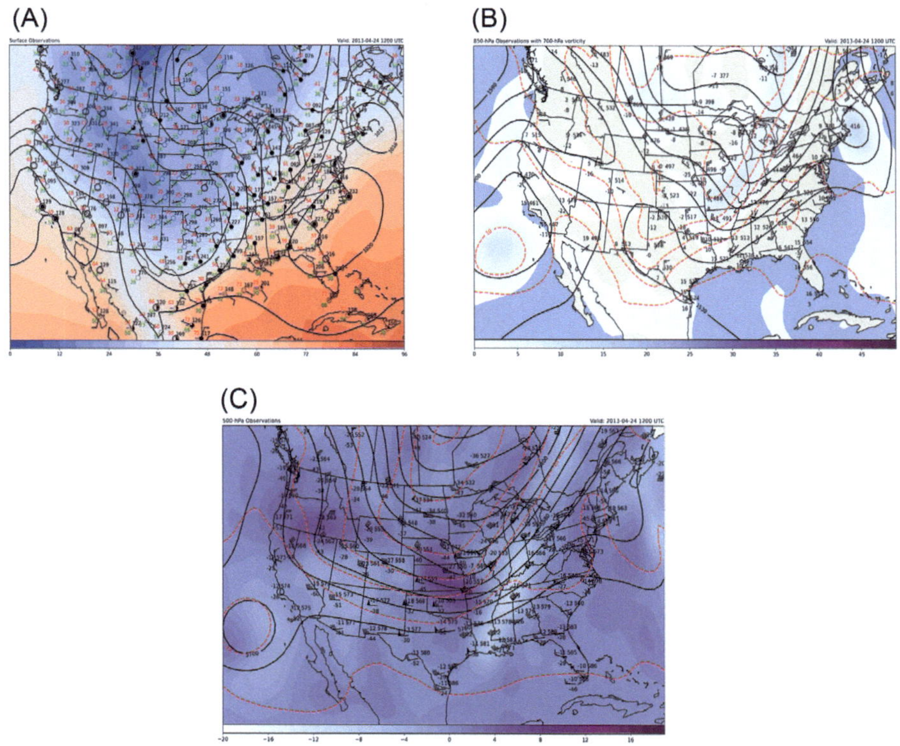

FIGURE 11.21 Case study of cold front on April 24, 2013

(A) Surface observations with isobars in solid black contours and isotherms contour filled. (B) 850-hPa observations with isohypses in solid black contours, isotherms in dashed red contours, and 700-hPa vorticity contour filled. (C) 500-hPa observations with isohypses in solid black contours and omega velocity in contour filled.

No Permission Required

13. Figure 11.21 provides surface observations, 850-hPa observations, and 500-hPa observations on April 24, 2013.

  a. Draw the cold front boundary on the surface chart. Based on the surface chart, determine whether frontogenesis is occurring along the cold front.

  b. Mark "WAA" in region of strongest warm advection and place a "CAA" in the region of strongest cold advection on the 850-hPa chart.

  c. Based on the 500-hPa chart, is the surface front in thermal wind balance? Defend your answer.

References

Markowski, Paul, Richardson, Yvette, 2010. Mesoscale Meteorology in Midlatitudes. John Wiley & Sons, Ltd.

Stull, Roland, 2017. Practical Meteorology: An Algebra-based Survey of Atmospheric Science. University of British Columbia, British Columbia. Available from. https://www.eoas.ubc.ca/books/Practical_Meteorology/.

CHAPTER

Midlatitude cyclones

12

Chapter outline

The single most common weather system in the midlatitudes is the frontal cyclone, also known as the **midlatitude cyclone**. As shown in Figure 12.1, a midlatitude cyclone is a synoptic-scale system in which winds rotate cyclonically (ie, counterclockwise in the Northern Hemisphere and clockwise in the Southern Hemisphere) around a closed low-pressure center. Near the surface, turbulent drag causes these cyclonic winds to spiral inward toward the center.

As shown in Figure 12.1, surface fronts are often associated with midlatitude cyclones. In the Northern Hemisphere, a cold front typically extends equatorward from the low-pressure center, while a warm front extends eastward. In this chapter, we provide an overview of the dynamics of midlatitude cyclones. Specifically, we examine the physical processes that govern their formation and intensification of midlatitude cyclones (a process known as **cyclogenesis**), as well as the mechanisms responsible for their eventual dissipation, a process known as **cyclolysis**.

An Introduction to Synoptic-Dynamic Meteorology. https://doi.org/10.1016/B978-0-443-24858-0.00012-5

12.1 Characteristics of midlatitude cyclones

Before providing an overview of the dynamics of midlatitude cyclones, we will first examine their basic observational features.

12.1.1 Structure of midlatitude cyclones

Figure 12.1 presents a two-dimensional depiction of a midlatitude cyclone as it typically appears on a surface weather chart. However, radiosonde networks and, more recently, satellite observations have revealed that a midlatitude cyclone is a three-dimensional system (Reed, 1990). As discussed in Ch. 11, the three-dimensional flow associated with a midlatitude cyclone can be broken down into the three distinct air streams, as illustrated in Figure 12.2:

1. **The warm conveyor belt (WCB):** This stream ascends from the warm sector of the cyclone (ie, ahead of the cold front). As it rises and intersects the warm front, it veers to the right (ie, eastward).
2. **The dry conveyor belt (DCB)**: Originating near the tropopause within the cold sector (ie, behind the cold front), this stream descends towards the surface. Upon encountering the cold front, it undergoes diffluence and stretching deformation.

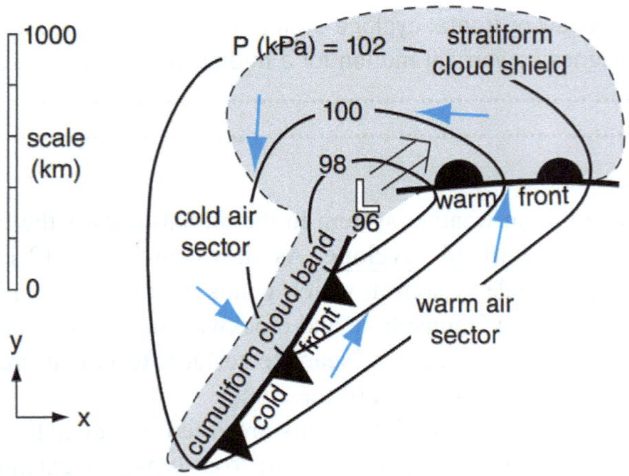

FIGURE 12.1 Schematic of a frontal cyclone

This figure provides the typical components of a midlatitude cyclone in the Northern Hemisphere. The light gray shading shows clouds. Cyan arrows show near-surface winds. Thin black lines are isobars in kPa, and thick black lines indicate fronts. The double-shaft arrow shows movement of the central low-pressure area.

From Stull, R., 2017: Practical Meteorology: An Algebra-Based Survey of Atmospheric Science—*version 1.02b. University of British Columbia. 940 pages.*

3. **The cold conveyor belt (CCB)**: This stream originates near the surface ahead of the warm front. It ascends and, upon interacting with the warm front, rotates eastward.

In addition to the airflow, the pressure-height structure of a midlatitude cyclone exhibits distinct features. Figure 12.3 shows the 1000-hPa isohypses superimposed with 300-hPa observations for January 8, 2024. At 1000 hPa, the isohypses indicate a low-pressure center located in the Texas Panhandle. The sharp gradient, or "kinking," of the isohypses over southwestern Texas reveals the presence of a cold frontal boundary extends equatorward from the surface low pressure, consistent with the frontal cyclone structure described previously.

At 300 hPa, the isohypses reveal a long-wave trough with its axis centered over central New Mexico. The trough lies to the west of the surface low-pressure system, indicating that the cyclone's vertical axis tilts westward with height. As will be discussed in Ch. 12.2, this westward tilt is a hallmark of intensifying and developing midlatitude cyclones. In this configuration, the cyclone's warm sector is located beneath a large upper-level ridge, while its cold sector lies beneath a deep upper-level trough. This alignment underscores the critical role of interactions between upper-tropospheric jet stream dynamics and lower-tropospheric thermodynamic processes in cyclone development and evolution.

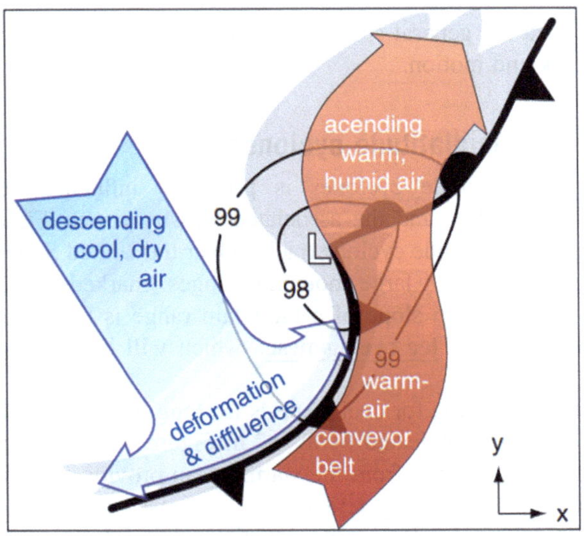

FIGURE 12.2 Three-dimensional atmospheric flow around a midlatitude cyclone

This figure provides a schematic of the ascending and descending atmospheric flow in a midlatitude cyclone. Thin black lines with numbers denote isobars in kPa, and thick black lines illustrate fronts.

From Stull, R., 2017: Practical Meteorology: An Algebra-Based Survey of Atmospheric Science—*version 1.02b. University of British Columbia. 940 pages.*

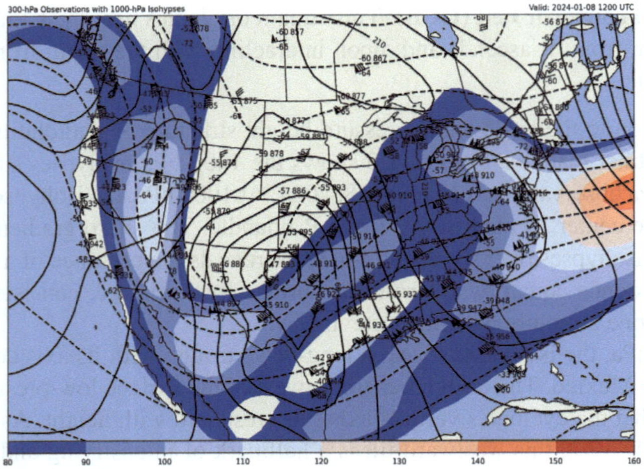

FIGURE 12.3 Vertical structure of a mature midlatitude cyclone

This figure shows the 300-hPa observations for January 8, 2024, at 12 UTC with 300-hPa wind speeds in contour color fill. Solid contours correspond to 1000-hPa isohypses, and dashed contours correspond to 300-hPa isohypses.

-No Permission Required

Having outlined the general structure of midlatitude cyclones, we now turn to their general track and motion.

12.1.2 Tracks of midlatitude cyclones

The motion of midlatitude cyclones is primarily influenced by the upper-tropospheric westerly flow and the associated Rossby waves. Figure 12.4 shows the climatology of midlatitude cyclone tracks. Over the United States, cyclogenesis is favored in regions east of large mountain ranges (marked with an “m” in the figure). Because the eastern slope of a mountain range is known as the lee side, this phenomenon is called **lee cyclogenesis**, which will be discussed in detail in Ch. 12.4.

Another preferred region for cyclogenesis is along the Gulf Stream off the east coast of North America (indicated by a “w” in Figure 12.4). As explained in Ch. 12.3, significant latent heating from the warm ocean provides additional energy to developing cyclones. The strong temperature contrast between the Gulf Stream and the adjacent continent enhances the jet stream through thermal wind balance, creating a feedback loop that promotes cyclogenesis.

As shown in Figure 12.4, midlatitude cyclones over the United States generally move eastward and poleward, consistent with the mean jet stream flow. In the Northern Hemisphere, they typically develop over a period of 2-7 days and travel at speeds of 12 to 15 m/s (Stull, 2017). Cyclone tracks are also influenced by climate

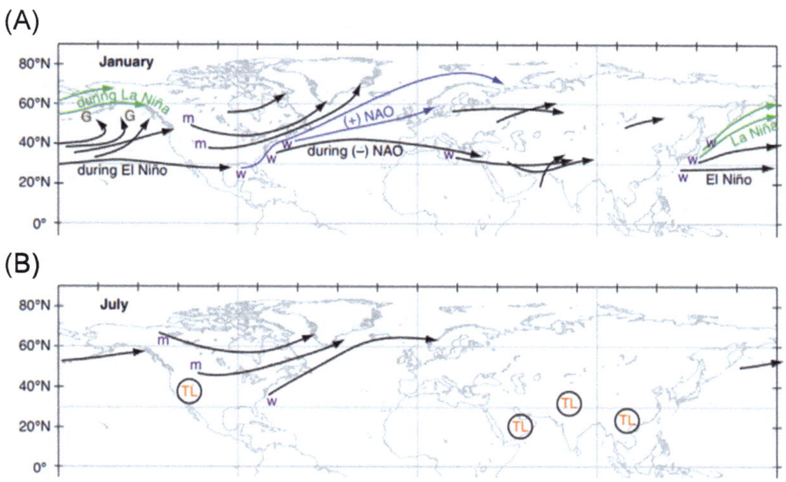

FIGURE 12.4 Climatology of midlatitude cyclone tracks

Climatology of midlatitude cyclone tracks (lines with arrows) for (A) January and (B) July. Other symbols represent genesis and decay regions, as explained in the text. Circled symbols indicate stationary cyclone.

From Stull, R., 2017: Practical Meteorology: An Algebra-Based Survey of Atmospheric Science—*version 1.02b. University of British Columbia. 940 pages.*

variability at different scales. As discussed in Ch. 9, the strength and position of the polar jet stream vary seasonally; it is strongest and located further equatorward during winter. As a result, midlatitude cyclones in winter tend to be stronger, faster, and located farther south compared with summer.

The El Niño Southern Oscillation (ENSO) also affects cyclone tracks by altering the jet stream. During a typical winter El Niño event, the Pacific jet stream intensifies over the eastern Pacific Ocean, while the polar jet stream shifts northward. Consequently, storm tracks are displaced toward the southern United States, as shown in Figure 12.4. Conversely, during a typical winter La Niña , the Pacific jet stream shifts into the Northern Pacific Ocean, while the polar jet stream extends southward into the United States. In these cases, cyclone tracks often originate in the Pacific Northwest and progress into the central United States.

Another source of variability in cyclone tracks is the **North Atlantic Oscillation (NAO)**, which reflects changes in the pressure difference between the Bermuda High and the Icelandic Low. During a positive NAO phase, the pressure difference strengthens, producing a larger poleward pressure gradient across the North Atlantic. This intensifies the westerlies, steering cyclones toward northern Europe, as shown in Figure 12.4. During a negative NAO phase, the pressure difference weakens, and storm tracks shift toward southern Europe.

12.2 Dynamics of cyclogenesis

Now that we have provided a general climatology of midlatitude cyclones, we turn to the physical processes that lead to cyclogenesis. Cyclogenesis is not a single event but a sequence of processes that produce wind, temperature, and pressure changes. It is typically characterized by rising motion, decreasing surface pressure, and increasing vertical vorticity. In this section, we focus on the dynamical processes involved in cyclogenesis.

12.2.1 The role of upper-tropospheric processes

As discussed in Chs. 9 and 10, upper-tropospheric dynamical processes are linked to the evolution of the jet stream, jet streaks, troughs, and ridges. Rising motion is connected to net divergence in an atmospheric column through the continuity equation:

$$\frac{\partial u}{\partial x}+\frac{\partial v}{\partial y}=-\frac{\partial \omega}{\partial p} \tag{12.1}$$

Eq. (12.1) shows that net horizontal divergence (horizontal outflow) is balanced by rising motion (vertical inflow). Integrating from the surface up to the pressure p yields:

$$\omega(p)=-\int_{p_{sfc}}^{p}\left(\frac{\partial u}{\partial x}+\frac{\partial v}{\partial y}\right)dp \tag{12.2}$$

Thus, significant horizontal divergence associated with the jet stream produces rising motion to conserve mass. Most of this rising motion occurs near the mid-troposphere (around 500 hPa), while strong stability near the tropopause suppresses vertical motion there.

Two primary processes cause horizontal divergence of the jet stream: Rossby waves and jet streaks. As discussed in Ch. 7, gradient wind flow is subgeostrophic in troughs and supergeostrophic in ridges. If the isohypse spacing remains constant, winds accelerate around ridges and decelerate through troughs, as shown in the trough-ridge couplet in Figure 12.5.

As explained in Ch. 10, this produces speed divergence downstream of the trough and speed convergence upstream of the trough. According to Eq. (12.1), the downstream divergence induces rising motion, which removes mass from the surface and lowers surface pressure. Consequently, jet-stream curvature tends to generate cyclogenesis east of the trough and anticyclogenesis (formation of anticyclones) west of it, as shown in Figure 12.5.

Jet streaks amplify these effects. As discussed in Ch. 9, divergence arises from the ageostrophic component of the flow:

$$\boldsymbol{V}_{ag}=\frac{\boldsymbol{k}}{f}\times\frac{d\boldsymbol{V}_H}{dt} \tag{12.3}$$

In a straight jet streak, acceleration toward the jet core produces strong divergence in the left-exit and right-entrance quadrants, making these regions favorable for cyclogenesis. The thermally indirect circulation in the exit region also weakens static stability, further enhancing rising motion in the left-exit quadrant. For a curved jet streak at the base of a trough (ie, a cyclonically curved jet streak), divergence is especially strong in the left-exit region, making this the most favorable area for cyclogenesis.

So far, we have shown how jet stream processes can produce rising motion and falling surface pressure. We now consider how they also increase vertical vorticity, or the "spin-up" of the cyclone. As shown in Ch. 10, vertical vorticity at synoptic scales changes through both vorticity advection and divergence/convergence. Advection only transports vorticity, whereas divergence and convergence can create or destroy it locally.

Returning to the trough-ridge couplet in Figure 12.5: the trough axis corresponds to cyclonic curvature, making its base a local maximum of positive vorticity. The ridge crest, by contrast, represents anticyclonic curvature and a local maximum of negative vorticity. As air parcels move through the trough-ridge couplet, positive vorticity is advected downstream from the trough into the region of rising motion. This positive vorticity advection enhances cyclone spin-up by transporting higher vorticity into the surface low.

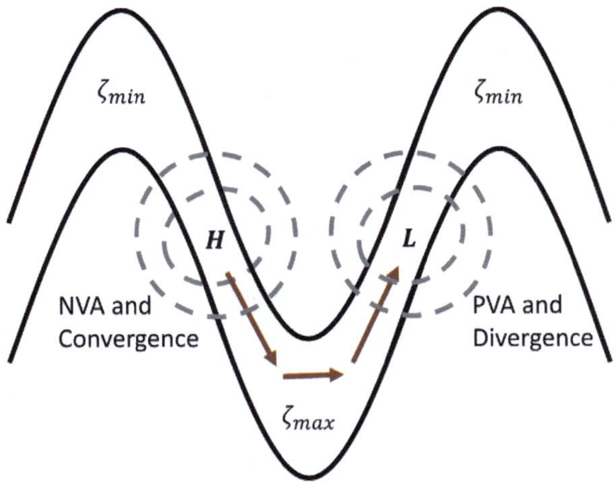

FIGURE 12.5 Schematic of the role of jet stream curvature

Trough-ridge couplet associated with 300-hPa jet stream. Brown arrows correspond to the wind through the trough in which the flow is subgeostrophic at the base of the trough and geostrophic away from the trough axis. The gray contours correspond to the position of surface cyclones and anticyclones. Notice that there is negative vorticity advection and convergence upstream of the trough and positive vorticity advection and divergence downstream of the trough.

-No Permission Required

Finally, we can see how upper-tropospheric divergence reinforces this process. As parcels decelerate into the trough and accelerate out of it, speed divergence develops aloft. By Dines' compensation principle, this produces low-level convergence, which further increases cyclonic vorticity. Thus, upper-tropospheric divergence generated by the jet stream and the jet streaks drives rising motion, lowers surface pressure, and enhances vertical vorticity.[1]

12.2.2 The role of lower-tropospheric processes

The primary dynamic process in the lower troposphere is **boundary-layer pumping, which refers to the upward motion generated by horizontal wind drag around a cyclone**. To understand this process, we must examine the winds associated with surface cyclones.

As discussed in Ch. 11, turbulent drag within the boundary layer causes the horizontal winds to spiral inward toward the cyclone's center. As illustrated in Figure 12.6, these winds have both a tangential component (ie, parallel to the isobars) and a radial component (i.e. directed inward). According to the continuity equation, this horizontal inflow must be balanced by vertical outflow in order to conserve mass.

Because turbulence in the planetary boundary layer (PBL) transfers surface friction throughout its entire depth, a net horizontal inflow develops into the surface low-pressure area. By continuity, this inflow induces rising motion above the PBL, producing boundary-layer pumping. At the same time, turbulent drag slows the winds within the PBL, which reduces vertical vorticity - a process often described as "spin-down" of the cyclone - and can ultimately lead to cyclolysis (the dissipation of the cyclone).

From this reasoning, it follows that turbulent drag tends to increase surface pressure as air spirals into the low-pressure center. In other words, PBL flow always acts to fill a surface low. This filling process occurs quickly, often eliminating a surface cyclone in less than a day, which is typical in the final stage of a midlatitude cyclone.

However, during cyclogenesis (the development phase), upper-tropospheric dynamics can offset the filling effect of PBL flow. For cyclogenesis to occur, upper-level divergence above the surface low must exceed the convergence within the PBL. This happens when the axis of low pressure tilts westward with height, as shown in Figure 12.3.

For example, when an upper-level trough is positioned west of the surface low, the downstream side of the trough - where upper-level divergence is strongest - aligns directly above the surface low, enhancing its development. By contrast, when the upper-level trough is vertically aligned with the surface low, the associated

[1] Because lower-tropospheric convergence will shrink the area of the air column and rising motion will stretch the height of the air column, this process is sometimes known as the stretching term for vertical vorticity.

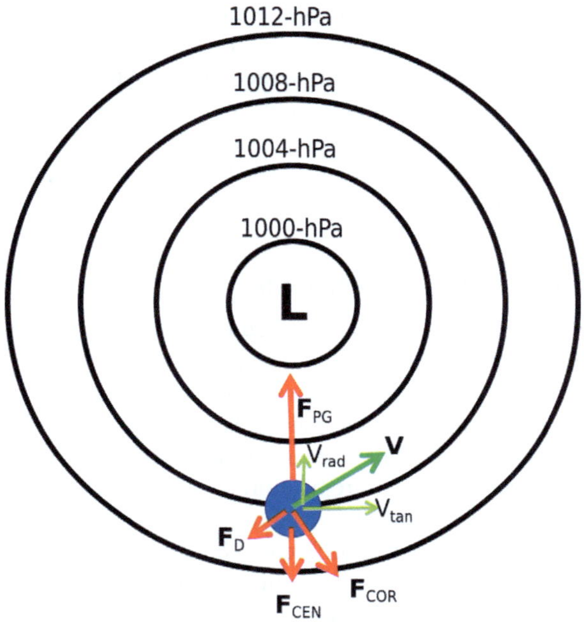

FIGURE 12.6 Schematic of an air parcel within a surface cyclone

An air parcel moving within a surface cyclone. The red arrows correspond to atmospheric forces, where F_{PG} is the pressure gradient force, F_{COR} is the Coriolis force, F_{CEN} is the centrifugal force, and F_D is the turbulent drag force. The dark green arrow corresponds to the velocity vector of the air parcel, which is decomposed into a tangential component V_{tan} and a radial component V_{rad}.

-No Permission Required

divergence is displaced to the east of the surface low. Without divergence aloft over the surface low, the PBL inflow dominates, leading to filling and cyclolysis.

This leads to a common rule of thumb when it comes to the evolution of midlatitude cyclones.

Westward-tilting low-pressure systems are associated with cyclogenesis, while ***vertically stacked low-pressure systems are associated with cyclolysis.***

12.3 Thermodynamics of cyclogenesis

Surface cyclones are associated with low surface pressure, so processes that further deepen this low pressure promote cyclogenesis. As discussed in the previous section, upper-tropospheric dynamics enhance rising motion, lower surface pressure, and increase positive vorticity near the surface. A deepening cyclone is also characterized by decreasing geopotential heights.

On an isobaric surface, the time rate of change of geopotential height is called **geopotential height tendency**, whereas on a constant height surface, the time rate of change of sea-level pressure is called **surface pressure tendency**. Because falling pressure and decreasing geopotential heights are closely related, it is important to examine the processes that control geopotential height tendency - many of which are thermodynamic. This section focuses on those thermodynamic processes and their role in cyclogenesis.

12.3.1 Differential thermal advection

As discussed in Chs. 5 and 8, the thickness equation establishes a general relationship between temperature and geopotential heights on isobaric maps. In many cases, thermal advection (equivalent to thickness advection), particularly in the lower troposphere, is the dominant physical mechanism driving large-scale ascent. Typically - but not always - the effect of thermal advection decreases with height because the atmosphere becomes more barotropic aloft, where isotherms and isohypses tend to run parallel.

In Ch. 8.5, we demonstrated that differential horizontal thermal advection (ie, the vertical gradient in thermal advection) modifies the lapse rate and thus affects static stability. For example, if cold advection increases with height, the lapse rate increases, reducing static stability. Similarly, if warm advection decreases with height, the lower troposphere warms more rapidly than the upper troposphere, also increasing the lapse rate within the air column.

To see how this works dynamically, we can examine how changes in the vertical gradient of thermal advection lead to vertical motion.

As shown in Figure 12.7, rising motion occurs east of the surface low, in advance of the warm front, where positive thickness advection is the strongest. Conversely, sinking motion occurs west of the surface low, in the region of strongest cold advection. Warm advection beneath the 500-hPa ridge (east of the surface low) increases the 1000-500-hPa thickness, which in turn bows the 500-hPa surface upward. This enhances anticyclonic vorticity at the ridge crest. In the absence of negative vorticity advection (NVA), the only way to decrease vertical vorticity is through horizontal divergence. By continuity, divergence aloft implies rising motion beneath the ridge axis and convergence at low levels. Thus, low-level warm advection (positive thickness advection) induces rising motion and rising geopotential heights aloft.

It is important to note that the key driver of vertical motion is the vertical *gradient* of thermal advection. Consider the schematics in Figure 12.8. In Figure 12.8A, cold advection is present at 700 hPa but absent aloft. This reduces thickness near 700 hPa, causing isohypses to bow upward below 700 hPa and downward above 700 hPa. The result is falling geopotential heights at 500 hPa and rising geopotential heights near the surface. Hence, cold advection beneath an upper-level trough tends to deepen the trough. If the cold advection at 700 hPa was equal to that at 300 hPa, the isohypses would remain evenly spaced, producing no geopotential height tendency.

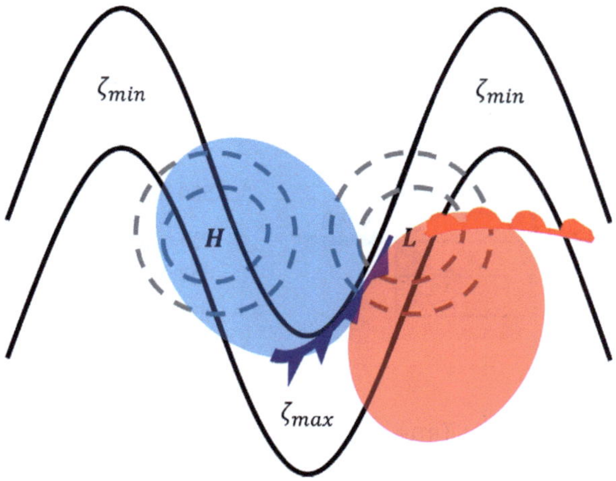

FIGURE 12.7 Schematic of the relationship between thermal advection and vertical motion

The relationship between lower-tropospheric warm advection and rising motion. The solid black contours correspond to 500-hPa isohypses and the dashed gray contours correspond to surface isobars. The region shaded in red is the location of lower-tropospheric warm advection and rising motion, whereas the region shaded in blue is the location of lower-tropospheric cold advection and sinking motion.

-No Permission Required

In Figure 12.8B, warm advection occurs at 700 hPa but not aloft. This increases thickness near 700 hPa, causing isohypses to bow upward above 700 hPa and downwards below 700 hPa. The result is rising geopotential heights at 500 hPa and falling geopotential heights near the surface. Hence, warm advection beneath an upper-level ridge tends to amplify the ridge.

Differential thermal advection is commonly observed near midlatitude cyclones. Figure 12.7 illustrates how low-level thermal advection shapes the geopotential height tendency field. Warm advection east of the cyclone, especially ahead of the warm front, is associated with falling pressure in the lower troposphere and ridge-building aloft. Meanwhile, cold advection west of the cyclone (behind the cold front), corresponds to rising surface pressure and falling upper-level geopotential height.

It is also important to recognize that vertical motion partially offset the direct effects of thermal advection. For example, thermal advection near fronts initiates ascent, but as discussed in Ch. 8, rising motion induces adiabatic cooling. These opposing processes restore thermal wind balance in response to thermal advection, as explained in Ch. 11 with regard to frontal circulations.

In summary, low-level warm advection that decreases with height and/or upper-tropospheric cold advection that increases with height generates rising motion and falling geopotential heights near the surface, which in turn drive changes in sea-

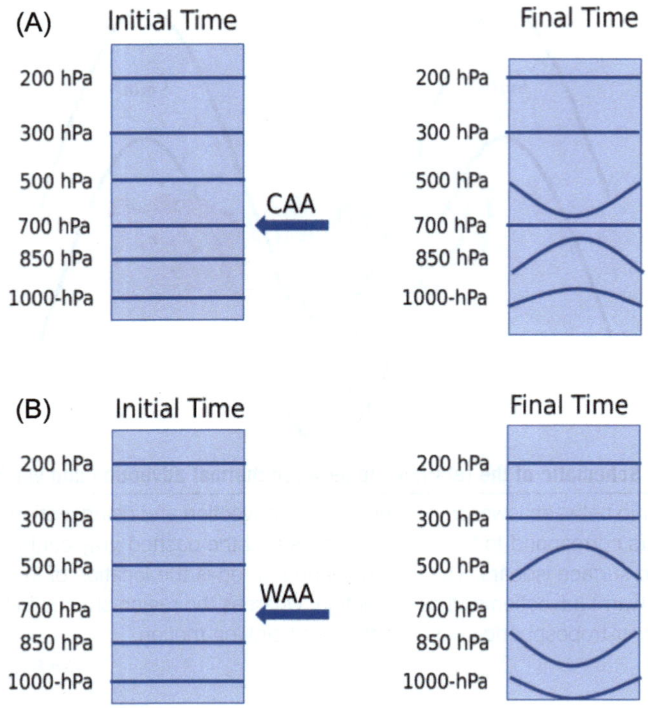

FIGURE 12.8 Differential thermal advection

(A) The effect of cold air advection (CAA) decreasing with height on geopotential height tendency. (B) Schematic of the effect of warm air advection (WAA) decreasing with height on geopotential height tendency.

-No Permission Required

level pressure. As discussed in Ch. 12.2.1, these thermodynamic also influence surface vorticity and thus play a central role in cyclogenesis.

12.3.2 Differential diabatic heating

As discussed in Ch. 8, diabatic heating processes (such as solar or infrared radiation, condensation, evaporation, etc.) can alter atmospheric thickness. For example, cloud formation releases latent heat, which warms the air column, whereas rain evaporation absorbs latent heat, cooling the air below cloud base. As noted earlier, Ch. 8 the precipitation that reaches the ground is directly tied to the net latent heating, and by the thickness equation, these temperature changes are linked to pressure changes.

Like thermal advection, the vertical *gradient* of diabatic heating influences geopotential height tendencies and vertical motion. Consider the schematic in Figure 12.9. When condensation occurs during cloud formation, latent heat is released, warming the air and increasing layer thickness. If maximum heating occurs near 500 hPa (as shown in Figure 12.9), isohypses bow upward above 500 hPa and bow downward below it. This produces falling geopotential heights beneath 500 hPa and rising geopotential heights above - consistent with the local increase in thickness. This process, called **differential diabatic heating,** is critical for cyclogenesis. Latent heat release during condensation injects additional energy into the cyclone and modifies its thermal structure.

Figure 12.10 illustrates this effect. In the warm sector of a frontal cyclone, the WCB enhances warming through both thermal advection and latent heat release. This warming sharpens the temperature gradient along the cold front, which further strengthens warm advection in the region. As discussed earlier, stronger low-level thermal advection promotes stronger ascent, deepening surface low pressure and intensifying the cyclone. In other words, **differential diabatic heating can establish a positive feedback loop in the warm sector of cyclone**.

This mechanism - where latent heat release reinforces cyclone development - is known as the **self-development paradigm**. The interaction between latent heating and the cyclone often produces the most rapid phases of intensification, referred to as **explosive cyclogenesis** (Martin, 2006). Research shows that a sea-level pressure drop of at least 24 hPa within 24 hours is the defining characteristic of such events (Sanders and Gyakum, 1980).

In the United States, explosive cyclogenesis events (sometimes called bomb cyclones) is commonly associated with one or more of the following conditions (Martin, 2006):

1. They frequently occur downstream of strong troughs and upstream of strong ridges, where upper-tropospheric divergence above the surface low drives intense rising motion.
2. They often develop near frontal zones, where warm advection and latent heat release are strongest in the warm sector.
3. They typically form in regions with weak static stability, which do not inhibit vertical motions (often linked to jet streak dynamics, particularly when the cyclone lies in the left-exit region).
4. They often occur where cold advection flows over warm water, such as along the Gulf Stream where evaporation supplies moisture and destabilizes the boundary layer.
5. They often produce heavy precipitation poleward and westward of the cyclone center.

12.4 Life cycle of midlatitude cyclone

Based on our analysis of the physical processes that drive cyclogenesis and frontogenesis, we can now present a model outlining the life cycle of a midlatitude cyclone. Much of our understanding of this life cycle comes from the conceptual framework developed by early 20th-century Norwegian meteorologists, known as

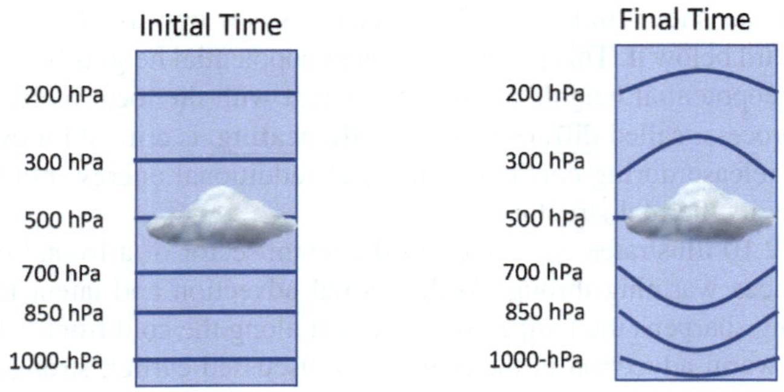

FIGURE 12.9 Schematic of differential diabatic heating

The effect of cloud condensation on geopotential height tendency. In this figure, diabatic heating increases with height up to 500 hPa and diabatic heating decreases with height above 500 hPa.

-No Permission Required

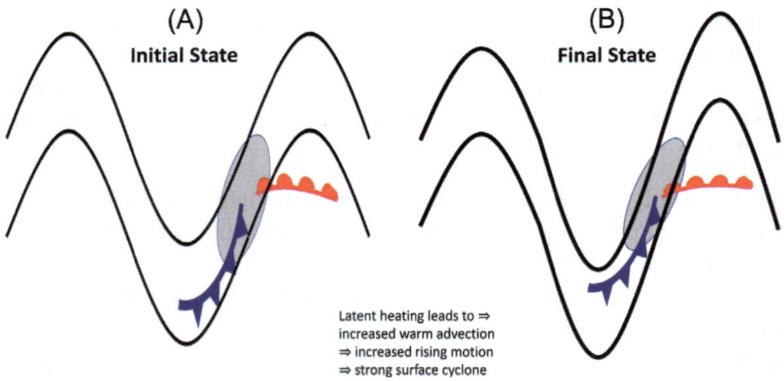

FIGURE 12.10 Schematic of self-development paradigm

The self-development paradigm in which differential diabatic heating generates a positive feedback loop in the warm sector of the cyclone. Latent heat release leads to enhanced lower-tropospheric warm advection, greater rising motion, and a deepening surface cyclone.

-No Permission Required

the **Norwegian Cyclone Model (NCM)**. In this section, we will summarize the NCM and use the dynamics and thermodynamics discussed earlier to address some of its limitations.

12.4.1 Overview of frontal cyclone evolution

An overview of the midlatitude cyclone life cycle, according to the NCM, is shown in Figure 12.11. The energy source for cyclone development lies in the polar front,

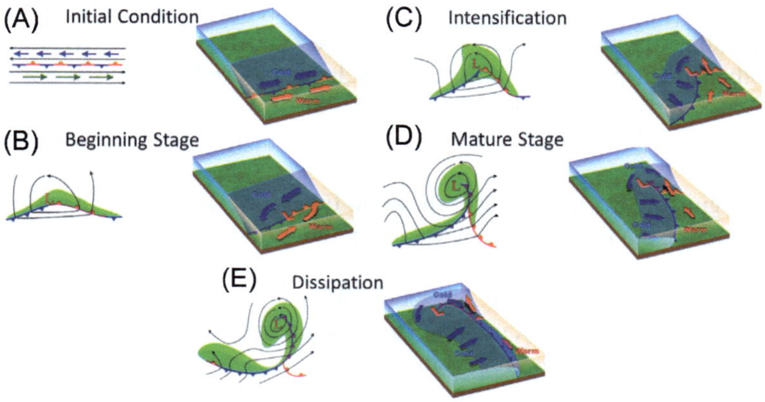

FIGURE 12.11 Overview of NCM

The life cycle of a midlatitude cyclone according to the Norwegian cyclone model (NCM). The five stages are (A) initial condition, (B) beginning stage, (C) intensification, (D) mature stage, and (E) dissipation.

Courtesy of the National Oceanic and Atmospheric Administration. -No Permission Required

which forms along the boundary between cold, polar air moving equatorward, and warm, tropical air moving poleward.[2]

At the initial stage, the polar front is depicted as a stationary front, where the temperature gradient serves as a reservoir of potential energy. Cyclone development begins when an instability along the front allows tropical air to be advected poleward and polar air equatorward.[3] During this process, deformation along the front increases, and central sea-level pressure continuously falls until the cyclone reaches its mature phase. In this phase, the cold front extends equatorward and the warm front extends eastward from the low-pressure center. Eventually, the cold front overtakes the warm front, cutting off the cyclone center from the warm sector. Within the NCM framework, this marks the beginning of the **post-mature phase**, characterized by the formation of an occluded front and the onset of cyclolysis.

The NCM, developed before the global radiosonde network and satellite observations, was a landmark achievement in synoptic meteorology. It successfully

[2]To be more specific, the polar front is located at the boundary of the polar cell and the Ferrel cell. The polar cell is a thermally direct circulation in which very cold air is transported equatorward near the surface. In contrast, the Ferrel cell is a thermally indirect circulation in which tropical air is transported poleward near the surface.

[3]The initial state of the polar front can be thought of as being in a state of unstable equilibrium within the NCM. Since neither air mass on both sides of the front are advancing, then polar front remains stationary. However, when small disturbances occur on the polar front, the disturbances grow over time.

explained the cloud and precipitation patterns associated with cyclones and their corresponding fronts. However, the NCM does not describe the dynamical and thermodynamical processes underlying cyclogenesis and cyclolysis, nor does it capture the vertical structure and evolution of midlatitude cyclones. With the framework established in this chapter, we can supplement and refine the original NCM.

We begin with the initial stage, shown in Figure 12.11A. Along the polar front, the temperature gradient points poleward, implying westerly gestrophic flow in thermal wind balance. Because this flow is uniform, there are no vertical vorticity gradients, and therefore no vorticity advection. Without vorticity advection, there is no geopotential height tendency, meaning that uniform westerly geostrophic flow alone cannot initiate cyclogenesis. For cyclogenesis to begin, an upper-level disturbance - manifested as a vorticity maximum - is required.

One common source of such disturbances is atmospheric flow over high mountain ranges. Consider uniform westerly geostrophic flow crossing the Rocky Mountains, as illustrated in Figure 12.12. As discussed in Ch. 10, in adiabatic and frictionless flow, potential vorticity (PV) must be conserved.

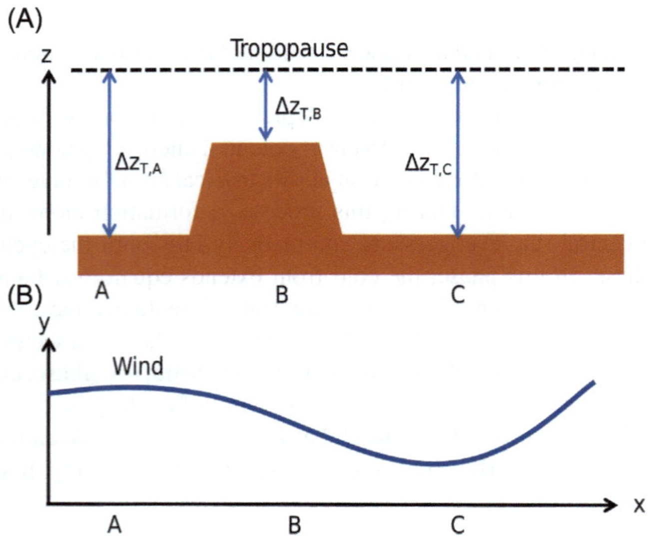

FIGURE 12.12 Schematic of lee cyclogenesis

Lee cyclogenesis relative to the Rocky Mountains. (A) A vertical cross-section with the positive *x*-direction pointing to the east and the positive *z*-axis pointing upward. (B) A horizontal cross-section with the positive *x*-direction pointing to the east and the positive *y*-direction pointing to the north.

-No Permission Required

$$PV = \frac{f_A}{\Delta z_{T,A}}$$

Here, f_A is the Coriolis parameter at point A, and $\Delta z_{T,A}$ is the tropospheric depth at point A. As the flow is forced between the mountain top and the tropopause at point B, the effective depth decreases such that $\Delta z_{T,B} < \Delta z_{T,A}$ where $\Delta z_{T,B}$ is the effective depth of the troposphere at point B. Assuming latitude remains approximately constant, this decrease in depth must be balanced by a reduction in absolute vorticity, producing anticyclonic vorticity as the flow passes over the mountain.This anticyclonic curvature rotates the wind equatorward, which lowers the Coriolis parameter. To conserve PV, the reduced Coriolis parameter must be offset by an increase in vertical vorticity.

As the air descends east of the Rockies at point C, the tropospheric depth returns to its original value, but the Coriolis parameter remains lower because the flow has shifted equatorward. Conservation of PV therefore requires even greater cyclonic vorticity. This sequence generates a Rossby wave pattern in the upper-tropospheric flow, which often serves as the disturbance that initiates cyclogenesis. This explains why cyclogenesis frequently occurs east of large mountain ranges.

Suppose such an upper-level disturbance develops, leading to the developing stage of the NCM as shown in Figure 12.11B. This stage can be represented as a trough-ridge couplet superimposed on westerly geostrophic flow. As explained in Ch. 8, geostrophic flow produces no vorticity advection along the axes of the trough or ridge, and thus cannot by itself strengthen them. However, positive vorticity advection (PVA) downstream of the trough induces ascent, while negative vorticity advection (NVA) upstream induces subsidence, as shown in Figure 12.5. The rising motion downstream of the trough lowers surface pressure, enhances low-level convergence, and increases cyclonic vorticity near the surface. The result is the development of a midlatitude cyclone that tilts westward with increasing height.

As discussed in Ch. 12.3, thermodynamic processes further intensify the cyclone. In westward-tilting system, the surface low lies beneath an upper-level ridge, while the surface high lies beneath an upper-level trough. The low-pressure circulation produces lower-tropospheric warm advection (decreasing with height) beneath the ridge, strengthening it, and low-level cold advection (increases with height) beneath the trough, strengthening it as well.

Thus, thermal advection raises mid-tropospheric geopotential heights east of the surface low and lower the heights to its west, as shown in Figure 12.7. This enhances the trough-ridge couplet and creates a positive feedback loop between upper-tropospheric vorticity advection and lower-tropospheric thermal advection.

In summary, during the **intensification phase** of a midlatitude cyclone (as shown in Figure 12.11C), upper-level processes (such as jet stream processes and trough-ridge interactions) and thermodynamic processes (including thermal advection and latent heating) strengthen the cyclone. In contrast, boundary-layer processes (such as surface friction and convergence) act to weaken it. When upper-level and thermodynamic processes dominate, the cyclone reaches its **mature phase**, as shown in Figure 12.11D, characterized by an equatorward-extending cold front and an eastward-extending warm front, enclosing a distinct warm sector, as shown in Figure 12.13.

Cyclolysis begins as the upper-level disturbance progresses eastward. The once westward-tilted system becomes vertically stacked, so that surface convergence no longer coincides with upper-level divergence. At this point, boundary-layer processes dominate, leading to rising surface pressure and dissipation of the system.

Finally, as the upper disturbance shifts eastward, the cold front overtakes the warm front, disconnecting the surface low from the warm sector, as shown in Figure 12.14. With weakened low-level warm advection, the thermodynamic support for the cyclone collapses. The formation of the occluded front marks the **post-mature phase**, ultimately leading to cyclone dissipation.

12.4.2 Diagnosing vertical motion for a midlatitude cyclone

The development of a surface cyclone depends on the position of vertical motion relative to the cyclone center. Throughout this chapter, we have discussed several diagnostic tools for identifying vertical motion, including upper-tropospheric vorticity advection and lower-tropospheric warm advection. These processes can be described within the framework of **quasi-geostrophic (QG) theory**, which is based on the observation that synoptic-scale flow is approximately geostrophic and hydrostatic. Accordingly, QG theory focuses on atmospheric motion above the boundary layer, assuming an approximate thermal wind balance.[4]

A central insight of QG theory is that vertical motion in the atmosphere can be related to geostrophic vorticity advection and the thermal wind vector. Although the full derivation of this relationship is beyond the scope of this text,[5] it can be shown that the midlevel vertical velocity can be expressed as:

$$\omega_{500} = -\frac{2\Delta p}{f}\left[\boldsymbol{V}_T \cdot \overline{\nabla_P\left(\zeta_g + \frac{f}{2}\right)}\right] \qquad (12.4)$$

In this equation, ω_{500} represents the omega velocity at 500 hPa, ζ_g is the geostrophic vorticity (ie, the vertical vorticity associated with the geostrophic wind), and $\boldsymbol{V}_T$ is the thermal wind vector between the 1000- and 500-hPa levels. The overbar indicates an average over the 1000–500 hPa layer. This formulation

[4] There are four assumptions associated with QG theory. First, it is assumed that ageostrophic flow is assumed to be less than approximately 10% of the geostrophic flow. Second it is assumed that atmospheric flow is adiabatic and frictionless. Third, horizontal variability in atmospheric stability is ignored. Fourth, the atmosphere is in hydrostatic balance.

[5] The derivation of the QG omega equation can be found in any standard textbook in atmospheric dynamics.

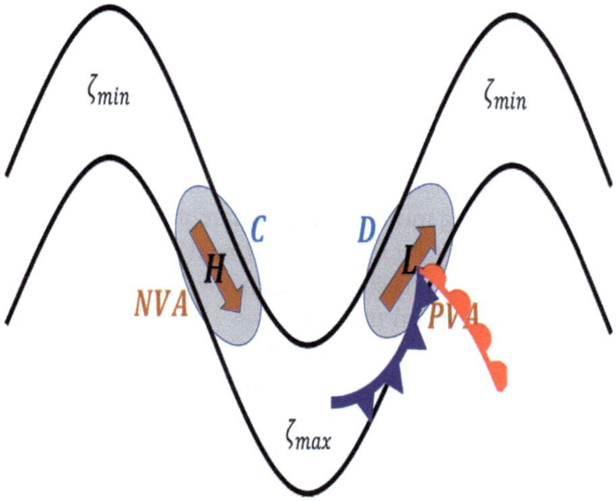

FIGURE 12.13 Schematic of the mature phase of a midlatitude cyclone

The mature phase of the midlatitude cyclone. The solid black contours correspond to 500-hPa isohypses in which there is PVA and divergence (D) downstream of the trough and NVA and convergence (C) upstream of the trough. The surface low pressure (L) along with its accompanying surface fronts is located east of the trough axis, whereas the surface high (H) pressure is located west of the trough axis.

-No Permission Required

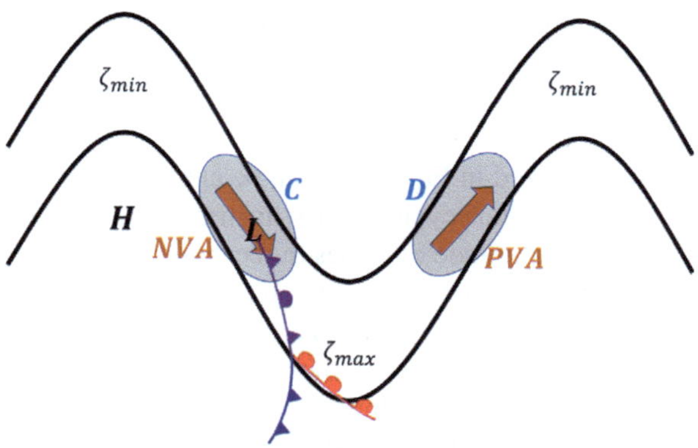

FIGURE 12.14 Schematic of the dissipation phase of a midlatitude cyclone

The dissipation phase of the midlatitude cyclone. The solid black contours correspond to 500-hPa isohypses in which there is PVA and divergence (D) downstream of the trough and NVA and convergence (C) upstream of the trough. During the dissipation phase, the surface low pressure (L) is upstream of the trough, which produces an occlusion.

-No Permission Required

is an application of the **Trenberth omega equation**, named after meteorologist Kevin Trenberth, who originally derived it. The term in square brackets represents the advection of absolute vorticity by the thermal wind.

To approximate the average geostrophic vorticity in the 1000-500 hPa layer, values from either the 850-hPa or 700-hPa are often used. Physically, Eq. (12.4) implies that the strongest rising motion occurs where the thermal wind vector indicates maximum PVA, while sinking motion occurs where there is maximum NVA. Strong upward motion favors cloud development and heavier precipitation. Additionally, rising motion lowers surface pressure beneath the ascent region, helping deepen the surface low and pull it toward that area.

Using Eq. (12.4), vertical motion can be diagnosed by overlaying 700-hPa absolute vorticity on a 1000–500 hPa thickness map. The direction and magnitude of the thermal wind vector can be inferred from the thickness contours. Vertical motion is strongest where tightly packed thickness contours align with a steep gradient in absolute vorticity.

This process is illustrated in Figure 12.15. On the left, the 700-hPa absolute vorticity field is superimposed on the 1000–500 hPa thickness map, with the maximum PVA associated with the thermal wind is located within the highlighted region. On the right, the omega velocity confirms this expectation: the strongest negative omega values (indicating ascent) are collocated with the region of maximum PVA.

The thickness field also provides insight into thermal advection, since the thermal wind flows parallel to isotherms representing the average temperature of the layer. Furthermore, because the thermal wind vector is defined as the vertical difference in the geostrophic wind across the layer, vorticity advection by the thermal wind corresponds to the vertical gradient of vorticity advection. This vertical gradient is directly tied to vertical motion.

The dynamics summarized in Eq. (12.4) can be applied to forecast cyclogenesis and analyze evolving midlatitude cyclones. To illustrate this, we now examine one of the most powerful U.S. midlatitude cyclones in recent decades: the North American blizzard of March 2019. This storm produced up to 2 feet of snow across the Northern Plains and Midwest.[6] Like the Great Blizzard of 1993, it was classified as a bomb cyclone after its central pressure dropped more than 24 hPa in 24 hours.

As shown in Figure 12.16, the cyclone formed in Arizona on March 12 before being steered northeastward by the upper-level flow, reaching Colorado on March 13. At 00Z on March 13, a deep trough was positioned over New Mexico, with a jet streak located downstream of the trough base, as shown in Figure 12.17. This placed eastern Arizona in the left-exit region of the jet streak, a configuration that enhanced upper-level divergence. At the same time, the 1000–500 hPa thickness chart revealed PVA by the thermal wind, suggesting synoptic-scale ascent. Together,

[6]The melting of this snow also caused historic flooding.

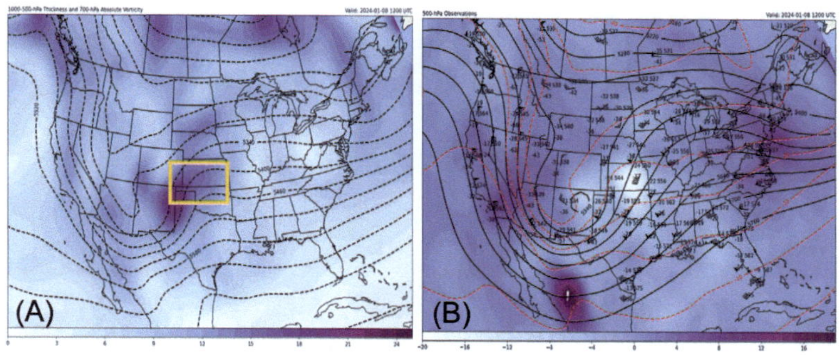

FIGURE 12.15 Example of the use of the QG-omega equation

(A) 1000–500 hPa thickness map with 700-hPa absolute vorticity on January 8, 2024, at 12Z. (B) 500-hPa observations with isohypses contoured as solid black contours, isotherms contoured as dashed red contours, and 500-hPa omega velocity as contour filled for January 8, 2024, at 12Z. The highlighted region corresponds to the location of maximum positive vorticity advection by the thermal wind.

-No Permission Required

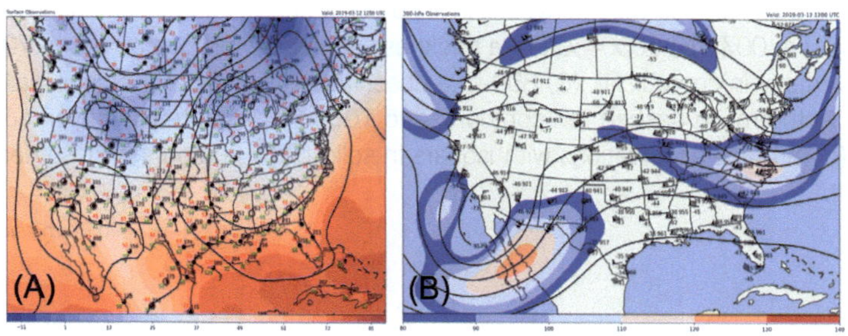

FIGURE 12.16 Observations of 2019 cyclone on March 12

(A) Surface observations for March 12 at 12Z with isobars in solid black contours and isotherms in contour color fill. (B) 300-hPa observations with isohypses in solid black contours and isotachs in contour color fill.

-No Permission Required

these features indicated that the surface cyclone would intensify as it moved toward Colorado.

By the time the cyclone reached Colorado, the low pressure deepened further, supported by strong upper-tropospheric forcing, as shown in Figure 12.18. At 300-hPa, northeastern Colorado lay beneath an area of jet stream diffluence and divergence, particularly in the left-exit region of a straight jet streak. The upper-level trough had acquired a negative tilt, signaling intensification of the surface cyclone. The 1000–500 hPa thickness map confirmed strong PVA by the thermal wind over northeast Colorado, directly above the surface low pressure. With the

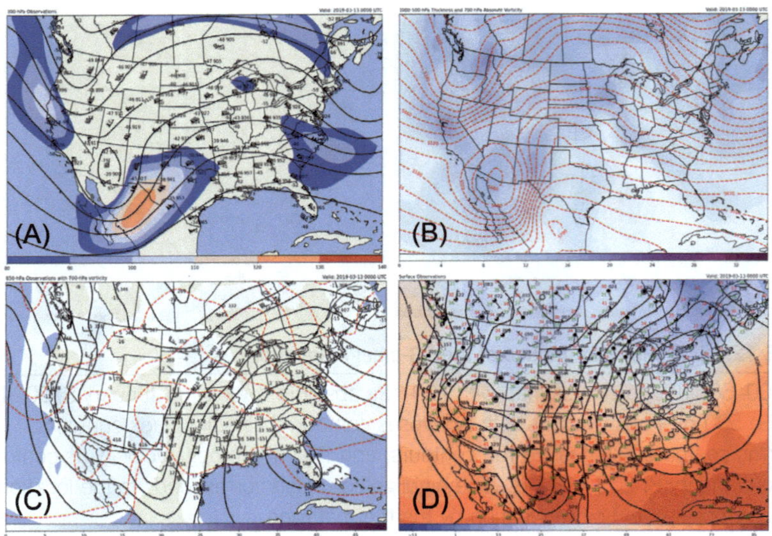

FIGURE 12.17 Observations of the 2019 cyclone on March 13 at 00Z

(A) 300-hPa observations with isohypses in solid black contours and isotachs in color fill for March 13 at 00Z. (B) 1000–500 hPa thickness in dashed red contours with 700-hPa absolute vorticity in contour fill. (C) 850-hPa observations with isohypses in solid black contours, isotherms in dashed red contours, and 700-hPa absolute vorticity in contour color fill. (D) Surface observations with isobars in solid black contours and isotherms in contour fill.

-No Permission Required

thermal wind directed northeastward around the cyclone, the system was expected to deepen further while tracking toward the northeast.

By March 13 at 12Z, the surface cyclone reached a minimum pressure below 984 hPa, producing hurricane-force wind gusts across the Midwest. From March 13 at 12Z through March 14 at 00Z, cyclogenesis and frontogenesis proceeded in tandem, as shown in Figure 12.19. Cold advection intensified behind the cold front, sharpening the frontal temperature gradient. As discussed in Ch. 11, this increase in cold advection strengthens the geostrophic wind near the front, enhancing cyclonic shear vorticity. The resulting shear vorticity further strengthened ascent along the frontal boundary, consistent with the 1000–500 hPa thickness chart shown in Figure 12.19. Because the surface cyclone still tilted westward with height, enhanced divergence aloft coincided with frontal ascent, reinforcing frontogenesis. In effect, cyclogenesis and frontogenesis fueled one another, driving the cyclone's rapid deepening to a central pressure near 968 hPa. Guided by the steering flow in Figure 12.19, the storm advanced toward the Great Lakes.

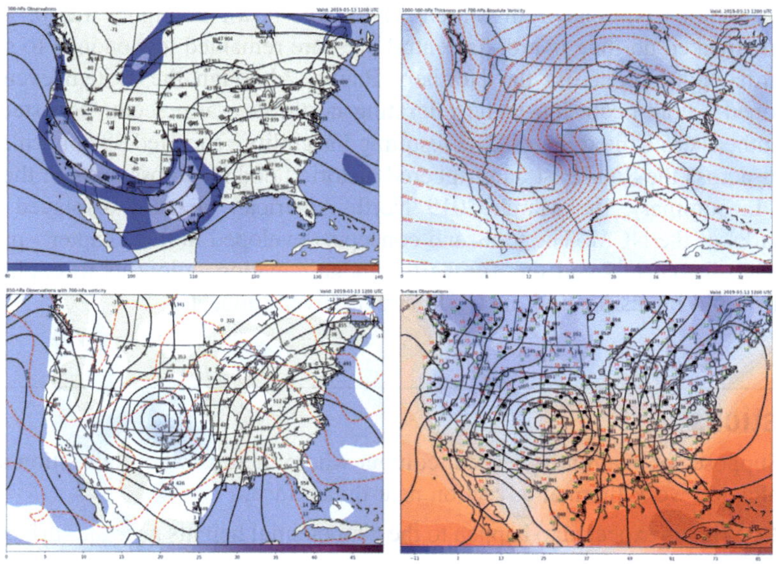

FIGURE 12.18 Observations of the 2019 cyclone on March 13 at 12Z

Same as Figure 12.17, except for March 13 at 12Z.

-No Permission Required

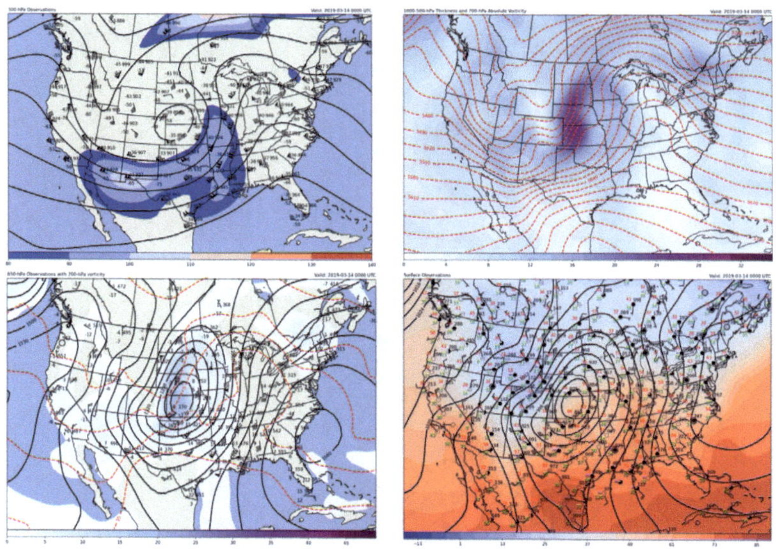

FIGURE 12.19 Observations of the 2019 cyclone on March 14 at 00Z

Same as Figure 12.17, except for March 14 at 00Z.

-No Permission Required

Surface observations from March 14-15 are shown in Figure 12.20. As the cyclone tracked northeastward, its central pressure remained low enough to sustain blizzard conditions across the Northern Plains. Although the system eventually developed a cold occlusion while migrating north, its warm sector persisted over the southeastern United States. The resulting instability generated 38 confirmed tornadoes across the Southeast and triggered historic flooding across parts of the Midwest. In summary, the March 2019 North American blizzard displayed many hallmarks of the NCM. The tools and concepts developed in this chapter - particularly QG diagnostics and the Trenberth omega equation -allow us to explain the storm's rapid intensification, its track, and its severe impacts.

Questions

1. Suppose we examine a vertical column where there is warm advection at 500 hPa and weak thermal advection near the surface.
   a. What is expected to happen to the geopotential heights near the surface over time? Please explain your answer.

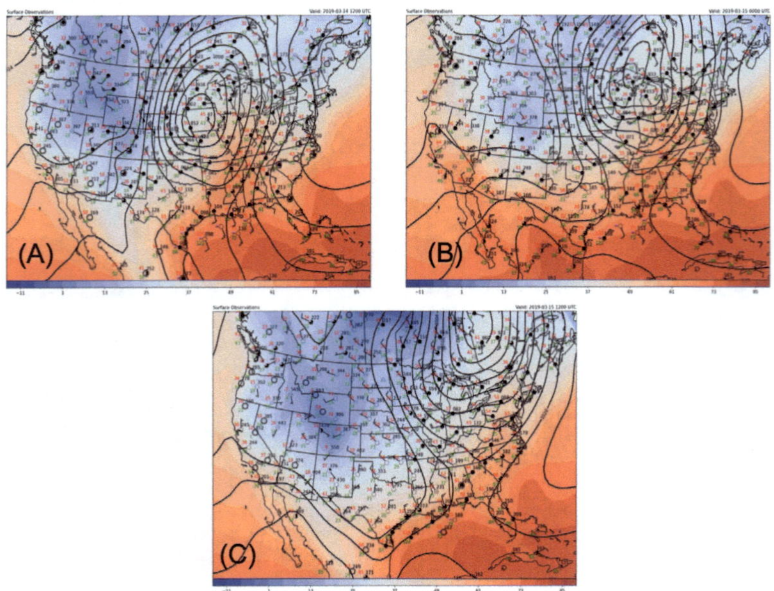

FIGURE 12.20 Observations of the 2019 cyclone from March 14 to 15

Surface observations with isobars in solid black contours and isotherms in contour fill for March 14 at 12Z (A), March 15 at 00Z (B), and March 15 at 12Z (C).

-No Permission Required

- **b.** What is expected to happen to the geopotential height near the jet stream over time? Please explain your answer.

2. Answer the following questions regarding midlatitude cyclones given in (a) and (b):

- **a.** Explain why cyclogenesis and frontogenesis often occur simultaneously.
- **b.** Explain why surface cyclones propagate poleward in the midlatitudes.

3. Answer the following questions regarding cyclogenesis given in (a) and (b):

- **a.** Explain why cyclogenesis is often favored downwind (lee) of the Rocky Mountains in the United States.
- **b.** Suppose a midlatitude cyclone is located on the lee side of the Rocky Mountains. Use the conservation of potential vorticity to explain why the cyclone is likely to move equatorward and downslope (ie, eastward).

4. The 20th-century British meteorologist R.C. Sutcliffe argued that if the vorticity in the air column is positively advected by the thermal wind, it must be associated with greater air divergence at the column top compared with the bottom.

- **a.** Explain how this statement is essentially equivalent to the meaning of the Trenberth omega equation.
- **b.** Use Sutcliffe's statement to argue that PVA by the thermal wind implies falling surface pressures.

5. In Figure 12.21, 1000–500 hPa thickness is shown in solid contours, while 700-hPa absolute vorticity is shown in color-filled contours. Identify the states over which sinking motion is the strongest. Identify the states over which rising motion is the strongest. Justify your answers.

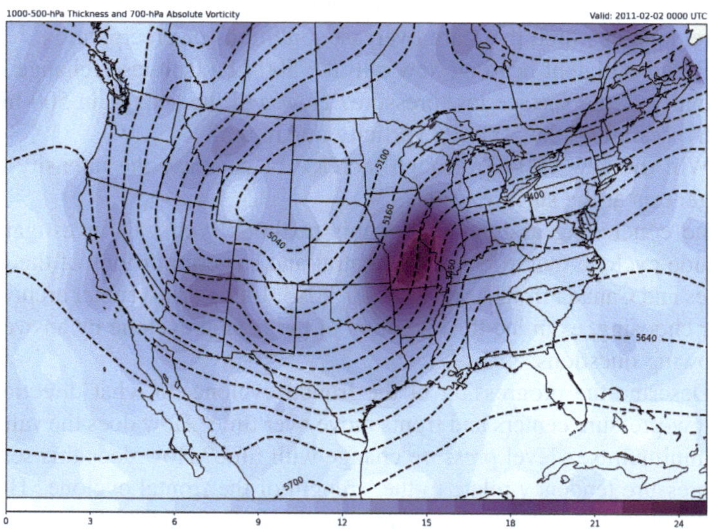

FIGURE 12.21 Case study on February 2, 2011

1000–500 hPa thickness map with 700-hPa absolute vorticity.

-No Permission Required

6. Figure 12.22 presents 850-hPa isohypses in solid contours and 1000–850 hPa thickness in dashed contours for a frontal cyclone on February 2, 2011.
   a. Use an arrow to indicate the direction of the 1000–850 hPa thermal wind. Justify your answer.
   b. Use a "+" to indicate the regions of synoptic-scale upward motion and use a "−" to indicate the regions of synoptic-scale downward motion. Justify your answer.
7. Figure 12.23 provides observations associated with a frontal cyclone on February 2, 2011. Based on the synoptic charts, determine whether the cyclone will undergo cyclogenesis or cyclolysis. Justify your answer using information from all three synoptic charts.
8. Figure 12.24 shows surface observations with isobars in solid black contours and isotherms in contour fill for February 24, 2012, at 12Z.
   a. Locate the frontal cyclone. Place an "L" next to the central low-pressure area, and draw the accompanying cold front and warm front.
   b. Analyze the cloud distribution around the frontal cyclone and shade the regions where overcast conditions or precipitation are occurring.
   c. How will the cloud distribution marked in part (b) reinforce the differential thermal advection? How should this frontal cyclone develop?
9. Figure 12.25 provides surface observations with isobars in solid black contours and isotherms in contour fill for February 29, 2012, at 00Z. 500-hPa isohypses have been superimposed on the surface chart in dashed black contours.
   a. Locate the frontal cyclone. Place an "L" next to the central low pressure and draw the accompanying cold front and warm front.
   b. Analyze the cloud distribution around the frontal cyclone and shade the regions in which there are overcast conditions or precipitation.
   c. Based on latent heating, how will the 500-hPa isohypses change in the vicinity of the surface low pressure? How will the change in 500-hPa isohypses affect the 500-hPa vorticity maximum?
   d. Will the new 500-hPa trough increase or decrease the intensity of cyclogenesis at the surface? Why?
10. In the concluding assignment for this textbook, you will investigate a midlatitude cyclone that affected the central and eastern portions of the United States and Canada from April 17 to 20, 2013. Using any weather archive site of your choosing, examine the evolution of the frontal cyclone by answering the following questions:
    a. Describe the progression of the frontal cyclone. In what direction do the low-pressure centers and fronts move over time? How does the value of the minimum sea-level pressure change with time? How does this sea-level pressure tendency relate to the strength of the frontal cyclone? How does the pressure gradient change around the surface cyclone?
    b. Describe the progression of the trough. Where is the position of the trough axis relative to the center of the surface cyclone? Based on the position of

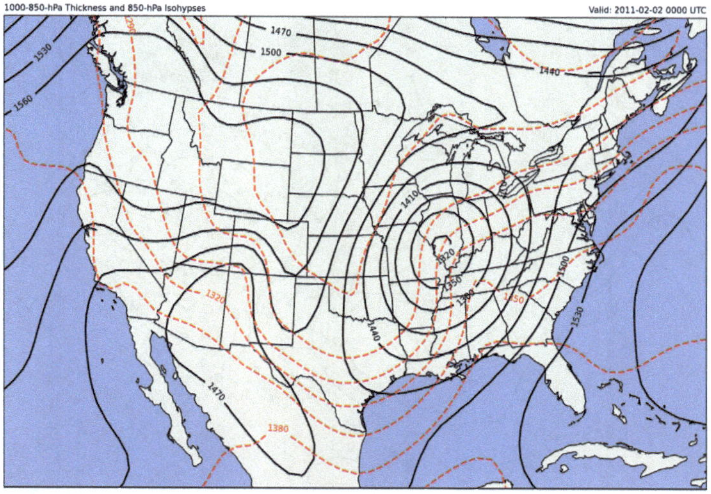

FIGURE 12.22 Case study on February 2, 2011

1000–850 hPa thickness map (in dashed red contours) with 850-hPa isohypses (in solid black contours).

-No Permission Required

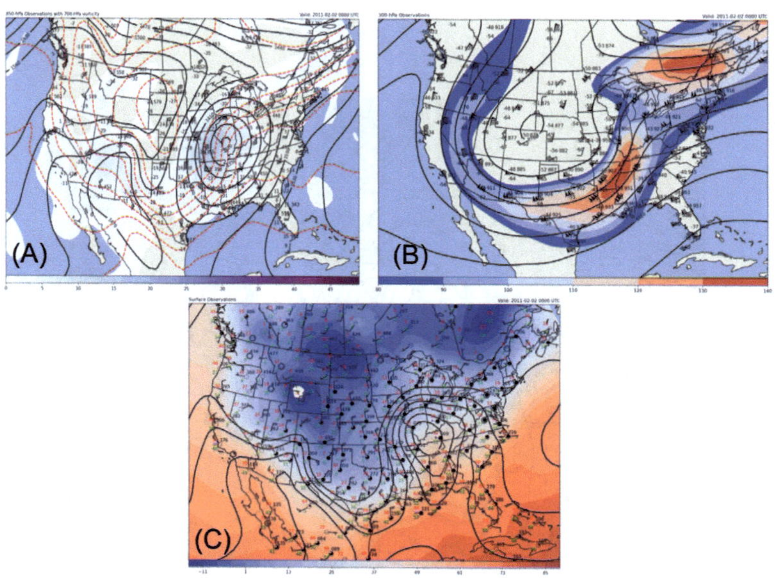

FIGURE 12.23 Case study on February 2, 2011, at 00Z

(A) 850-hPa observations with isohypses in solid black contours, isotherms in dashed red contours, and 700-hPa vorticity in contour fill. (B) 300-hPa observations with isotachs in contour fill. (C) Surface observations with isobars in solid black contours and isotherms in contour color fill.

-No Permission Required

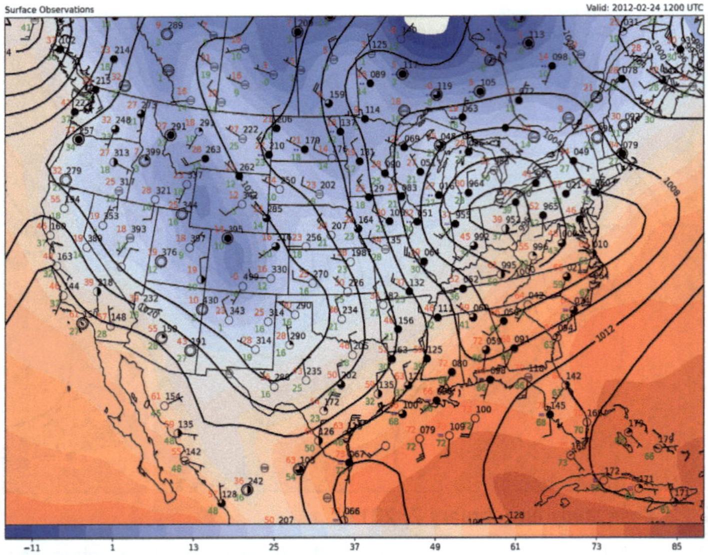

FIGURE 12.24 Case study on February 24, 2012, at 12Z

Surface observations with isobars in solid black contours and isotherms in contour color.

-No Permission Required

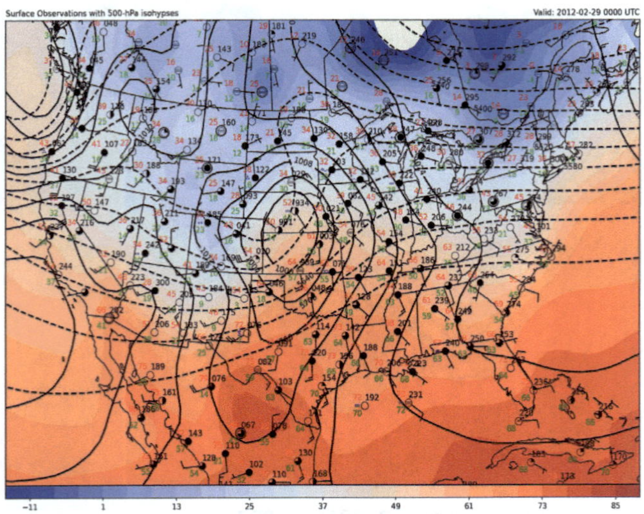

FIGURE 12.25 Case study on February 29, 2012, at 00Z

Surface observations with isobars in solid black contours and isotherms in contour color. 500-hPa isohypses are contoured in dashed black contours.

-No Permission Required

the trough and the surface cyclone, should there be cyclogenesis or cyclolysis?

c. Describe the progression of the jet stream and the development of jet streaks. Based on the position of the jet stream and jet streaks, should there be cyclogenesis or cyclolysis?

d. Describe the cloud distribution around the frontal cyclone. How do the clouds correspond with the surface and upper-level features? Based on the cloud distribution, should there be cyclogenesis or cyclolysis?

References

Martin, J., 2006. Mid-Latitude Atmospheric Dynamics: A First Course, 1st ed., John Wiley & Sons Inc.

Sanders, F., Gyakum, J.R., 1980. Synoptic-dynamic climatology of the "bomb". Mon. Weather Rev. 108, 1589–1606. https://doi.org/10.1175/1520-0493(1980)108<1589:SDCOT>2.0.CO;2.

Stull, R., 2017. Practical Meteorology: An Algebra-Based Survey of Atmospheric Science. University of British Columbia, British Columbia University of British Columbia, British Columbia.

Reed, R.J., 1990. Advances in knowledge and understanding of extratropical cyclones during the past quarter century: an overview. In: Newton, C.W., Holopainen, E.O. (Eds.) Extratropical Cyclones. American Meteorological Society, Boston, MA. https://doi.org/10.1007/978-1-944970-33-8_3.

Index

Note: Page numbers followed by *f* indicate figures and *t* indicates tables.

D

E

F

G

H

N

O

P

Z